Lecture Notes in Physics

For information about Vols. 1–99, please contact your bookseller or Springer-Verlag.

Vol. 100: Einstein Symposion Berlin. Proceedings 1979. Edited by H. Nelkowski et al. VIII, 550 pages. 1979.

Vol. 101: A. Martin-Löf, Statistical Mechanics and the Foundations of Thermodynamics. V, 120 pages. 1979.

Vol. 102: H. Hora, Nonlinear Plasma Dynamics at Laser Irradiation. VIII, 242 pages. 1979.

Vol. 103: P. A. Martin, Modèles en Mécanique Statistique des Processus Irréversibles. IV, 134 pages. 1979.

Vol. 104: Dynamical Critical Phenomena and Related Topics. Proceedings, 1979. Edited by Ch. P. Enz. XII, 390 pages. 1979.

Vol. 105: Dynamics and Instability of Fluid Interfaces. Proceedings, 1978. Edited by T. S. Sørensen. V, 315 pages. 1979.

Vol. 106: Feynman Path Integrals, Proceedings, 1978. Edited by S. Albeverio et al. XI, 451 pages. 1979.

Vol. 107: J. Kijowski, W. M. Tulczyjew, A Symplectic Framework for Field Theories. IV, 257 pages. 1979.

Vol. 108: Nuclear Physics with Electromagnetic Interactions. Proceedings, 1979. Edited by H. Arenhövel and D. Drechsel. IX, 509 pages. 1979.

Vol. 109: Physics of the Expanding Universe. Proceedings, 1978. Edited by M. Demiański. V, 210 pages. 1979.

Vol. 110: D. A. Park, Classical Dynamics and Its Quantum Analogues. VIII, 339 pages. 1979.

Vol. 111: H.-J. Schmidt, Axiomatic Characterization of Physical Geometry. V, 163 pages. 1979.

Vol. 112: Imaging Processes and Coherence in Physics. Proceedings, 1979. Edited by M. Schlenker et al. XIX, 577 pages. 1980.

Vol. 113: Recent Advances in the Quantum Theory of Polymers. Proceedings 1979. Edited by J.-M. André et al. V, 306 pages. 1980.

Vol. 114: Stellar Turbulence. Proceedings, 1979. Edited by D. F. Gray and J. L. Linsky. IX, 308 pages. 1980.

Vol. 115: Modern Trends in the Theory of Condensed Matter. Proceedings, 1979. Edited by A. Pekalski and J. A. Przystawa. IX, 597 pages. 1980.

Vol. 116: Mathematical Problems in Theoretical Physics. Proceedings, 1979. Edited by K. Osterwalder. VIII, 412 pages. 1980.

Vol. 117: Deep-Inelastic and Fusion Reactions with Heavy Ions. Proceedings, 1979. Edited by W. von Oertzen. XIII, 394 pages. 1980.

Vol. 118: Quantum Chromodynamics. Proceedings, 1979. Edited by J. L. Alonso and R. Tarrach. IX, 424 pages. 1980.

Vol. 119: Nuclear Spectroscopy. Proceedings, 1979. Edited by G. F. Bertsch and D. Kurath. VII, 250 pages. 1980.

Vol. 120: Nonlinear Evolution Equations and Dynamical Systems. Proceedings, 1979. Edited by M. Boiti, F. Pempinelli and G. Soliani. VI, 368 pages. 1980.

Vol. 121: F. W. Wiegel, Fluid Flow Through Porous Macromolecular Systems. V, 102 pages. 1980.

Vol. 122: New Developments in Semiconductor Physics. Proceedings, 1979. Edited by F. Beleznay et al. V, 276 pages. 1980.

Vol. 123: D. H. Mayer, The Ruelle-Araki Transfer Operator in Classical Statistical Mechanics. VIII, 154 pages. 1980.

Vol. 124: Gravitational Radiation, Collapsed Objects and Exact Solutions. Proceedings, 1979. Edited by C. Edwards. VI, 487 pages. 1980.

Vol. 125: Nonradial and Nonlinear Stellar Pulsation. Proceedings, 1980. Edited by H. A. Hill and W. A. Dziembowski. VIII, 497 pages. 1980.

Vol. 126: Complex Analysis, Microlocal Calculus and Relativistic Quantum Theory. Proceedings, 1979. Edited by D. Iagolnitzer. VIII, 502 pages. 1980.

Vol. 127: E. Sanchez-Palencia, Non-Homogeneous Media and Vibration Theory. IX, 398 pages. 1980.

Vol. 128: Neutron Spin Echo. Proceedings, 1979. Edited by F. Mezei. VI, 253 pages. 1980.

Vol. 129: Geometrical and Topological Methods in Gauge Theories. Proceedings, 1979. Edited by J. Harnad and S. Shnider. VIII, 155 pages. 1980.

Vol. 130: Mathematical Methods and Applications of Scattering Theory. Proceedings, 1979. Edited by J. A. DeSanto, A. W. Sáenz and W. W. Zachary. XIII, 331 pages. 1980.

Vol. 131: H. C. Fogedby, Theoretical Aspects of Mainly Low Dimensional Magnetic Systems. XI, 163 pages. 1980.

Vol. 132: Systems Far from Equilibrium. Proceedings, 1980. Edited by L. Garrido. XV, 403 pages. 1980.

Vol. 133: Narrow Gap Semiconductors Physics and Applications. Proceedings, 1979. Edited by W. Zawadzki. X, 572 pages. 1980.

Vol. 134: $\gamma\gamma$ Collisions. Proceedings, 1980. Edited by G. Cochard and P. Kessler. XIII, 400 pages. 1980.

Vol. 135: Group Theoretical Methods in Physics. Proceedings, 1980. Edited by K. B. Wolf. XXVI, 629 pages. 1980.

Vol. 136: The Role of Coherent Structures in Modelling Turbulence and Mixing. Proceedings 1980. Edited by J. Jimenez. XIII, 393 pages. 1981.

Vol. 137: From Collective States to Quarks in Nuclei. Edited by H. Arenhövel and A. M. Saruis. VII, 414 pages. 1981.

Vol. 138: The Many-Body Problem. Proceedings 1980. Edited by R. Guardiola and J. Ros. V, 374 pages. 1981.

Vol. 139: H. D. Doebner, Differential Geometric Methods in Mathematical Physics. Proceedings 1981. VII, 329 pages. 1981.

Vol. 140: P. Kramer, M. Saraceno, Geometry of the Time-Dependent Variational Principle in Quantum Mechanics. IV, 98 pages. 1981.

Vol. 141: Seventh International Conference on Numerical Methods in Fluid Dynamics. Proceedings. Edited by W. C. Reynolds and R. W. MacCormack. VIII, 485 pages. 1981.

Vol. 142: Recent Progress in Many-Body Theories. Proceedings. Edited by J. G. Zabolitzky, M. de Llano, M. Fortes and J. W. Clark. VIII, 479 pages. 1981.

Vol. 143: Present Status and Aims of Quantum Electrodynamics. Proceedings, 1980. Edited by G. Gräff, E. Klempt and G. Werth. VI, 302 pages. 1981.

Lecture Notes in Physics

Edited by H. Araki, Kyoto, J. Ehlers, München, K. Hepp, Zürich
R. Kippenhahn, München, H. A. Weidenmüller, Heidelberg
and J. Zittartz, Köln

184

Stochastic Processes Formalism and Applications

Proceedings of the Winter School
Held at the University of Hyderabad, India
December 15 – 24, 1982

Edited by G. S. Agarwal and S. Dattagupta

Springer-Verlag
Berlin Heidelberg GmbH 1983

Editors

G. S. Agarwal
S. Dattagupta
School of Physics, University of Hyderabad
Hyderabad 500 134, India

ISBN 978-3-540-12326-2 ISBN 978-3-540-40923-6 (eBook)
DOI 10.1007/978-3-540-40923-6

Originally published by Springer-Verlag Berlin Heidelberg New York Tokyo in 1983

2153/3140-543210

PREFACE

A winter school on STOCHASTIC PROCESSES - FORMALISM AND APPLICATIONS was held at the University of Hyderabad December 15 - 24, 1982. These proceedings are based on lectures delivered at the School. The aim was to introduce research workers at the pre- and post-doctoral levels to the basic concepts, techniques and applications of stochastic theories. Some of the discussions, such as those on the properties of Gaussian and Markov processes and on Langevin and Fokker-Planck equations, have by now become standard textbook material. It was decided, however, to include them in this volume in order to make it a self-contained source book or even a graduate level reference book on the subject. On the other hand, certain other topics, not normally dealt with in the context of stochastic processes (e.g. random matrices) also find place in the proceedings. Similarly, the important problem of localization of electronic states in disordered metals is included in view of its natural connection to diffusion, a paradigm of stochastic processes.

The lectures can be divided into two broad categories : (i) basic concepts and techniques, and (ii) applications. Under (i), in addition to fundamental ideas in stochastic processes, certain formal techniques involving projection operators in both stochastic differential equations and master equations, system-size expansions, continuous-time random-walk models, decay of metastable states, etc. are treated in detail. Two lectures on numerical techniques, an often neglected but useful aspect, are also presented here. The basic formalism covered here forms a natural setting for the applications discussed later. It is hoped that the reader will find the right kind of balance between techniques on the one hand and physical applications on the other.

The applications, of course, constitute the central motivating theme of the school. Here, such diverse and topical subjects as Brownian motion, nonequilibrium phase transitions, disordered systems, wave propagation in random media, line shapes in optics and other spectroscopy problems are discussed. For the sake of conciseness, only an outline of the mathematical steps is provided in some lectures. This lacuna is, however, filled-in with the aid of numerous references which the interested reader can look into.

It would not have been possible to organize the school without the financial and material support from the Department of Science and Technology, Government of India and the University of Hyderabad. We would like to record our gratitude to both these institutions. We thank Mr. A. Neela Kantam and Mr. P.C.P. Reddy for skilful typing of the manuscripts. We are greatly indebted to Ms. B. Aruna, Mr. N.V. Mukteswara Rao and Mr. D. Sridhara Rao for their invaluable assistance in the preparation of the proceedings.

G.S. AGARWAL
S. DATTAGUPTA

April 1983
Hyderabad, India

TABLE OF CONTENTS

			Page
1.	R. Vasudevan	SOME BASIC CONCEPTS IN THE THEORY OF STOCHASTIC PROCESSES AND INTRODUCTION TO MARKOV PROCESSES	2
2.	S. Chaturvedi	GAUSSIAN STOCHASTIC PROCESSES	19
3.	G.S. Agarwal	FOKKER-PLANCK EQUATIONS FOR STOCHASTIC PROCESSES	30
4.	R. Vasudevan	STOCHASTIC DIFFERENTIAL EQUATIONS	37
5.	R. Jagannathan	ON SOME NEW CONCEPTS IN PROBABILITY THEORY	54
6.	S. Dattagupta and S.R. Shenoy	DECAY OF METASTABLE STATES: KRAMERS, FIRST PASSAGE TIME AND VARIATIONAL APPROACHES	61
7.	V. Srinivasan	INSTANTONS IN THE DYNAMICAL EVOLUTION OF FOKKER-PLANCK SYSTEMS	77
8.	G.S. Agarwal	PROJECTION OPERATOR TECHNIQUES IN STOCHASTIC PROCESSES	83
9.	S. Chaturvedi	PROJECTION OPERATOR METHODS IN LINEAR STOCHASTIC DIFFERENTIAL EQUATIONS	88
10.	V. Balakrishnan	CONTINUOUS-TIME RANDOM WALK THEORY AND NON-EXPONENTIAL DECAYS OF CORRELATION FUNCTIONS	96
11.	G. Ananthakrishna	ON THE APPROXIMATE SOLUTIONS OF THE NONLINEAR LANGEVIN EQUATIONS	104
12.	M.C. Valsakumar	SOLUTION OF FOKKER-PLANCK EQUATIONS USING TROTTER'S FORMULA	112
13.	K.P.N. Murthy	MONTE CARLO METHODS : AN INTRODUCTION	116
14.	R. Indira	NUMERICAL SOLUTION FOR THE NONLINEAR FOKKER-PLANCK EQUATION	122
15.	G.V. Anand	STABILITY OF STOCHASTIC SYSTEMS	125
16.	G.S. Agarwal	OPTICAL RESONANCE IN PARTIALLY COHERENT FIELDS	134
17.	S. Dattagupta	STOCHASTIC MODELLING OF RELAXATION EFFECTS IN LINE SHAPES	147
18.	N. Kumar	CLASSICAL AND QUANTUM DIFFUSION	166
19.	D. Kumar	RELAXATION OF SINGLE DOMAIN MAGNETIC PARTICLES	186
20.	K.N. Pathak	LANGEVIN EQUATION : APPLICATION TO LIQUID STATE DYNAMICS	197
21.	J.H. Freed	STOCHASTIC MODELLING OF MOLECULAR DYNAMICS	220
22.	G.Venkataraman and K. Neelakantan	NONEQUILIBRIUM PHASE TRANSITIONS - A REVIEW	228

23.	S.R. Shenoy	ANALOGUE OF OPTICAL BISTABILITY IN DRIVEN JOSEPHSON JUNCTIONS	238
24.	S. Chaturvedi	NONLINEAR PHENOMENA IN CHEMICAL KINETICS	245
25.	V. Srinivasan	GOLDSTONE MODES IN NONEQUILIBRIUM PHASE TRANSITIONS	253
26.	S.V. Lawande	PHASE TRANSITIONS IN A SYSTEM OF ATOMS INTERACTING WITH A COHERENT FIELD	257
27.	T.V.Ramakrishnan	LOCALIZATION AND DIFFUSION	270
28.	V. Srivastava	CONTINUOUS-TIME RANDOM-WALK IN DISORDERED SYSTEMS	279
29.	C.K. Majumdar	RANDOM MATRICES IN CONDENSED MATTER PHYSICS	286
30.	D. Dhar	STOCHASTIC EVOLUTION IN ISING MODELS	300
31.	D. Kumar	RELAXATION DYNAMICS OF SPIN GLASSES NEAR TRANSITION TEMPERATURE	314
32.	G.S. Agarwal	WAVE PROPAGATION IN RANDOM MEDIA	319

BASIC CONCEPTS AND TECHNIQUES IN THE THEORY OF STOCHASTIC PROCESSES

SOME BASIC CONCEPTS IN THE THEORY OF STOCHASTIC PROCESSES AND INTRODUCTION TO MARKOV PROCESSES

R. Vasudevan
The Institute of Mathematical Sciences
Madras - 600113, India

Basic concepts in probability theory

The central role of modern probability theory in unravelling the mysteries of nature whether for explanation or prediction can be appreciated when one realises that few mathematical disciplines have contributed to such a wide spectrum of areas from physics to number theory, from biology to econometrics. Very few disciplines have gone so deep into our thinking at the conceptual level . Games of chance attracted the attention of Pascal and Fermat. The efforts of Huyghens, Laplace, Bernoulli and Quetelet and others and even earlier studies of Darwin laid the foundations of statistical methods followed later by the works of Galton, Weldon and more quantitative studies of Fisher. The gigantic work of the members of the Russian school, Tschebychev, Markov, Bernstein, Kintchine and Kolmogorov enhanced the concepts of statistical mechanics in physics. The works of Lindberg, Feller, Long, Doob and others and also the econometricians like Slutzky and Yule introduced new important tools like sequential sampling, time-series analysis etc.

One knows very well the difference between statics and dynamics. Much of economic theory and those relating to classical statistical mechanics are erected on equilibrium structure. Now they model their studies to represent closely our dynamic changing world. Thus stochastic theory in common parlance may mean a wider theory of statistics of change.

The basis of all statistical theories is probability theory governing random processes the study of which usually begins with coin tossing experiments.

According to the <u>frequency approach</u> of Von Mises if A is the outcome of an experiment occurring k times in n trials then the limit

$$\lim_{n\to\infty} \frac{k}{n} \sim P(A) \tag{1}$$

is the probability for A determined by the relative frequency of A. How large n should be is not a question that can be answered with mathematical rigor.

As an illustration let us consider the distribution of N indistinguishable Bose particles in C_j cells of phase space having energy parameter E_j. Since any number of particles can be allotted to each cell the generating function of such an arrangement is $G(u) = 1/(1-u)$ and for C_j cells $G(u,C_j) = (1-u)^{-C_j}$. Hence the coefficient of the term u^{N_j} will yield $[(N_j+C_j-1)!/(N_j!(C_j-1)!)]$ which is the number of ways of distributing N_j particles. This has to be minimised along with the conditions that the total number of particles is fixed and also the total energy i.e. $\sum_j N_j = N$

and $\sum_j N_j E_j = E$. Under such conditions this yields the Bose-Einstein distribution. Derivations for Maxwell-Boltzman and Fermi-Dirac statistics can be arrived at similarly (cf. e.g. Ramakrishnan [1] for details).

In the axiomatic approach all the outcomes of an experiment constituting a set of events is described as the sample space Ω of events. Subsets of Ω, say A,B,C,.. etc. composed of elementary events $\omega \in \Omega$ may be the events we are interested in. Any field $\mathcal{B}$ on Ω is called a Borel field (or σ field) $\mathcal{B}$ on Ω if it is closed under denumerable unions and intersections of the subsets A_1, A_2, A_3, ... etc. i.e.

$$\begin{aligned} &(1) \quad \text{if } A_1, A_2, \ldots, A_n \in \mathcal{B} \quad \text{then } \bigcup_{N=1}^{\infty} A_n \in \mathcal{B} \\ &(2) \quad \text{if } A_1, A_2, \ldots, A_n \in \mathcal{B} \quad \text{then } \bigcap_{n=1}^{\infty} A_n \in \mathcal{B} \end{aligned} \tag{2}$$

Also the compliments $\bar{A}_i$ for all i and unions and intersections of $\bar{A}_i \in \mathcal{B}$ for i = 1,2, ..., n... The members of a given Borel field are called $\mathcal{B}$ measurable. Let the collections a include all A_i's and $\bar{A}_i$'s and their unions and intersections. Then we assign a probability P(A) which is a set function for each member of the collection and thus we have a triplet $(\Omega, \mathcal{B}, P)$ defining our probability space. The P(A)'s satisfy the following:

(1) $0 \leqslant P(A)$ (2) $P(\Omega) = 1$ (3) If A_i's are mutually independent then $P(A_1A_2 \ldots A_n) = P(A_1) + P(A_2) + \ldots + P(A_n)$ while if A_1 and A_2 are not mutually exclusive we have

$$(4) \quad P(A_1 \cup A_2) = P(A_1) + P(A_2) - P(A_1 \cap A_2) \tag{3}$$

For illustration let us consider a single die with six faces. The space Ω is the totality of out comes $\Omega = \{1,2,3,4,5,6\}$ and the class given by $0, \Omega, \{1,3,5\}$ and $\{2,4,6\}$ may be a chosen field $\mathcal{B}$ on Ω. A number P(A) is assigned to each event in the class a of events.

If P(AB) is the joint probability of both A and B occuring and P(A|B) is the conditional probability that A occurs given that B has occurred then P(AB)=P(A|B)P(B) =P(B|A)P(A). Hence we can arrive at the generalized Bayes' theorem that for any set of $\mathcal{B} \subset \Omega$ [If (1) $\bigcup A_i = \Omega$ and (2) $A_iA_j = \emptyset$ for $i \neq j$ where $\emptyset$ is null] as:

$$P(A_j|B) = \frac{P(A_j)P(B|A_j)}{\sum_i P(A_i)P(B|A_i)} \tag{4}$$

Examples of discrete distributions are:

(a) $P(x=k) = \binom{n}{k} p^k q^{n-k}$ (Binomial distribution-k denotes the number of successes in n trials)

(b) $P(x=n) = [\exp(-\lambda)] \left(\frac{\lambda^n}{n!}\right)$ (Poisson distribution giving probability for n events) (5)

Examples of continuous distributions are

(a) $f(x) = \frac{1}{\pi(1+x^2)}$; $-\infty < x < \infty$ (Cauchy distribution)

(b) $f(x) = \frac{1}{\sigma\sqrt{2\pi}} \exp\left[-(x-\mu)^2/2\sigma^2\right]$ (Gaussian distribution) (6)

There are a large number of discrete and continuous distributions used in statistical modelling of phenomena. In the so called Pearsonian systems the frequency function f(x) satisfied the equation

$$\frac{df}{dx} = \frac{x+a}{c_o+c_1x+c_2x^2} \tag{7}$$

with different values for the constants c_o, c_1 and c_2. In many applications we are interested only in the statistical features like mean and moments of the distributions. Hence we describe a few transformations of the frequency function which yeild the desired information in a simple fashion

(1) The probability generating function is

$$G(s) = \sum_n P_n s^n \quad \text{so that} \quad P_n = \frac{1}{n!} \left.\frac{d^nG(s)}{ds^n}\right|_{s=0} \tag{8}$$

$\forall n = 0,1,2,$ ...etc.

With E denoting expectation

$$\left.\frac{d^kG(s)}{ds^k}\right|_{s=1} = E\left[n(n-1)\cdot\cdot\cdot\cdot(n-k+1)\right] \tag{9}$$

are called the factorial moments which are related to the actual moments $E(n^k) = \sum_{r=1}^{k} C_k^r G^r(1)$ where C_k^r are the stirling numbers of the second kind

(2) If we put $s=e^t$ in (8) we get the moment generating function

$$G(e^t) = G_m(t) = \sum_{r=0}^{\infty} m_r \frac{t^r}{r!} \tag{10}$$

where m_r is the r-th moment

(3) For continuous distributions

$$G_m(x;t) = \int e^{tx}\, dF_X(x) \tag{11}$$

is the moment generating function for the random variable X and the central moment generating function is given by

$$C_m(t) = e^{-t\mu} G_m(x;t) = 1+\sum \bar{\mu}_\ell \frac{t^\ell}{\ell} \tag{12}$$

where

$$\bar{\mu}_\ell = E\left\{\left[x-\langle x\rangle\right]^\ell\right\}$$

(4) The characteristic function for the distribution function $F_X(x)$ is

$$\phi_X(t) \quad = \quad \int dx\ f(x)e^{itx} \tag{13}$$

and this is well defined since $|e^{itx}| \leqslant 1$ and ϕ_X is uniformly continuous in t with $\phi_X(0) = 1$. The characteristic function for $z = x_1+x_2$ where x_1 and x_2 are independent random variables is $\phi_z(t) = \phi_{X_1}(t)\phi_{X_2}(t)$ the product of the two characteristic functions ϕ_{X_1} and ϕ_{X_2}.

(5) The cumulant generating function K(t) is given by

$$K(t) \quad = \quad \log \phi_X(t) \quad = \quad \sum_{s=1}^{\infty} K_s \frac{(it)^s}{s!} \tag{14}$$

$\{K_s\}$ being called the cumulants which for the Poisson distribution are given by $K(t) = \lambda(e^{it}-1) = \lambda \sum_{s=1}^{\infty} \frac{(it)^s}{s!}$ and all the cumulants are the same λ the mean of the Poisson distribution.

Many limit theorems can be easily obtained using the concept of characteristic or generating functions. For example the generating function for the binomial distribution in $G_n(u) = [q+pu]^n$ where p is the success probability. If as n becomes large p is small enough so that $np = \lambda$ = a constant then $G_n(u) = [1+\frac{\lambda}{n}(u-1)]^n \xrightarrow[n\to\infty]{} \exp[\lambda(u-1)]$ which is the generating function for the Poisson distribution. Similar limiting distributions as contemplated in the central limit theorem, the law of large numbers etc. can be arrived at using characteristic functions.

If a random variable $X(\omega) = x$ corresponding to an element Ω of the sample space and if $Y = g(x) = g(X(\omega))$ is a function of the random variable and g is a mapping and $f_X(x)$ is the probability frequency function of the random variable X, the frequency function of the random variable Y is

$$f_Y(y) \quad = \quad \frac{f_X(x_1)}{|g'(x_1)|} + \frac{f_X(x_2)}{|g'(x_2)|} + \ldots + \frac{f_X(x_n)}{|g'(x_n)|} \tag{15}$$

where $x_1, x_2 \ldots$ are the zeroes of the function g(x) expressed in terms of y. It is easy to see that if x is normally distributed with mean m and variance σ and if $y = \exp(x)$ the frequency function for Y variable is called log normal distribution

$$F_Y(y) \quad = \quad \frac{1}{\sqrt{2\pi}\sigma y} \exp\left[-(\log y-m)^2/2\sigma^2\right] \tag{16}$$

To develop a calculus of random variables we briefly sketch three types of convergences relating to sequences of random variables

(1) Convergence in probability:

Let $X_1, X_2, \ldots X_n$ be a sequence of random variables and this sequence is said to converge in probability to the random variable X if

$$\lim_{n\to\infty} \{P_r\ |X_n - X| \geqslant \epsilon\} \quad = 0; \qquad X_n \xrightarrow[n\to\infty]{P} X \tag{17}$$

Similarly one can define convergence in distribution

(2) Almost certain convergence:

The sequence of variables X_n converges to X if in the limit $n \longrightarrow \infty$, almost certainly if

$$Pr(X_n = X) = 1 \text{ as } n \to \infty \quad \text{and} \quad \lim_{n \to \infty} X_n(\omega) = X(\omega) \tag{18}$$

for almost all $\omega \in \Omega$. Otherwise stated $(Pr[\,|X_n - X| > \epsilon\,])$ as $n \to \infty$ is zero for atleast one m, m being greater than n.

(3) Mean square convergence:

The sequence X_n of the random variables is said to converge in mean square (m.s) to the random variable X if

$$E(X_n^2) < \infty \;;\; E(X^2) < \infty \quad \text{and if}$$

$$\lim_{n \to \infty} E[(X_n - X)^2] = 0 \tag{19}$$

because of Chebyshev inequality

$$Pr[\,|X_n - X| \geqslant \epsilon\,] \leqslant E[(X_n - X)^2]/\epsilon^2 \text{ for any } \epsilon > 0 \tag{20}$$

For m.s. convergence the right hand side of (20) tends to zero in the limit $n \to \infty$. Hence m.s. convergence implies convegence in probability. Almost certain convergence also implies convergence in probability.

The subject matter detailed till now is contained in the literature cited in [1-7] at the end.

Stochastic Processes

Let Ω be a set whose Borel field of subsets generates events. $\{X(\omega,t);\ \omega \in \Omega, t \in T\}$ is called a stochastic process which depends on two variables ω (event space) and t (the index set). The index t denotes time when the process evolves in time. X may be a real valued function for each ω with its domain on T or a random variable on the probability space (Ω, F, P) at each t. For every finite set of t-values $t_1, t_2, \ldots, t_n$ the corresponding random variables $X_1(\omega_1, t_1)$, $X_2(\omega_2, t_2)$, ... $X_n(\omega_n, t_n)$ have an n-dimensional distribution function

$$F_n(x_1, x_2, \ldots, x_n; t_1, t_2, \ldots, t_n) = Pr[X_n(t) \leqslant x_n, X_{n-1}(t) \leqslant x_{n-1}, \ldots, X_1(t) \leqslant x_1] \tag{21}$$

These F_n's should satisfy

(1) The symmetry condition i.e. it is symmetric in all pairs of (x_j, t_j) values which means F_n's are invariant for the same permutations of x_j and t_j.

(2) The Kolmogorov consistency condition which leads to the following

$$F_n(x_1, x_2, \ldots, x_n; t_1, t_2, \ldots, t_n) = F_{n-1}(x_1, x_2, \ldots, x_{n-1}; t_1, t_2, \ldots, t_{n-1}) \tag{22}$$

when $x_n \longrightarrow \infty$

Since both variables x and t can be either discrete or continuous there are four categories of stochastic processes.

Stationary stochastic processes are characterised by the fact that the finite-dimensional joint distributions are invariant under translations of the origin. It is often illustrated that

$$\mathrm{Cov}\,[X(t)X(t+h)] = E[X(t)X(t+h)] - E[X(t)]\;E[X(t+h)] \qquad (23)$$

is only a function of h for all $t \in T$. If X(t) process is such that

$$E\{[x(t) - E(x(t))]\,[x(s) - E(x(s))]\} = R(t-s) \qquad (24)$$

then the process is wide sense stationary or second order stationary. Then (1) if $R(\tau)$ in (24) is continuous for all τ and has second derivatives, X(t) is differentiable in the m.s. sense (2) if $R(\tau)$ is nonnegative definite for all τ, (τ being real) the process is stationary (3) Also $R(\tau)$ for a mean square continuous stationary process has the mathematical form of the characteristic function (Fourier transform) of a distribution function (Bochner's theorem) [2]. Stationarity and normality of X(t) vis a vis the nature of $R(\tau)$ will be dealt with later while discussing gaussian Markov processes.

The power spectrum of a process X(t) is given by the Fourier transform

$$F_X(\omega) = \int_{-\infty}^{\infty} X(t)\,\exp(-i\omega t)\,dt \qquad (25)$$

$F_X(\omega)$ being a generalised function. If $G(\omega_1,\omega_2) = \langle F_X(\omega_1)F_X(\omega_2)\rangle$ and if it so happens that $G(\omega_1,\omega_2) = G(\omega)\,\delta(\omega_1-\omega_2)$ then $G(\omega)$ is the power spectrum of the stationary process and its different frequency components are uncorrelated.

A stochastic process defined over the interval (0,T) can be expanded in terms of a complete set of orthogonal functions $\phi_i(t)$ such that [10]

$$X(t) \underset{N\to\infty}{=} \sum_{i=1}^{N} \chi_i\phi_i(t), \qquad \int_0^T \phi_i(t)\phi_j(t)dt = \delta_{ij} \qquad (26)$$

The coefficients χ_i are given by $\chi_i = \int_0^T X(t)\phi_i(t)dt$. If we want the coefficients χ_i to be uncorrelated then we should find the function ϕ_i such that

$$\int_0^T R(t,t')\phi_j(t')dt' = \mu_j\phi_j, \quad 0 \leqslant t \leqslant T \qquad (27)$$

assuming that $\langle X(t)\rangle = 0$. This is the main content of the <u>Karhunan-Levy expansion of a stochastic process</u> X(t). Some of the interesting properties of these μ's are

$$\sum_{i=1}^{\infty} \mu_i = \int_0^T G_x(t,t)\,dt; \quad \sum_{i=1}^{\infty} \mu_i^2 = \iint |G_x(t_1,t_2)|^2 dt_1dt_2$$

$$G_x(t_1,t_2) = \sum_{i=1}^{\infty} \mu_i\phi_i(t_1)\phi_i(t_2) \quad \text{etc.} \qquad (28)$$

and if X(t) is stationary then in the limit of large T, much larger than the width of $G_x(\tau)$ we find that

$$\phi_n \sim \frac{1}{\sqrt{T}} \exp(i\omega_n t) \quad \text{where} \quad \omega_n = \frac{2\pi n}{T} \tag{29}$$

Point processes [1,8-12]

A process whose realisations are related to a series of point events occuring in a continuous one-dimensional parameter space (such as time etc.) are point processes which are distinguished by their times of occurrence. The Poisson events and the renewal processes have been studied even before a theory of point process was developed. Studies of assemblage of particles distributed in phase space are also a type of point process. A powerful tool for the study of point processes is the product density technique of Ramakrishnan [2,12]. The central quantity of interest is $dN(x,t)$ denoting the number of entities with parametric values between x and $x+dx$ at time t. This is proportional to dx. It is assumed that the probability that there is more than one particle in that range is of order $O(dx)$ and hence the probability that there are n particles in that range is given by

$$P(1) = f_1(x,t)dx, \quad P(n) = 0, \quad n>1, \quad P(0) = (1-f_1(x,t)dx) \tag{30}$$

Hence $E[n^m] = E[dN(x,t)^m] = \sum P(n)n^m$

$$= E[dN(x,t)] + O(dx) = f_1(x,t)\,dx \tag{31}$$

If $dN(x,t)$ takes the value unity if the variable x happens to lie between x and $x+dx$ then $E[dN(x,t)]$ is the probability that an entity with parametric values between x and $x+dx$ occurs at t. The regularity condition is that $\sum_{n\geq 2} P(n,x;x+\Delta) = O(\Delta)$. Though ideas of this type for first order densities existed in earlier works [13,14] this concept has been deeply analysed and perfected as a sophisticated tool for the study of point processes by Ramakrishnan [12] and many applications and development of these ideas have been made by Ramakrishnan and his group involving higher order correlation functions [1,11].

$f_1(x,t)dx$ is called the first order product density and the expectation of the number in a given interval $[a,b]$ is

$$E[N(a,t)-N(b,t)] = \int_a^b f_1(x,t)\,dx \tag{32}$$

The second order product density and higher order correlations are given by

$$E[dN(x_1,t)dN(x_2,t)] = f_2(x_1,x_2,t)\,dx_1dx_2 \quad \text{provided } x_1 \text{ and } x_2 \text{ do not overlap}$$

$$E[dN(x_1,t)dN(x_2,t)\ldots dN(x_n,t)] = f_n(x_1,x_2,\ldots,x_n,t)\,dx_1dx_2\ldots dx_N \tag{33}$$

For the second moment of the number of points we find

$$E\left\{[N(a,t) - N(b,t)]^2\right\} = \int_a^b\!\!\int E[dN(x_1,t)dN(x_2,t)] \tag{34}$$

and due to the singular behaviour if x_1 and x_2 coalesce the integral in the equation splits into two regions to yield

$$E\left\{[N(a,t)-N(b,t)]^2\right\} = \int_a^b f_1(x,t)\,dx + \int_a^b\int_a^b f_2(x_1,x_2\ t)\,dx_1 dx_2 \tag{35}$$

Similarly any r-th moment of the number of entities in a given region of x-space is given by

$$E\left\{[N(a,t)-N(b,t)]^r\right\} = \sum_{s=1}^{r} C_s^r \int\int_{x_1 x_2 \ldots x_s}\ldots\int f_s(x_1,x_2\ldots x_s;t)\,dx_1 dx_2\ldots dx_s \tag{36}$$

where C_s^r denotes the number of (r-s) fold degeneracy of an r-fold product and $\left[C_s^r\right]$ are the Stirling numbers of the second kind. So the moments are related to the factorial moments as

$$E[N^r] = \sum_{s=1}^{r} C_s^r \langle N(N-1)\ \ldots.(N-s+1)\rangle \tag{37}$$

The product densities are related to what are called the Janossy densities $J_n(x_1,x_2,\ldots;t)$ which represent the probability that there are exactly n particles distributed in the parametric space at x_i's, $i = 1,2,\ldots,n$. Since particles are indistinguishable the probability for the occurence of n particles is

$$P(n,t) = \frac{1}{n!}\int_{\Omega} J_n(x_1,x_2\ldots,x_n;t)\ dx_1 dx_2\ldots dx_n \tag{38}$$

and the product density of order h which expresses the probability of finding particles or entities in the ranges (x_1,x_1+dx_1), (x_2,x_2+dx_2)....etc. irrespective of what happens in other ranges is given by

$$f_h(x_1,x_2,\ldots x_h;t) = \sum_{n=h}^{\infty}\frac{1}{(n-h)!}\int\int\ldots\int J_n(x_1,x_2,\ldots x_h,x_{h+1},\ldots x_n;t)$$

$$dx_{h+1}dx_{h+2}\ldots dx_n \tag{39}$$

It follows from (38) and (39) that [15,16]

$$P(n,t) = \frac{1}{n!}\sum_{k=n}^{\infty}\frac{(-1)^{k-n}}{(k-n)!}\int\int\ldots\int f_k(x_1,x_2\ldots,x_k;t)\,dx_1\ldots dx_k \tag{40}$$

The moment generating function $\theta(u) = \sum_{n=0}^{\infty} P_n \exp(inu)$ in view of the above relations can be expressed in terms of the product density functions as follows

$$\theta(u;t) = \sum_{s=0}^{\infty}\frac{(e^{iu}-1)^s}{s!}\int\int_{\Omega}\ldots\int f_s(x_1,x_2,\ldots x_s;t)\,dx_1 dx_2\ldots dx_s \tag{41}$$

Since moments can be obtained from $\theta(u;t)$ we easily see from (36) that the C_s^r coefficients are given by

$$C_s^r = \left\{\frac{1}{i^r}\frac{\partial^r}{\partial u^r}\left[\frac{(e^{iu}-1)^s}{s!}\right]\right\}_{u=0} = \frac{1}{s!}\sum_{k=0}^{s} k^r\binom{s}{k}\ (-1)^{s-k} \tag{42}$$

The product densities have separable parts and nonseparable cluster terms called irreducible or connected parts as given below

$$f_1(x;t) = g_1(x;t) \; ; \; f_2(x_1,x_2) = g_1(x_1)g_1(x_2) + g_2(x_1,x_2) \; ;$$

$$f_3(x_1,x_2,x_3) = g_1(x_1)g_1(x_2)g_3(x_3) + 3\{g_1g_2\} + g_3(x_1,x_2,x_3) \tag{43}$$

Herein we have suppressed the t dependence and the symbol $\{ \ \}$ means proper symmetrisation. Similar expression for higher order correlation functions can be explicitly written down.

The product density generating functional can be written down as

$$D_\Omega(v) = 1 + \sum_{s=1}^{\infty} \frac{1}{s!} \iint_\Omega \cdots \int f_s(x_1,x_2,\ldots,x_s)v(x_1)v(x_2)\ldots v(x_s) \quad dx_1dx_2..dx_s \tag{44}$$

and f_s can be obtained as the sth order functional derivative of D_Ω with respect to $v(x_1),v(x_2)\ldots.v(x_s)$ and it can also be shown that

$$D_\Omega(v) = \exp\left\{ \sum_{s=1}^{\infty} \frac{1}{s!} \iint_\Omega \cdots \int g_s(x_1,x_2,\ldots x_s)dx_1dx_2\ldots dx_s \right\} \tag{45}$$

where $g_s(x_1,x_2,\ldots x_s)$ is the sth order cluster function. The probability generating function $G(z) = \sum z^n P(n,t) = D_\Omega(z-1)$ and hence for the Furry distribution with $f_n(x_1,x_2,\ldots x_n) = n!f_1(x_1)f_1(x_2)\ldots.f_1(x_n)$ we have

$$G(z) = (1-z \int_\Omega f_1(x_1)dx_1)^{-1} \tag{46}$$

If a light beam incident on a photodector with sensitivity α ejects electrons proportional to the intensity $I(t)$ of the beam the product density of getting an electron in the time interval $(t,t+dt)$ is given by $f_1(t)dt = \langle \alpha I(t) \rangle dt$ where $\langle I(t) \rangle = \langle V(t)V^*(t) \rangle$ V being the analytic signal characteristic of the light which has a gaussian distribution for chaotic light. The product density generating functional

$$D_\Omega(v) = \left\langle e^{\alpha \int_0^T I(t)u(t)dt} \right\rangle \tag{47}$$

for this case. This, by (45), is given in terms of the connected parts as

$$D_\Omega(u) = \exp\left\{ \sum_{s=1}^{\infty} \frac{\alpha^s}{s!} \iint_0^T \cdots \int \langle I(t_1)I(t_2)\ldots I(t_s) \rangle_c dt_1dt_2..dt_s \right\} \tag{48}$$

The cluster part is given by

$$\alpha^s \langle I(t_1)I(t_2)\ldots I(t_s) \rangle_c = \Gamma^s(t_1,t_2,\ldots t_s) \tag{49}$$

where Γ^s is the s-th order coherence function. We have seen that the probability generating function $G(z)$ can be obtained by simply getting $D_\Omega(z-1) = G(z)$ and hence

$$G(z) = \exp \sum_{s=1}^{\infty} \frac{(z-1)^s}{s!} \iint_0^T \cdots \int \Gamma^s(t_1,t_2,\ldots,t_s)dt_1dt_2\ldots dt_s \tag{50}$$

$\Gamma^s(t_1,t_2,\ldots t_s)$ can be taken as $(s-1)!(\alpha IT)^s$ for the case of chaotic light, I being a constant. Then we have [11,17,18]

$$G(z) = (1+\alpha IT-\alpha zT)^{-1} \tag{51}$$

Hence inverting we get

$$P(n,T) = \frac{1}{(1+\bar{n})} \frac{1}{(1+\frac{1}{\bar{n}})^n} \tag{52}$$

which is a Boson distribution. For mixing of different types of light beams etc., the reader is referred to [17,18] .

Multiple product densities [19] where a twin or an i-tuple can constitute a point process with parametric values between x and x+dx can be formulated. In this case $f_1^i(x) = E[dN^i(x)]$ and $f_2^{i_1,i_2}(x_1,x_2) = E\, dN^{i_1}[(x_1)dN^{i_2}(x_2)]$ where f_1^i is the product density of first order for the occurrence of an i-tuple between x and x+dx and the mixed density $f_2^{i_1,i_2}(x_1,x_2)$ denotes the probability that there is an i_1-tuple in (x_1,x_1+dx_1) and an i_2-tuple in (x_2,x_2+dx_2). These are useful in the birth of pions etc., and in neutron multiplication problems [10,19,23].

A sequent product density such as $F_2(E_1,E_2;t_1,t_2|E_o,o)dE_1,dE_2$ represents the probability of occurrence of particles in the energy ranges (E_1,E_1+dE_1) and (E_2,E_2+dE_2) at times t_1 and t_2 respectively, given that there was a particle at t=0 with energy E_o. These find great use in cosmic ray shower theories. When $t_1 < t_2$

$$F_2(E_1,E_2;t_1,t_2|E_o) = \int f_2(E_1,E_2'|E_o,t_1)\, f_1(E_2|E_2';t_2-t_1)\, dE_2' + f_1(E_1|E_o;t_1)\, f_1(E_2|E_1,t_2-t_1) \tag{53}$$

and when $t_1 \rightarrow t_2$ the limiting case is represented as

$$F_2(E_1,E_2|t_2,t_2) = f_2(E_1,E_2;t_2) + f_1(E_1,t_2)\,\delta(E_2-E_1) \tag{54}$$

These were formulated in [20] and evolutionary product densities in product space Ω of both x and t have been widely used in cosmic ray theories [21].

Starting with probability generating function $F(u) = \sum_{n=o}^{\infty} u^n P(n)$ we can express

$$F(u) = \exp\left\{\sum_k (u^k-1)c_k\right\} \tag{55}$$

where c_k's are called 'combinants'. They are a measure of the deviation of the probability function P(n) from a Poisson. Assuming that $P(0) \neq 0$ for the Poisson case $F(u) = \exp[\bar{n}(u-1)]$. Hence only one combinant $c_1 = \bar{n}$ exists in this case. For a general correlated process [22]

$$\log F(u) - \log P(0) = \sum_{k=1}^{\infty} c_k u^k,\quad P(0) = \exp\left\{-\sum_{k=1}^{\infty} c_k\right\} \tag{56}$$

c_k's are expressible in terms of Pr's upto the same order r = k and conversely if c's are known upto any order probabilities upto that order can be found. Also it can be shown by using this concept and the relations in equations (59,56 etc.) that the cumulants of a probability distribution can be arrived at as

$$K_r = \sum_s c_s^r \tau_s \tag{57}$$

where τ_s are integrals over cluster functions

$$\tau_s = \iint \cdots \int g_s(x_1, x_2, \ldots, x_s) dx_1 dx_2 \ldots dx_s \tag{58}$$

Also we can show that

$$K_r = \sum_{\ell=1}^{\infty} \ell^r c_\ell \tag{59}$$

where c_ℓ's are the combinants. For more details see references [22,(a),(b)].

Markov Processes[1,23-27]

Let a physical system occupy one of the finite and discrete set of states $\{X_n\}$, n = 0,1,2,...etc. (which are random variables) at discrete set of times $\{n\Delta\}$. The sequence of $\{X_n\}$ constitute a Markov chain if the random variables X_{n_m} have a conditional probability which depend only on the value of the random variable at one step earlier only i.e.

$$\Pr\left\{X_{n_m} = x_{n_m} \mid X_{n_{m-1}} = x_{n_{m-1}}, X_{n_{m-2}} = x_{n_{m-2}}, \ldots, X_{n_o} = x_{n_o}\right\}$$

$$= \Pr\left\{X_{n_m} = x_{n_m} \mid X_{n_{m-1}} = x_{n_{m-1}}\right\} \tag{60}$$

This is a discrete one step Markov chain. Simplifying the notation let us consider $P_i(m)$ the probability that the system is in the state i at the time step m and formulating the one step transition probability by

$$P_{ij} = \Pr\left\{X_{m+1} = j \mid X_m = i\right\} = P_{ij}(1) \tag{61}$$

we can see that $P_{ij}(m)$, the transition probability for m steps can be found as

$$P_{ij}(m) = P_{ij}^m \tag{62}$$

Hence for a homogeneous Markov chain $\{X_n\}$ the (n+m) step transition probability satisfies the Chapman-Kolmogarov relation

$$P_{ij}(n+m) = \sum_k P_{ik}(n) P_{kj}(m) \tag{63}$$

Also $\sum_j P_{ij} = 1$. Such matrices with elements $[P_{ij}]$ summing upto unity along the rows (or columns) are called stochastic matrices.

A state is called a recurrent or persistent state if starting from it the ultimate return to it is a certainty. The time of first return to that state is a random variable called recurrence time. A state is called positive recurrent or null recurrent according as the mean recurrence time is finite or infinite. A periodic recurrent state is one which can be reached only at t^{th}, $2t^{th}$, $3t^{th}$....time steps.

A recurrent, nonnull, aperiodic state is called an ergodic state.

If we take an initial unconditional probability vector $\vec{\Pi}(0)$ for a finite state system and obtain the probability vector $\vec{\Pi}(n)$ at the n-th time step then we would have

$$\vec{\Pi}(n) = (\bar{P})^n \vec{\Pi}(0) \qquad (64)$$

where $\bar{P}_{ij} = P_{ji}$ i.e. $\bar{P}$ is the transpose of the transition probability matrix $[P_{ij}]$ described above. If $\vec{\Pi}(n) \underset{n\to\infty}{\longrightarrow} \vec{\Pi}$ independent of n then the chain is ergodic. Hence an irreducible and aperiodic Markov chain is ergodic if we can find a non-null solution of the equation

$$\vec{X} = \bar{P}\vec{X} \text{ for } \sum_i |X_i| < \infty \qquad (65)$$

Markov processes in continuous time like the Poisson process $P(n,t)$ or the birth-death processes represent evolution processes without memory. Given the state of the system at time t the development of the system between t and $t+\Delta$ depends only on the state of the system at time t and independent of the states prior to t. This is the probabilistic analogue of the deterministic trajectories in mechanics. An elementary one dimensional random walk where the state at n-th step $Y(t_n) = \sum_{i=1} X_i$ with X_i's as independent random variables corresponding to the sizes of the steps $\{i\}$, is a Markov process.

Let us take discrete states and continuous time and consider the transition probability, $\Pi(X_i|X_k,t)$, that the system goes to the state X_i from the state X_k in the time interval $(0,t)$. Then by the Chapman-Kolmogorov relation

$$\Pi(X_i|X_k;t+\Delta) = \sum_j \Pi(X_j|X_k;t)\,\Pi(X_i|X_j;\Delta) \qquad (66)$$

Let us further assume that $\lim_{\Delta\to 0} \Pi(X_i\ X_j;\Delta) = R_{ij}\Delta$. Hence taking account of the probabilities that can exhaust the possibilities that can happen in the interval Δ we have

$$\Pi(X_i|X_k;t+\Delta) = (1-\sum_{j\neq i} R_{ji}\Delta)\,\Pi(X_i|X_k;t) + \sum_j \Pi(X_j|X_k;t)R_{ij}\Delta \qquad (67)$$

In the limit $\Delta \longrightarrow 0$, this leads us to the forward differential equation

$$\frac{\partial\Pi(X_i|X_k;t)}{\partial t} = -\sum_{j\neq i} R_{ji}\,\Pi(X_i|X_k;t) + \sum_{j\neq i} R_{ij}\,\Pi(X_j|X_k;t) \qquad (68)$$

By relating $\Pi(X_i|X_j;t)$ to $\Pi(X_i|X_k;t-\Delta)$ taking into consideration to possibilities in the first interval $(0,\Delta)$ the backward differential equation can be obtained as

$$\frac{\partial\Pi(X_i|X_k;t)}{dt} = -\Pi(X_i|X_k;t)\sum_{j\neq k} R_{jk} + \sum_{j\neq k}\Pi(X_i|X_j;t)R_{jk} \qquad (69)$$

It is an interesting tool to derive equations governing the evolution of the moment generating function for a Markov continuous process. This can be done in a genera-

generalized fashion. The moment generating function at time $t+\Delta t$ for the $X(t)$ process is

$$M(\theta,t+\Delta t) = \underset{t+\Delta t}{E} \left\{ \exp\left\{(\theta X(t+\Delta t))\right\}\right\} \tag{70}$$

where $E_{t+\Delta t}$ is the expectation with respect to the random variable $X(t+\Delta t)$. Let us call E_t, the expectation with respect to the variable $X(t)$, and $E_{\Delta t/t}$ the expectation of the variable $X(t)$ at the end of time Δt due to fluctuations in ΔX given that at time t the variable has a value $X(t)$. Then

$$M(\theta,t+\Delta t) = \underset{t}{E}\left\{ \exp(\theta X(t) \underset{\Delta t/t}{E} \left[\exp(\theta \Delta X(t))\right]\right\} \tag{71}$$

Hence

$$\lim_{\Delta t\to o} \frac{M(\theta,t+\Delta t)-M(\theta,t)}{\Delta t} = \underset{t}{E}\left\{ \exp(\theta X(t)) \lim_{\Delta t\to o} \underset{\Delta t/t}{E} \left[\exp\ (\theta\Delta X(t))-1\right]\right\} \tag{72}$$

and hence

$$\frac{\partial M}{\partial t} = \underset{t}{E}\left\{\exp(\theta X(t))\ \psi(\theta,t,X)\right\} = \psi(\theta,t,\frac{\partial}{\partial\theta})M(\theta,t) \tag{73}$$

where $\psi(\theta,t,X) = \underset{\Delta t/t}{E} \left\{\exp\ (\theta\Delta X(t))-1\right\}$

For a finite state process in continuous time let

$$\Pr\left\{\Delta X(t) = j|X(t)\right\} = f_j(X)\,\Delta t;\ j \neq o$$

$$\Pr\left\{\Delta X(t) = 0|X(t)\right\} = 1-\sum_j f_j(X)\,\Delta t \tag{74}$$

Hence $\psi(\theta,t,X) = \sum_{j\neq o} f_j(X)\left[\exp\ (\theta_j)-1\right]$. Thus for a pure birth and death process with $f_{+1} = \lambda X$ and $f_{-1} = \mu X$ we have the partial differential equations for M as

$$\frac{\partial M(\theta,t)}{\partial t} = \left[\lambda(\exp[\theta]-1) + \mu(\exp[-\theta]-1)\right]\frac{\partial M(\theta,t)}{\partial\theta} \tag{75}$$

with $M(\theta,o) = \exp\ (a\theta)$ if $X(0) = a$. This equation can be arrived at also by the usual Chapman-Kolmogorov techniques.

If we take continuous set of states and continuous time

$$P(y_3|y_1;t+\tau) = \int_\Omega P(y_3|y_2;\tau)\cdot P(y_2|y_1;t)dy_2 \tag{76}$$

If for small τ we take $P(y_2|y_1;\tau) = (1-a_o(y_1)\tau)\ \delta(y_2-y_1)+\tau W(y_2|y_1)$ where $a_o(y_1) = \int_\Omega W(y_2|y_1)dy_2$ we have the socalled Master equation which after suppressing the intial state symbol y runs as

$$\frac{\partial P(y,t)}{\partial t} = \int_\Omega\left[W(y|y')P(y';t) - W(y'|y)P(y;t)\right]dy' \tag{77}$$

The solution of these types of equations have been dealt with in [26].

Product densities and electromagnetic cascades:

There are two important processes which are key factors in the development of the shower of electrons and photons: (1) Bremstrahlung process in which an electron of energy E_o emits a photon of energy (E_o-E') and proceeds with energy E' in its passage through a medium and the cross section for such a process is $R_1(E'|E_o)dE'$ per unit thickness; (2) Pair production process in which a photon of energy E_o produces an electron of energy E' and a positron with the rest of the energy. The cross section or probability for such a process is $R_2(E'|E_o)\ dE'$ per unit depth. With an initial particle we use the method of invariant imbedding [29] to obtain a complete set of equations for the product densities for finding an electron or a photon of given energy at a given depth. The philosophy of this method is to study what happens at the initial depth Δ and then to express the process by the same function for the rest of the depth $t-\Delta$ with new initial conditions. Thus we imbed the original problem in a class of similar problems leading to functional equations.

To this end we define first order densities $\{f_1^i(E,t|E_o)\}$ which denote the probability that there is an electron with energy in the range $(E,E+dE)$ at a depth t, the shower being initiated by a particle i with energy E_o. If $i = 1$ it is an electron initiated shower and $i = 2$ means a photon initiated shower. The imbedding equations are

$$\frac{\partial f_1^i(E,t|E_o)}{\partial t} = - f_1^1(E,t|E_o)\int_0^{E_o} R_1(E'|E_o)dE' + \int_0^{E_o} R_1(E'|E_o)\,[f_1^1(E,t|E') + f_1^2(E,t|E_o-E')]\ dE'$$

$$\frac{\partial f_1^2(E,t|E_o)}{\partial t} = 2\int_0^{E_o} R_2(E'|E_o)\ f_1^1(E,t|E')dE' - f_1^2(E,t|E_o)\int_0^{E_o} R_2(E'|E_o)dE' \qquad (78)$$

With proper initial conditions these equations can be solved using Mellins transforms. We can also formulate equations for higher order product densities. In every order one gets after taking the transforms, an equation of the type

$$\frac{d}{dt}\vec{F}_n = [A]\,\vec{F}_n + \vec{\Psi} \qquad (79)$$

By this imbedding approach one always gets A to be a matrix of dimensions 2 X 2 and $\vec{\Psi}$ is also two component vector. Thus the dimensionality of the problem is very much reduced [30]. In the usual methods [28a-d] A will be a matrix of order $2^n x 2^n$, etc. Also there are no mixed densities as those occuring in earlier methods. These equations describe the evolution of the correlations at $t+\Delta$ in terms of the spectrum at time t and do not yield higher correlations at later times. In this sense these equations, though of the same type as the Chapman-Kolmogorov equations, can be called sub-markovian. Many more processes of this type of branching processes can be treated by imbedding methods [28b,1].

Fokker-Planck equations[1,23,24,26,31]

Let us start with Chapman-Kolmogorov equations for the continuous process in continuous time. Then

$$P(x|y;t+\Delta) = \int_{-\infty}^{\infty} P(z\ y;t)\ P(x|z,\Delta)dz \tag{80}$$

and calling $\theta(u,z) = \int \exp[iu(x-z)]\ P(x|z;\Delta)dx$ we can rewrite (80) as

$$P(x|y;t+\Delta) = \frac{1}{2\pi}\iint \exp[-iu(x-z)]\ \theta(u,z)\ P(z|y;t)dzdu \tag{81}$$

We call the conditional moments $m_s = E[(x-z)^s]_{P(x|z;\Delta)}$ and note that

$\frac{1}{2\pi}\int \exp[-iu(x-z)]du(iu)^s = \left(-\frac{\partial}{\partial x}\right)^s \delta(x-z)$. We can expand the r.h.s of (81) in terms of m_s since $\theta(u,z) = 1+\sum_{s=1}^{\infty}\frac{(iu)^s}{s!}m_s$. Then we find that

$$P(x|y;t+\Delta) = P(x|y;t) + \sum_{s=1}^{\infty}\left(-\frac{\partial}{\partial x}\right)^s \frac{1}{s!}[m_s(x)P(x|y;t)] \tag{82}$$

If $m_s = k_s\Delta$ for each s, we obtain as $\Delta \longrightarrow o$ the following generalised Fokker-Planck equation

$$\frac{\partial P}{\partial t} = -\frac{\partial}{\partial x}[k_1(x)P(x|y;t)] + \frac{1}{2!}\frac{\partial^2}{\partial x^2}[k_2(x)P(x|y;t)] + \frac{1}{3!}\frac{\partial^3(k_3P)}{\partial x^3} + \ldots \tag{83}$$

For white noise, jump moments higher than k_2 are zero. Then if $k_1 = \beta x$ and $k_2 = 2D$ we get the equation

$$\frac{\partial P}{\partial t} = \beta\frac{\partial}{\partial x}(xP) + D\frac{\partial^2 P}{\partial x^2} \tag{84}$$

the solution of which is the Uhlenbech-Ornstein process with

$$P(x|y;t) = \frac{1}{\sqrt{2\pi\sigma^2}}\exp[-(x-\bar{y})^2/2\sigma^2] \tag{85}$$

with $\bar{y} = y\exp(-\beta t)$ and $\sigma^2 = (D/\beta)(1-\exp[-2\beta t])$. Instead of (83) we can also get the backward Fokker-Planck differential equation

$$\frac{\partial P(x\ y;t)}{\partial t} = K_1(y)\frac{\partial P(x|y;t)}{dy} + \frac{1}{2}k_2(y)\frac{\partial^2 P(x|y;t)}{\partial y^2} \tag{86}$$

If the Fokker-Planck equation has boundaries a and b and if $b = -\infty$, we can define the first passage time T of the process to reach the boundary a as [27(a)]

$$T = \inf\{t \mid x(t) > a\} \tag{87}$$

The distribution function for the first passage time $T(X_o)$ is

$$G(x_o;a;t) = \Pr\{T(x_o) \leqslant t\} \tag{88}$$

and $\frac{dG}{dt} = g(x_o;a;t)$ the density function for the first passage time to be in (t,t+dt). Let $f(x_o;x;t)$ be the unbounded solution of the Fokker-Planck equations, which is the density function for the particle to lie in the interval (x,x+dx) at time in (t,t+dt) starting from initial value x_0. If $f^*(x_0;x,\theta)$ and $g^*(x_0;a;\theta)$ are Laplace transforms with respect to time t and if $x_0 < a < x$ we have [16,23].

$$f(x_0;x;t) = \int_0^t g(x_0;a;\tau)\, f(a;x;t-\tau)\,d\tau \tag{89}$$

and

$$g^*(x_0;a;\theta) = f^*(x_0;x;\theta)/f^*(a;x;\theta) \tag{90}$$

First passage times for jump processes can be found for special types of exponential jumps by introducing compensation functions (to take care of the boundaries) in the equation for the transition probability $f(x;t|x_0)$ and treating the equation as free boundary problem. These lead to interesting results for the first passage densities in terms of the free Green's functions. These methods can also be applied to moving boundaries (For details cf.[32-34/27].

References

1. A. Ramakrishnan, Handbuch der Physik III/2, 413 (Springer-Verlag, 1959).
2. M.S. Bartlett, "An introduction to stochastic processes" Cambridge University Press, (1955).
3. A. Papoulis, "Probability, random variables and stochastic processes" McGraw Hill, New York.
4. Yu.A. Prophorov and Yu.A. Rozanov, "Probability Theory" Springer-Verlag (1969).
5. E. Lukas, "Characteristic functions", Griffin Monographs No.5, London (1960).
6. S.K. Srinivasan and K.M. Mehta, "Probability and random processes" end Edn. Tata McGraw Hill, New Delhi, (1981).
7. W. Feller, "An introduction to probability theory and applications" Vols. I and II, John Wiley, New York (1972).
8. N.J. Bailey, "The elements of stochastic processes" John Wiley, New York, (1964).
9. D.L. Snyder, "Random point processes", Wiley Interscience, New York (1975).
10. B. Saleh, "Photoelectron statistics" Springer Series in Optical Sciences, (1978).
11. S.K. Srinivasan, "Stochastic point processes", Griffin statistical monographs, London (1974).
12. A. Ramakrishnan, Proc. Camb. Phil. Soc. 46, (1950) 595.
13. D.G. Kendall, J. Roy. Stat. Soc. B11 (1949) 230.
14. J. Yuon, "Fluctuations en densite actualites scientifique et industriells" Paris.
15. P.L. Kuznetzov, R.L. Stratanovitch and V.I. Tikhonov, "Nonlinear transformations of stochastic processes" Pergamon press, London (1965) Vol.I.
16. R.L. Stratonovich "Topics in the theory of random noise" Gorden Breach, New York (1963).
17. R. Vasudevan and S.K. Srinivasan, Nuovo Cimento Ser. V 47 (1967) 183.
18. R. Vasudevan and S.K. Srinivasan, Nuovo Cimento Ser. B 8 (1972) 278.
19. A. Ramakrishnan and S.K. Srinivasan, Bull. Math. Biophys. 20 (1958) 288.
20. A. Ramakrishnan and T.K. Radha, Proc. Camb. Phil. Soc. 57 (1961) 843.
21. S.K. Srinivasan, "Stochastic theory and Cascade processes", Elsivier, New York (1969).
22a. R. Vasudevan, P.R. Vittal and K.V. Parthasarathy (to be published) Matscience Preprint.
22b. Kauffman S.K. and Gyulassy M., J. Phy A (1979) 11 p.p 715
23. A.T. Barucha Reid, "Elements of the theory of Markov processes" McGraw Hill, New York (1960).
24. Nelson Wax, "Selected papers on noise and stochastic processes" Dover, New York (1956).
25. N.V. Prabhu, "Stochastic processes", John Wiley, New York (1970).
26. N.V. van Kampen, "Stochastic processes in physics and chemistry" North Holland (1981).

27a. D.R. Cox and H.D. Miller, "Theory of stochastic processes" 2nd edn., Metheun, London (1965).
27b. T.E. Harris, "Theory of branching processes", Springer-Verlag (1963).
28a. S.K. Srinivasan, "Stochastic theory and Cascade processes" Elsivier, New York (1969).
b. A. Ramakrishnan, "Elementary particles and Cosmic rays" Pergamon (1962).
c. A. Ramakrishnan and S.K. Srinivasan, "Proc. Ind. Acad. Sci. A 44 (1956) 26.
d. A. RAmakrishnan and P.M. Mathres, Prog. Theor. Phys. 11 (1954) 95.
29. R.E. Bellman and M. Wing, "Introduction to the theory of invariant imbedding", Acad. Press (1976).
30. R. Vasudevan, A. RAmakrishnan and S.K. Srinivasan, J. Math. Anal. Appl. 11 (1965) 278.
31. R. Vasudevan and S.K. Srinivasan, "An introduction to the theory of random differential equations" Elsivier, New York (1971).
32. J. Kielson, "Green's function methods in probability theory" Griffin Monograph London (1965).
33. R. Vasudevan and P.R. Vittal, J. Theor. Neurobiology 1 (1982) 219.
34. R. Vasudevan and P.R. Vittal, J. Neurological Research Vol.4 (1982) 1/2, p.63-87.

GAUSSIAN STOCHASTIC PROCESSES

S. Chaturvedi
School of Physics, University of Hyderabad
Hyderabad - 500134, India

Introduction

Among all possible stochastic processes, Gaussian stochastic processes constitute a very important class. These occur in many areas of physics. A historically important example of a Gaussian stochastic process is that of Brownian motion. The intensity of the light emitted by a thermal source is another example of such a process. The main reason why Gaussian stochastic processes have been studied so extensively is that they are completely specified by the first two moments. This makes them particularly easy to handle.

We shall begin our discussion on Gaussian stochastic processes by studying Gaussian random variables. This, as we shall see later, will enable us to define Gaussian stochastic processes and to discuss some of their important properties.

Gaussian Random variables

Let us briefly recapitulate what a Gaussian random variable is.

A random variable X is defined by specifying

(i) the range of values x it can take and

(ii) a probability distribution over this range.

A random variable is said to be Gaussian if the range of values it can take extends from $-\infty$ to $+\infty$ and if the probability distribution over this range is a Gaussian distribution

$$P(x) = \frac{1}{\sqrt{2\pi}\sigma} \exp\left(-\frac{(x-\langle X\rangle)^2}{2\sigma^2}\right) \tag{1}$$

If instead of a single variable we have a vector $\underline{X}$ having n components then (1) generalizes to

$$P(x_1 \ldots x_n) = \frac{(\det A)^{\frac{1}{2}}}{(2\pi)^{n/2}} \exp\left[-(\underline{x}-\langle\underline{X}\rangle)^T A(\underline{x}-\langle\underline{X}\rangle)\right] \tag{2}$$

where A is a positive definite symmetric matrix. The probability distribution (2) is known as a multivariate Gaussian distribution. We shall now discuss some of its properties.

(a) Let us first check that $\langle\underline{X}\rangle$ appearing on the R.H.S. of (2) is indeed the mean value of $\underline{X}$ and that the distribution is correctly normalised.

$$\langle\underline{X}\rangle = \frac{(\det A)^{\frac{1}{2}}}{(2\pi)^{n/2}} \int_{-\infty}^{\infty} d\underline{x}(\underline{x}) \exp\left[-\frac{1}{2}(\underline{x}-\langle\underline{X}\rangle)^T A\,(\underline{x}-\langle\underline{X}\rangle)\right]$$

$$= \frac{(\det A)^{\frac{1}{2}}}{(2\pi)^{n/2}} \int_{-\infty}^{\infty} d\underline{y}\left[\underline{y}+\langle\underline{X}\rangle\right] \exp\left[-\frac{1}{2}\underline{y}^T A\,\underline{y}\right]$$

Since A is a symmetric matrix, it can be diagonalized by an orthogonal matrix.

$$\Lambda = S^T A S \; ; \; S^T S = 1$$

Putting $\underline{y} = S\underline{z}$ in the expression above and making use of the fact that the Jacobian of the transformation is unity we obtain

$$\langle \underline{X} \rangle = \frac{(\det A)^{\frac{1}{2}}}{(2\pi)^{n/2}} \int_{-\infty}^{\infty} d\underline{z} \left[S\,\underline{z} + \langle \underline{X} \rangle \right] \exp\left[-\frac{1}{2} \underline{z}^T \Lambda\, \underline{z} \right]$$

The first term, being odd in $\underline{z}$, vanishes so that

$$\langle \underline{X} \rangle = \langle \underline{X} \rangle \frac{(\det A)^{\frac{1}{2}}}{(2\pi)^{n/2}} \int_{-\infty}^{\infty} dz_1 \ldots dz_n \exp\left[-\frac{1}{2} \sum_{i=1}^{n} \Lambda_i z_i^2 \right]$$

$$= \langle \underline{X} \rangle \frac{(\det A)^{\frac{1}{2}}}{(2\pi)^{n/2}} \frac{(2\pi)^{n/2}}{(\Lambda_1 \ldots \Lambda_n)^{\frac{1}{2}}} = \langle \underline{X} \rangle$$

Here Λ_i are the eigenvalues of A .

This also shows that (2) is correctly normalised.

(b) We now want to show that

$$\langle (X_i - \langle X_i \rangle)(X_j - \langle X_j \rangle) \rangle = (A^{-1})_{ij} \qquad (3)$$

$$\langle (X_i - \langle X_i \rangle)(X_j - \langle X_j \rangle) \rangle = \frac{(\det A)^{\frac{1}{2}}}{(2\pi)^{n/2}} \int_{-\infty}^{\infty} d\underline{y}\, (y_i y_j) \exp\left[-\frac{1}{2} \underline{y}^T A\, \underline{y} \right]$$

$$= \frac{(\det A)^{\frac{1}{2}}}{(2\pi)^{n/2}} \int_{-\infty}^{\infty} d\underline{z}\, (S\underline{z})_i (S\underline{z})_j \exp\left[-\frac{1}{2} \underline{z}^T \Lambda\, \underline{z} \right]$$

$$= \frac{(\det A)^{\frac{1}{2}}}{(2\pi)^{n/2}} \sum_{\alpha\beta} S_{i\alpha} S_{j\beta} \int_{-\infty}^{\infty} d\underline{z}\, z_\alpha z_\beta \exp\left[-\frac{1}{2} \sum_{i=1}^{n} \Lambda_i z_i^2 \right]$$

$$= \sum_{\alpha\beta} S_{i\alpha} S_{j\beta}\, \delta_{\alpha\beta} \frac{1}{\Lambda_\alpha} = (S \Lambda^{-1} S^T)_{ij} = (A^{-1})_{ij}$$

We thus see that a Gaussian distribution is completely determined by the mean values of the variables and the second moments.

Henceforth we shall assume for convenience that the variables X_i have zero mean i.e. $\langle \underline{X} \rangle = 0$. All the results for the case $\langle \underline{X} \rangle \neq 0$ can easily be obtained by replacing $\underline{X}$ in the following by $\underline{X} - \langle \underline{X} \rangle$.

(c) We now establish a very useful property of multivariate Gaussian distribution

$$\langle X_i f(\underline{X}) \rangle = \sum_j \langle X_i X_j \rangle \left\langle \frac{\partial f(\underline{X})}{\partial X_j} \right\rangle \qquad (4)$$

where $f(\underline{X})$ is a polynomial in the X_i's.

To prove (4) we rewrite it using (3) as

$$\langle X_i f(\underline{X})\rangle = \sum_j (A^{-1})_{ij} \langle \frac{\partial f(\underline{X})}{\partial X_j}\rangle$$

or

$$\sum_j A_{ij} \langle X_j f(\underline{X})\rangle = \langle \frac{\partial f(\underline{X})}{\partial X_i}\rangle$$

Now

$$\sum_j A_{ij} \langle X_j f(\underline{X})\rangle = \frac{(\det A)^{\frac{1}{2}}}{(2\pi)^{n/2}} \int_{-\infty}^{\infty} d\underline{x}\ f(\underline{x}) \sum_j A_{ij}\ x_j \exp[-\frac{1}{2} \sum_{l\ m} x_l A_{lm} x_m]$$

$$= -\frac{(\det A)^{\frac{1}{2}}}{(2\pi)^{n/2}} \int_{-\infty}^{\infty} d\underline{x}\ f(\underline{x}) \frac{\partial}{\partial x_i} \exp[-\frac{1}{2}\underline{x}^T A \underline{x}]$$

and on integrating by parts

$$= \langle \frac{\partial f(X)}{\partial X_i}\rangle$$

Hence the proof.

(d) Repeated use of (4) enables us to show that all the even moments of a Gaussian distribution with zero mean factorize pairwise into the second moments. (The odd moments of such a distribution of course vanish as is easily seen.) Consider for instance the fourth moment $\langle X_i X_j X_k X_l\rangle$. From (4) we have

$$\langle X_i X_j X_k X_l\rangle = \sum_\alpha \langle X_i X_\alpha\rangle \langle \frac{\partial}{\partial X_\alpha} (X_j X_k X_l)\rangle$$

$$= \langle X_i X_j\rangle\langle X_k X_l\rangle + \langle X_i X_k\rangle\langle X_j X_l\rangle + \langle X_i X_l\rangle\langle X_j X_k\rangle \qquad (5)$$

The R.H.S. of (5) can be written compactly as

$$= \sum_{\text{pairs}} \langle X_p X_q\rangle\langle X_r X_s\rangle$$

where the indices p,q,r,s etc. are the same as i,j,k,l and the summation extends over all different ways in which i,j,k,l can be divided into pairs.

Proceeding in a similar fashion, we have, in general

$$\langle X_i X_j X_k X_l \cdots\cdots\rangle = \sum_{\text{pairs}} \langle X_p X_q\rangle\langle X_r X_s\rangle \cdots\cdots \qquad (6)$$

Thus the even moments of a Gaussian with zero mean factorise as in (6). For a moment of order 2k, there are $(2k!/2^k\ k!)$ terms on the R.H.S. of (6). Conversely one can show that if the moments of a probability distribution factorise as in (6) then the distribution is a Gaussian. It therefore follows that (6) is both necessary and sufficient for a distribution to be a Gaussian and is called the moment theorem for Gaussian distributions.

(e) A convenient way of calculating the moments of a probability function is to work out its characteristic function

$$C(\underline{h}) = \langle \exp(i\, \underline{h}.\underline{X}) \rangle$$

$$= \sum_{m_i = 0}^{\infty} \frac{(ih_1)^{m_1}}{m_1!} \frac{(ih_2)^{m_2}}{m_2!} \;\ldots\ldots \langle X_1^{m_1} \; X_2^{m_2} \;\ldots\ldots \rangle \qquad (7)$$

Given the characteristic function, an arbitrary moment can be worked out by differentiating it an appropriate number of times w.r.t. the h_i's and then setting $\underline{h} = 0$. For Gaussian distribution with zero mean $C(\underline{h})$ has a very simple form

$$C(\underline{h}) = \exp[-\tfrac{1}{2}\, \underline{h}^T A^{-1} \underline{h}] \qquad (8)$$

as is easily seen:

$$C(\underline{h}) = \frac{(\det A)^{\frac{1}{2}}}{(2\pi)^{n/2}} \int d\underline{x} \;\; \exp[-\tfrac{1}{2}\, \underline{x}^T A \underline{x} + \tfrac{i}{2}\, \underline{h}^T \underline{x} + \tfrac{i}{2}\, \underline{x}^T \underline{h}]$$

$$= \frac{(\det A)^{\frac{1}{2}}}{(2\pi)^{n/2}} \int d\underline{x} \;\; \exp[-\tfrac{1}{2}(\underline{x} - iA^{-1}\underline{h})^T A (\underline{x} - iA^{-1}\underline{h}) - \tfrac{1}{2}\, \underline{h}^T A^{-1} \underline{h}]$$

$$= \exp(-\tfrac{1}{2}\, \underline{h}^T A^{-1} \underline{h}) \frac{(\det A)^{\frac{1}{2}}}{(2\pi)^{n/2}} \int d\underline{y} \exp[-\tfrac{1}{2}\, \underline{y}^T A\, \underline{y}]$$

$$= \exp(-\tfrac{1}{2}\, \underline{h}^T A^{-1} \underline{h})$$

Using (3) we may write (8) as

$$C(\underline{h}) = \exp[-\tfrac{1}{2} \sum_{ij} h_i \langle X_i X_j \rangle h_j] \qquad (9)$$

Similarly for a Gaussian with $\langle \underline{X} \rangle \neq 0$, $C(\underline{h})$ is found to be

$$C(\underline{h}) = \exp[-\tfrac{1}{2}\, \underline{h}^T A^{-1} \underline{h} + i\, \underline{h}^T \langle \underline{X} \rangle]$$

$$= \exp[i \sum_i h_i \langle X_i \rangle - \tfrac{1}{2} \sum_{ij} h_i \langle\langle X_i X_j \rangle\rangle h_j] \qquad (10)$$

Another useful quantity is the logaritham of the characteristic function - the cumulant generating function:

$$K(\underline{h}) = \ln C(\underline{h})$$

$$= \sum_{m_i = 0}^{\infty} \frac{(ih_1)^{m_1}}{m_1!} \frac{(ih_2)^{m_2}}{m_2!} \;\ldots\ldots \langle\langle X_1^{m_1} \; X_2^{m_2} \;\ldots\ldots \rangle\rangle \qquad (11)$$

For a Gaussian we find that the cumulant generating function

$$K(\underline{h}) = i\, \underline{h}^T \langle \underline{X} \rangle - \tfrac{1}{2}\, \underline{h}^T A^{-1} \underline{h}$$

$$= i \sum_i h_i \langle X_i \rangle - \tfrac{1}{2} \sum_{ij} h_i \langle\langle X_i X_j \rangle\rangle h_j \qquad (12)$$

is at most quadratic in the auxiliary variables h_i's and hence all cumulants higher than the second vanish.

We mention here a theorem due to Marcinkiewicz [1-3] which states that the cumulant generating function of a probability distribution can not be a polynomial in the h_i's of degree greater than 2. In other words either $K(\underline{h})$ is at best quadratic in $\underline{h}$ or contains all of powers of $\underline{h}$. This in turn implies that either all but the first two cumulants of a probability distribution vanish or there are an infinite number of non vanishing cumulants.

With this background on Gaussian random variables we now go over to defining Gaussian stochastic processes.

Gaussian Stochastic Processes

A stochastic process is a function of two variables t, the time and Ω a random variable.

$$X_{\Omega}(t) \quad = \quad f(t,\Omega) \tag{13}$$

We may look upon (13) in two ways:

(i) For each value ω the random variable Ω takes, $X_{\omega}(t)$ is just an ordinary function of time and is called a realisation of the stochastic process $X_{\Omega}(t)$. The stochastic process is thus an ensemble of all such realisations.

(ii) For a fixed t, $X_{\Omega}(t)$ is a sotchastic variable - being a function of a random variable X. The stochastic process $X_{\Omega}(t)$ may be regarded as a continuum of random variables one for each t.

From the second point of view it therefore logically follows that in order to define a stochastic process completely we need to specify an infinite number of joint probabilities,

$$P_1(x,t)$$
$$P_2(x_2,t_2;x_1,t_1)$$
$$P_3(x_3,t_3;x_2,t_2;x_1,t_1)$$
$$\cdots\cdots\cdots\cdots\cdots\cdots$$

which say what is the probability that $X_{\Omega}(t)$ has a value x at t, what is the probability that $X_{\Omega}(t)$ has a value x_1 at t_1 and x_2 at t_2 etc. Given this infinite (over complete) set of joint probabilities the stochastic process is completely defined.

A stochastic process is said to be Gaussian if all these joint probabilities are Gaussian.

$$P_n(x_n,t_n; \;\ldots\ldots\; ; x_2,t_2; x_1,t_1)$$
$$= \frac{(\det A)^{\frac{1}{2}}}{(2\pi)^{n/2}} \exp\Big[-\frac{1}{2}\sum_{ij} (x_i-\langle X(t_i)\rangle)\, A_{ij}\, (x_j-\langle X(t_j)\rangle)\Big] \tag{14}$$

where the matrix A_{ij} is the inverse of the matrix A^{-1} with elements

$$(A^{-1})_{ij} \quad = \quad \langle (X(t_i)-\langle X(t_2)\rangle)(X(t_j)-\langle X(t_j)\rangle)\rangle$$
$$\equiv \quad \langle\langle X(t_i)\; X(t_j)\rangle\rangle \tag{15}$$

in analogy with (3). $\langle\langle X(t_i)\ X(t_j)\rangle\rangle$ is known as the auto correlation function.

Thus a Gaussian stochastic process is completely characterized by $\langle X(t)\rangle$ and the auto correlation function. All the formulae we had derived previously for Gaussian distributions can now be generalised to stochastic processes by replacing partial derivatives by functional derivatives, summation over i by integration over t etc. We list them below.

(a) Novikov's Theorem: For a functional f[X] of the stochastic process X(t) we have

$$\langle X(t)\ f[X]\rangle = \int dt'\ \langle X(t)\ X(t')\rangle \left\langle \frac{\delta f[X]}{\delta X(t')}\right\rangle \tag{16}$$

(b) Moment theorem:

For a Gaussian stochastic process with $\langle X(t)\rangle = 0$, the odd moments vanish and the even moments factorise pairwise.

$$\langle X(t_i)X(t_j)\ldots\rangle = \sum_{\text{pairs}} \langle X(t_p)\ X(t_q)\rangle \langle X(t_r)\ X(t_s)\rangle \ldots\ldots \tag{17}$$

(c) The characteristic functional

$$\begin{aligned} C\,[h] &= \langle \exp i \int_{-\infty}^{\infty} dt\ h(t)\ X(t)\rangle \\ &= \sum_{m=0}^{\infty} \frac{i^m}{m!} \int dt_1 \ldots \int dt_m\ h(t_1)\ldots h(t_m)\ \langle X(t_1)\ldots X(t_m)\rangle \end{aligned} \tag{18}$$

for a Gaussian stochastic process with zero mean is given by

$$C[h] = \exp\left[-\frac{1}{2} \int dt_1 \int dt_2\ h(t_1)\ h(t_2)\ \langle X(t_1)\ X(t_2)\rangle\right] \tag{19}$$

and for $\langle X(t)\rangle \neq 0$ by

$$C[h] = \exp\left[\ i\int dt_1 h(t_1) \langle X(t_1)\rangle - \frac{1}{2}\int dt_1 \int dt_2 h(t_1) h(t_2) \langle\langle X(t_1) X(t_2)\rangle\rangle\right] \tag{20}$$

where

$$\langle\langle X(t_1)\ X(t_2)\rangle\rangle = \langle (X(t_1) - \langle X(t_1)\rangle)(X(t_2) - \langle X(t_2)\rangle)\rangle \tag{21}$$

(d) The cumulant generating functional

$$\begin{aligned} K[h] &= \ln C[h] \\ &= \sum_{m=0}^{\infty} \frac{i^m}{m!} \int dt_1 \ldots \int dt_m\ h(t_1)\ldots h(t_m) \langle\langle X(t_1)\ldots X(t_m)\rangle\rangle \end{aligned} \tag{22}$$

for a Gaussian distribution with $\langle X(t)\rangle = 0$ reads

$$K[h] = -\frac{1}{2} \int dt_1 \int dt_2\ h(t_1)\ h(t_2) \langle x(t_1)\ x(t_2)\rangle \tag{23}$$

and for $\langle X(t)\rangle \neq 0$ as

$$K[h] = i\int dt_1 h(t_1) \langle X(t_1)\rangle - \frac{1}{2}\int dt_1 \int dt_2\ h(t_1) h(t_2) \langle\langle X(t_1) X(t_2)\rangle\rangle \tag{23a}$$

implying that all the cumulants of a Gaussian stochastic process higher than the second vanish. The Marcinkiewicz theorem holds for stochastic processes as well.

Of special interest to physicists and mathematicians are a class of stochastic process known as Markov processes [4]. A Markov process is fully determined by a single time distribution $P_1(x,t)$ and a conditional probability defined as

$$P(x_1,t_1|x_2t_2) \equiv \frac{P_2(x_1,t_1;x_2,t_2)}{P_1(x_2,t_2)} \tag{24}$$

satisfying

(i) the Chapman-Kolmogorov equation

$$P(x_3,t_3|x_1,t_1) = \int dx_2 P(x_3,t_3|x_2,t_2)\ P(x_2,t_2|x_1,t_1) \text{ for } t_3>t_2>t_1 \tag{25}$$

and

(ii) $$P_1(x_2,t_2) = \int dx_1\ P(x_2,t_2|x_1,t_1)\ P(x_1,t_1) \tag{26}$$

Another class of stochastic processes which are of special relevance to physics are the stationary processes. A stochastic process is stationary if all the joint probabilities are invariant under a shift in time.

$$P_n(x_n,t_n+\tau;\ \dots;x_2,t_2+\tau;\ x_1,t_1+\tau) = P_n(x_n,t_n;\dots;x_2,t_2;x_1,t_1) \tag{27}$$

This necessarily implies that the single time probability $P_1(x,t)$ is independent of time. Equation (27) in turn implies that

$$\langle X(t_n+\tau)\ \dots X(t_2+\tau)\ X_1(t_1+\tau)\rangle = \langle X(t_n)\ \dots\ X(t_2)X(t_1)\rangle \tag{28}$$

Having thus defined these three important classes of stochastic processes viz. Gaussian, Markovian and stationary stochastic processes, a natural question to ask is as follows: Among all Gaussian stochastic processes what characterizes those which have the additional attributes of being stationary and Markovian? The answer to this question is provided by Doob's Theorem:

A stationary Gaussian process is Markovian only if the auto correlation function is an exponential.

$$\langle\langle X(t_1)X(t_2)\rangle\rangle \propto \exp-\gamma|t_1-t_2| \tag{29}$$

(For a multicomponent stochastic process (29) is to be replaced by its obvious generalisation

$$\langle\langle \underline{X}(t_1)\underline{X}^T(t_2)\rangle\rangle \propto \exp-\Gamma|t_1-t_2| \tag{30}$$

where Γ is a constant matrix.)

We now briefly outline the proof of this important theorem.

For a Gaussian stochastic process the joint probabilities have the form given in (14). (For simplicity we shall consider Gaussian processes with $\langle X(t)\rangle = 0$). Substituting for $P_2(x_1,t_1;x_2,t_2)$ and $P_1(x_2,t_2)$ in (24), we find that the conditional probability for a Gaussian process has the following general form

$$P(x_1,t_1|x_2,t_2) = \frac{1}{\sqrt{2\pi\sigma^2(t_1)(1-\varrho^2(t_1,t_2))}}$$

$$\exp\left[-\frac{1}{2}\frac{1}{\sigma^2(t_1)(1-\varrho^2(t_1,t_2))}(x_1-\frac{\varrho(t_1,t_2)\sigma(t_1)}{\sigma(t_2)}x_2)^2\right] \quad (31)$$

where

$$\sigma^2(t) = \langle X(t)X(t)\rangle \quad (32)$$

and

$$\varrho(t_1,t_2) = \frac{\langle X(t_1)X(t_2)\rangle}{\sigma(t_1)\sigma(t_2)} \quad (33)$$

$\varrho(t_1,t_2)$ is known as the correlation coefficient.

From (31) it follows that the conditional average of $X(t)$ at time t_1 given that it had a value x_3 at time t_3

$$\langle X(t_1)\rangle_{X(t_3)=x_3} \equiv \int dx_1\, x_1\, P(x_1,t_1|x_3,t_3) \quad (34)$$

is given by

$$\langle X(t_1)\rangle_{X(t_3)=x_3} = \frac{\varrho(t_1,t_3)\,\sigma(t_1)}{\sigma(t_3)}\, x_3 \quad (35)$$

For a Gauss-Markov process, we have, on using (35) and the Chapman Kolmogorov equation (25)

$$\begin{aligned}\langle X(t_1)\rangle_{X(t_3)=x_3} &= dx_1\, x_1\, P(x_1,t_1\, x_3,t_3)\\ &= \iint dx_1 dx_2\, x_1\, P(x_1,t_1\,|\,x_2,t_2)\, P(x_2,t_2\,|\,x_3t_3); t_1 > t_2 > t_3\\ &= \frac{\varrho(t_1,t_2)\,\sigma(t_1)}{\sigma(t_2)}\int dx_2 x_2\, P(x_2,t_2\,|\,x_3,t_3)\\ &= \frac{\varrho(t_1,t_2)\,\sigma(t_1)}{\sigma(t_2)}\,\frac{\varrho(t_2,t_3)\,\sigma(t_2)}{\sigma(t_3)}\, x_3 \end{aligned} \quad (36)$$

From (35) and (36) we have

$$\varrho(t_1,t_3) = \varrho(t_1,t_2)\,\varrho(t_2,t_3) \quad . \quad t_1 \geqslant t_2 \geqslant t_3 \quad . \quad (37)$$

Thus we find that a necessary condition for a Gaussian process to be Markovian is that the correlation coefficients must satisfy (36). In fact this condition turns out to be both necessary and sufficient [5].

Let us now consider a stationary Gaussian process. Stationarity implies that $\sigma(t)$ is independent of time and that $\langle X(t_1)\, X(t_2)\rangle$ and hence $\varrho(t_1,t_2)$ depends only on t_1-t_2. From (36) it follows that for such a process to be Markovian we must have

$$\rho(t_1-t_3) = \rho(t_1-t_2)\,\rho(t_2-t_3)\ . \tag{38}$$

This functional equation is satisfied only if $\rho(t_1-t_3)$ is an exponential

$$\rho(t_1-t_3) = \exp - \gamma\,(t_1-t_3)$$

i.e.

$$\langle X(t_1)\;X(t_2)\rangle \propto \exp - |t_1-t_2| \tag{39}$$

Hence the proof.

Examples of Gaussian Stochastic Processes

In conclusion we list some important Gaussian stochastic processes which one frequently encounters in physics. It contains examples of Gaussian stochastic processes which have either the Markov property or stationarity or both or neither of them.

1. Gaussian White noise : The Gaussian "stochastic process" $\xi(t)$ characterized by

$$\langle \xi(t)\rangle = 0 \tag{40}$$

$$\langle \xi(t)\quad \xi(t')\rangle = \delta(t-t') \tag{41}$$

is usually referred to as Gaussian white noise. Such a stochastic process was first introduced by Langevin in the context of Brownian motion. Gaussian white noise is not a stochastic process in a strict mathematical sense. However, in physics it is often used as a model for very rapid fluctuations.

2. Wiener Process: Wiener process W(t) is an example of a Gaussian Markovian non-stationary stochastic process and is characterised by

$$\langle W(t)\rangle = 0 \tag{42}$$

$$\langle W(t)\;W(t')\rangle = \min\,(t,t') \tag{43}$$

That it is a non stationary process is clear from (43). From (43) it also follows that $\sigma^2(t)$ and $\rho(t_1,t_2)$ for this process are given by

$$\sigma^2(t) = \langle W(t)W(t)\rangle = t \tag{44}$$

$$\rho(t_1,t_2) = \frac{\langle W(t_1)W(t_2)\rangle}{\sigma(t_1)\sigma(t_2)} = \sqrt{\frac{t_2}{t_1}}\ ;\ t_1>t_2 \tag{45}$$

The single time probability $P_1(w,t)$ is therefore given by

$$P_1(w,t) = \frac{1}{\sqrt{2\pi t}}\exp\left[-\frac{w^2}{2t}\right] \tag{46}$$

Substituting from (44) and (45) in (31) we find that the conditional probability $P(w_1,t_1|w_2,t_2)$ for this process is given by

$$P(w_1,t_1|w_2,t_2) = \frac{1}{\sqrt{2\pi(t_1-t_2)}}\exp\left[-\frac{1}{2}\,\frac{(w_1-w_2)^2}{(t_1-t_2)}\right] \tag{47}$$

That this process is a Markov process is easily checked by verifying that $\rho(t_1,t_2)$ satisfies (37).

We can regard Wiener process as an integral of the Gaussian white noise

$$W(t) = \int_{-\infty}^{t} dt \;\xi(t) \tag{48}$$

in the sense that (48) together with (40) and (41) reproduce (42) and (43). We can also write (48) formally as a differential equation

$$\frac{dW}{dt} = \xi(t) \tag{49}$$

3. Ornstein Uhlenbeck Process: This is an example of a Gaussian Markovian stationary stochastic process. It is characterised by

$$\langle Y(t)\rangle = 0 \tag{50}$$

$$\langle Y(t_1)Y(t_2)\rangle = \exp[-|t_1-t_2|] \tag{51}$$

From (50) and (51) we can construct $P_1(y,t)$ and $P_1(y_1,t_1|y_2,t_2)$ just as in the case of Wiener process. These are given by

$$P_1(y,t) = \frac{1}{\sqrt{2\pi}} \exp\left[-\frac{1}{2}y^2\right] \tag{52}$$

$$P(y_1,t_1|y_2,t_2) = \frac{1}{\sqrt{2\pi(1-e^{-2(t_1-t_2)})}} \exp\left[-\frac{(y_1-y_2e^{-(t_1-t_2)})^2}{2(1-e^{-2(t_1-t_2)})}\right] \tag{53}$$

Again, as in the case of Wiener process, We may express the Ornstein Uhlenbeck process in terms of Gaussian white noise as follows

$$Y(t) = \int_{-\infty}^{t} dt' \; e^{-(t-t')} \;\xi(t') \tag{54}$$

in the sense that (54) reproduces (50) and (51). Equation (54) may be written as a differential equation

$$\frac{dY}{dt} = -Y + \xi(t). \tag{55}$$

4. An example of a Gaussian stochastic process which is stationary but non Markovian is easy to construct. Any Gaussian stationary process with a non exponential auto correlation function is, according to Doob's Theorem, non Markovian.

5. Finally an example of a Gaussian stochastic process which is neither Markovian nor stationary is the stochastic process defined to be the integral of Ornstein Uhlenbeck process

$$Z(t) = \int_{0}^{t} Y(t) \; dt' \tag{56}$$

For this process we can deduce using (50) and (51) that

$$\langle Z(t)\rangle = 0 \tag{57}$$

$$\langle Z(t_1)Z(t_2)\rangle = e^{-t_1}+e^{-t_2}-1-e^{-|t_1-t_2|}+2\min(t_1,t_2) \tag{58}$$

This process is not stationary as is evident from (58). It is also easy to check that (37) is not satisfied for this process and hence it is non Markovian.

We may write (55) as a differential equation

$$\frac{dZ}{dt} = Y(t) \tag{59}$$

where Y(t) obeys (54). With Z(t) and Y(t) identified with the position and velocity of a Brownian particle, (59) and (55) are the Langevin equation for a free Brownian particle. Although Z(t) is non Markovian Z(t) and Y(t) together constitute a Markov process.

The material presented above can be found in some form or another in any good text book on stochastic processes. See for instance [6] and [7].

References

1. J. Marcinkiewicz, Math. Z. 44, 612 (1939).
2. D.W. Robinson, Comm. Math. Phys. 1, 89 (1965).
3. A.K. Rajagopal and E.C.G. Sudarshan, Phys. REv. A10, 1852 (1974).
4. See Lectures by R. Vasudevan in these proceedings.
5. W. Feller, "Introduction to Probability Theory and its Applications", Vol.2 Wiley, New York (1966).
6. N.G. Van Kampen, "Stochastic Processes in Physics and Chemistry", North Holland, Amsterdam, New York, Oxford (1981).
7. C.W. Gardiner "A Hand book of Stochastic Methods for Physics, Chemistry and Natural Sciences". To be published, Springer (1983).

FOKKER-PLANCK EQUATIONS FOR STOCHASTIC PROCESSES

G.S. Agarwal
School of Physics, University of Hyderabad
Hyderabad-500134, INDIA

In this lecture I will discuss the various properties of the Fokker-Planck equations and the different methods, approximate and exact which are used to solve such equations.

1. Forward and Backward Equations for the Conditional Probability

Let $P(x,t|x_o,t_o)$ be the transition probability for a Markov process $x(t)$. This transition probability obeys CHAPMAN-KOLMOGOROV equation

$$P(x_3,t_3|x_1,t_1) = \int dx_2\, P(x_3,t_3|x_2,t_2)\, P(x_2,t_2|x_1,t_1),\; t_3 \geqslant t_2 \geqslant t_1 . \tag{1}$$

Using (1) , a dynamical equation for $P(x,t|x_o,t_o)$ can be obtained by introducing the conditional moments $M_n = \langle (x(t+\Delta t) - x(t))^n \rangle$ and by writing (1) as

$$P(x,t+\Delta t|x_o,t_o) = \int dx_1\, P(x,t+\Delta t\,|\,x_1,t)\, P(x_1,t\,|\,x_o,t_o) \tag{2}$$

and by using the formal relation

$$P(x,t+\Delta t\,|\,x_1,t) = \sum_{n=o}^{\infty} \frac{m_n(x_1)}{n!}\left(-\frac{\partial}{\partial x}\right)^n \delta(x-x_1) \tag{3}$$

The resulting dynamical equation, which is called KRAMER-MOYAL expansion is

$$\frac{\partial P}{\partial t} = \sum_1^{\infty}\left(-\frac{\partial}{\partial x}\right)^n [D_n(x)\, P(x,t|x_o,t_o)] , \tag{4}$$

where

$$D_n(x) = \lim_{t\to o} \frac{1}{tn!} \langle (x(t)-x)^n \rangle \tag{5}$$

Similarly by using CHAPMAN-KOLMOGOROV relation in the form

$$P(x,t|x_o,t_o-\Delta t_o) = \int dx_1 P(x,t|x_1 t_o)\, P(x_1,t_o|x_o,t_o-\Delta t_o) \tag{6}$$

and by expanding $P(x,t|x_1,t_o)$ in a Taylor series around x_o, we find the KOLMOGOROV or backward equation

$$-\frac{\partial P}{\partial t_o} = \sum_1^{\infty} D_n(x_o) \left(\frac{\partial}{\partial x_o}\right)^n P(x,t|x_o,t_o). \tag{7}$$

The backward equation (7) is the adjoint of the forward equation, with all the derivatives being taken with respect to x_o and t_o. The foregoing assumes that the limit (5) exists.

Let us now examine the meaning of the various terms in (4). Suppose that $D_n = 0 \;\forall\; n>1$, then (4) is a differential equation involving first order derivatives and hence

$$P(x,t_o|x_o,t_o) = \delta(x-x_o) \Rightarrow P(x,t|x_o,t_o) = \delta(x-x_o(t)),$$

$$\dot{x}_o(t) = D_1(x_o),\ x_o(t_o) = x_o. \tag{8}$$

In such a case the motion of the system is deterministic i.e. there are no fluctuations. The term D_1 is also known as the drift term. The fluctuations in the system arise due to the nonvanishing of the coefficients like D_2 etc. For example if $D_1 = 0$, D_2 = Constant and $D_n = o\ \forall\ n>2$, then (4) leads to

$$\langle x(t)\rangle = x_o,\quad \langle x^2(t)\rangle - x_o^2 = 2\,D_2 t, \tag{9}$$

which corresponds to the very familiar diffusion. Using the positiveness of P and the Schwarz inequality $(\int \psi(x)\phi(x)P(x)\,dx)^2 \leqslant (\int \psi(x)P(x)dx)(\int \phi^2(x)P(x)\,d\,x)$ and by choosing $\psi = (x(t+\tau)-x(t))^n$, $\phi = (x(t+\tau)-x(t))^{n+m}$, we prove the inequality

$$[(2n+m)!\ D_{2n+m}(x,t)]^2 \leqslant 2n!\ (2n+2m)!\ \ D_{2n}(x,t)D_{2n+2m}(x,t) \tag{10}$$

The inequality (10) can be shown to lead to the important result [1] - If KRAMER-MOYAL expansion terminates, then it can have at the most two terms D_1 and D_2. When KRAMER-MOYAL expansion terminates, then the resulting dynamical equation is called the FOKKER-PLANCK equation (FPE)

$$\frac{\partial P}{\partial t} = -\frac{\partial}{\partial x}[D_1(x)P] + \frac{\partial^2}{\partial x^2}[D_2(x)P], \tag{11}$$

whose multidimensional generalization is obviously

$$\frac{\partial P}{\partial t} = -\sum_i \frac{\partial}{\partial x_i}[D_i^{(1)}(\{x\})\,P] + \sum \frac{\partial^2}{\partial x_i \partial x_j}[D_{ij}^{(2)}(\{x\})P], \tag{12}$$

with

$$D_i^{(1)} = \lim_{t\to o} \frac{1}{t}\langle x_i(t)-x_i\rangle,\quad D_{ij}^{(2)}(x) = \lim_{t\to o}\frac{1}{2t}\langle (x_i(t)-x_i)(x_j(t)-x_j)\rangle \tag{13}$$

The diffusion matrix $D^{(2)}$ is obviously positive definite. The conservation of the probability follows form (12)

$$\frac{\partial P}{\partial t} + \vec{\nabla}.\vec{J} = 0,\quad J_i = D_i^{(1)}P - \frac{\partial}{\partial x_j}[D_{ij}^{(2)}P] \tag{14}$$

2. Examples of FPE

The FOKKER-PLANCK equations have been traditionally used in the context of the Brownian motion. However in the meanwhile, one has discovered that the behavior of such diverse systems as lasers [2], Josephson junctions [3], chemically reacting [4] species could be described by FPE. The Brownian motion of a particle in a potential V(x) is given by the dynamical equations

$$\dot{x} = p/m,\quad \dot{p} = -\frac{\partial V}{\partial x} - \gamma p + f(t) \tag{15}$$

where f(t) is a delta correlated Gaussian random process with zero mean

$$\langle f(t)\ f(t')\rangle = 2D\delta(t-t'), \tag{16}$$

The FPE corresponding to (15) is

$$\frac{\partial P(x,p,t)}{\partial t} = -\frac{\partial}{\partial x}\left(\frac{p}{m} P\right)+\frac{\partial}{\partial p}\left[\left(\gamma p+\frac{\partial v}{\partial x}\right)P\right] + D\frac{\partial^2 P}{\partial p^2}\ , \tag{17}$$

which in the limit of large damping leads to the SMOLUCHOWSKI equation for $P(x,t) = \int P(x,p,t)dp$

$$\frac{\partial P}{\partial t} = \frac{\partial}{\partial x}\left[\frac{V'}{m\gamma} P\right] + \frac{D}{m^2\gamma^2}\frac{\partial^2 P}{\partial x^2} \tag{18}$$

which in the case of a free particle reduces to the famous EINSTEIN diffusion equation

$$\frac{\partial P}{\partial t} = \left(\frac{D}{\gamma^2 m^2}\right)\frac{\partial^2 P}{\partial x^2}\ . \tag{19}$$

The microscopic theory [2] of the single mode laser operating in the threshold region shows that the distribution P of the laser photons is given by

$$\frac{\partial P}{\partial t} = -\frac{\partial}{\partial b}\left[(p-b^*b)bP\right] + 2\frac{\partial^2 P}{\partial b\partial b^*} + \text{c.c.}\ , \tag{20}$$

where b is the complex laser field amplitude and p is the pump parameter. A laser operating at threshold implies p = o. Very many other examples of the FPE and the contexts in which they arise will be discussed by other speakers in this school.

3. Solutions of the FPE

The FPE (12) is a partial differential equation of the parabolic type. We have to specify the boundary conditions on P(x,t) before one can solve even a one dimensional FPE. For a reflecting boundary one obviously should have the current vanish at the boundary i.e.

$$D_1 P - \frac{\partial}{\partial x}\left[D_2 P\right] = 0 \tag{21}$$

The case of an absorbing boundary is more involved. For the equation like (18), the condition at the absorbing boundary usually imposed is P = 0, though other forms have also been recently proposed [5]. In the case of natural boundaries at $\pm\infty$ one might impose $P(\pm\infty,t) = 0$. We now consider the various classes for which the FPE can be solved.

(a) $D_2 = 0$

In such a case as already discussed above, the conditional probability is exactly known provided the equations for the classical trajectories can be solved and then the probability distribution at time t will be

$$P(x,t) = \int dx_o\ P(x_o,t_o)\ \ \delta(x-x_o(t)) \tag{22}$$

Equation (22) yields information on how the initial fluctuations change with time. It may be added that this type of situation has arisen in a large number of diverse problems [6] such as lasers, superradiance, chamical reactions etc.

(b) $D_i^{(1)} = \sum a_{ij} x_j$, $D_{ij}^{(2)} =$ Quadratic function of x_i

In such a case it is obvious that the equations for the moments $\langle x_i(t) \rangle$, $\langle x_i(t) x_j(t) \rangle$ etc. form a closed set of linear equations and hence can be solved by standard matrix methods. Multiplicative stochastic processes, for example a randomly modulated harmonic oscillator, lead to the above type of drift and diffusion coefficients. Recently considerable progress has been made in the field of optical resonance in fluctuating fields [7] because of this realization.

(c) $D_i^{(1)} = \sum a_{ij} x_j$, $D_{ij}^{(2)} =$ Constant

This is a very important class of FPE. The underlying stochastic process is called the ORNSTEIN-UHLENBECK process. Brownian motion by a free particle, harmonic oscillator fall under this class. The FPE is exactly soluble. The solution, which has been obtained by the method of characteristics [8] , is

$$P[x,t|x_o,o] = [(2\pi)^N \det\sigma(t)]^{-\frac{1}{2}} \exp\left\{ -\frac{1}{2}\widetilde{(x-e^{-at}x_o)}\,\sigma^{-1}(t)\,(x-e^{-at}x_o) \right\}$$

$$\sigma_{ij}(t) = \langle (x_i(t) - \langle x_i(t)\rangle)(x_j(t) - \langle x_j(t)\rangle)\rangle \ ,$$

$$\sigma(t) = \sigma(\infty) - e^{-at}\sigma(\infty)\, e^{-\tilde{a}\, t},$$

$$a\sigma(\infty) + \sigma(\infty)\,\tilde{a} = 2D, \quad \langle x(t)\rangle = e^{-at}\, x_o. \tag{23}$$

The stochastic process in this case is Gaussian and markovian. The solution (23) has turned out to be of tremendous value in the study of the systems where $D^{(1)}$ is not a linear function of x_i but $D^{(2)} = g\, d^{(2)}$, where g is a infinitisimal parameter. In such a case one can calculate the behavior of the system to various orders in g. This is similar to VAN KAMPEN'S system size expansion [9] . We present the result to the lowest order in g. Write the stochastic process x(t) as

$$x(t) = x_o(t) + \sqrt{g}\; Y(t) + \ldots\ldots$$

$$\dot{x}_{oi} = D_i^{(1)}(\{x_o\}) \ , \quad x_o(0) = x_o \ ,$$

$$D^{(2)}(\{x\}) = g\, d^{(2)}(\{x\}) \approx g\, d^{(2)}(\{x_o\}) \ ,$$

$$D_i^{(1)}(\{x\}) = D_i^{(1)}(\{x_o\}) + \sqrt{g} \sum_j Y_j \frac{\partial}{\partial x_{oj}} D_i^{(1)}(\{x_o\}) + \ldots. \tag{24}$$

Then the variables $Y_i(t)$ form an ORNSTEIN-UHLENBECK process

$$\frac{\partial P(\{y\})}{\partial t} = -\sum_i \frac{\partial}{\partial y_i}\Big[\sum_j y_j \frac{\partial}{\partial x_{oj}} D_i^{(1)}(\{x_o\}) P \Big] + \sum_{ij} d_{ij}^{(2)} \frac{\partial^2 P}{\partial y_i \partial y_j} \tag{25}$$

but the drift and diffusion now depend on time t because of the time dependence of the solution of (24a). The steady state situation is much simpler and the solution given by (23) (in the limit $t \to \infty$) can be used. For the non-stationary problem, the resulting moment equations from (25) can be numerically integrated.

(d) One dimensional case-arbitrary $D^{(1)}$ and $D^{(2)}$

The one dimensional FPE can be solved by quadratures with the result

$$P(x) \sim e^{2\Phi} = \exp\left\{ \int^{x} \frac{dx}{D_2} (D_1 - D_2') \right\} \tag{26}$$

assuming that the current is zero at $\pm\infty$. The structure (26) of the steady state solution has been of great importance in the study of nonequilibriun phase transitions [10]. The time dependent problem is much harder. However it can be transformed into a Sturm-Liouville system

$$\dot{P} = \mathcal{L}P, \ \mathcal{L}\psi_j = -\lambda_j \psi_j, \quad \psi = e^{\Phi}\xi, \ L\xi_j = -\lambda_j \xi_j,$$

$$L = \frac{\partial}{\partial x} D_2 \frac{\partial}{\partial x} - (e^{-\Phi} \frac{\partial}{\partial x} D_2 \frac{\partial}{\partial x} e^{\Phi}) = L^{+}. \tag{27}$$

The conditional probability can be expressed in terms of the eigenfunctions ξ_j as

$$P(x,t|x_o,o) = \sum_i e^{-\lambda_i t} \xi_i(x) \xi_i^{*}(x_o) e^{\Phi(x)-\Phi(x_o)} \tag{28}$$

Thus the solution to $L\xi = -\lambda\xi$ determines the stochastic behavior of the system. Since one has a Sturm-Liouville system, various perturbative methods like WKB and variational, can be used to estimate at least some of the low lying eigenvalues and eigenfunctions. It may be added that the most general exactly soluble class will correspond to the case where $L\xi = -\lambda\xi$ can be reduced to an equation for hypergeometric or confluent hypergeometric functions.

(e) Multi-dimensional FPE

In the general case of a multi-dimensional FPE, one knows very little. Even it is not possible to obtain the steady state solution. There is however a subclass where steady state can be obtained by quadratures [11] as discussed by GRAHAM and HAKEN. On imposing the detailed balance criteria i.e. invariance of the multitime joint distributions under time reversal $x_i \to \tilde{x}_i$, $t_i \to -t_i$,

$$P_n[\{x_1\},0,\{x_2\},t_2 \ldots \{x_n\},t_n] = P_n[\{\tilde{x}_1\},0,\{\tilde{x}_2\},-t_2 \ldots \{\tilde{x}_n\},-t_n] \tag{29}$$

and on using the forward and backward equations, it is possible to show that in the steady state (i) the divergence of the reversible part of the current is zero

$$\sum_i \frac{\partial}{\partial x_i} [D_i^{(1)rev} P_{st}] = 0, \quad D_{ij}^{(2)}(\{x\}) = \epsilon_i \epsilon_j D_{ij}^{(2)}(\{\tilde{x}\}) \tag{30}$$

and (ii) that the irreversible current is zero

$$- D_i^{(1)irr} P_{s+} + \sum_j \frac{\partial}{\partial x_j} D_{ij}^{(2)}(\{x\}) P_{st} = 0 \tag{31}$$

Thus P_{st} can be obtained by quadratures provided that the integrability conditions and (30) are satisfied. This was a major step in the field and many important FPE such as those occuring in the theory of lasers, absorptive optical bistability, optical parametric oscillators etc. were solved for the stationary distribution function. Inspite of this achievement, the time dependent problem is still a major challenge in the field basically because of our inability to solve non-hermitean partial differential equations for eigenfunctions and eigenvalues.

(f) Approximate Methods

We have already mentioned that if the Fokker-Planck equation can be transformed into a hermitean problem, then the standard variational methods can be used. The other possibility which has been extensively used [12] by Risken's group and Lambropoulos's group and others is to transform the Fokker-Planck equation into a set of difference equations which can be solved numerically by using continued fractions. For example if we transform the original Fokker-Planck equation into the form

$$\frac{\partial P}{\partial t} = L_o P + L_1 P, \quad L_o^+ = L_o \tag{32}$$

and where the eigenfunctions of L_o are known $L_o \psi_n = - \lambda_n \psi_n$, then on expanding $P = \sum_n c_n \psi_n$ we obtain from (32)

$$\dot{c}_n = - \lambda_n c_n + \sum_m (L_1)_{nm} c_m \tag{33}$$

Now depending on the structure of L_1, we will get a finite term recursion relation. A three term recursion relation is easily converted into a continued fraction. It should be borne in mind that in general one has a matrix continued fraction, the methods of solution for which are available.

References

1. M. Lax in Statistical Physics, Phase Transitions and Superfluidity eds. M. Chretien et al (Gordon and Breach, N.Y. 1968) Vol.II, p.419
2. H. Haken, Handbuch der Physik, Vol. XXV/2C (Springer-Verlag, Berlin, 1970)
3. S.R. Shenoy and G.S. Agarwal, Phys. Rev. B23, 1977 (1981)
4. G. Nicolis and I. Prigogine, "Self Organization in Non-Equilibrium Systems" (Wiley, N.Y. 1977)
5. K. Razi Naqvi, K.J. Mork and S. Waldenstrom Phys. Rev. Lett. 49, 304 (1982)
6. J.P. Gordon and E.W. Aslaksen, IEEE J. Quant. Electron. QE 7, 428 (1970); F. Haake and R.J. Glauber, Phys. Rev. A5, 1457 (1972); E.W. Montroll in "Some Mathematical Problems in Biology" Vol. IV 1972 (Am. Math. Soc.; Providence) p.101
7. G.S. Agarwal, Phys. Rev. A18, 1490 (1978)
8. M.C. Wang and G.E. Uhlenbeck, Rev. Mod. Phys. 17, 323 (1945)
9. N.G. van Kampen, in "Advances in Chemical Physics" eds. I. Prigogine and S.A. Rice (Wiley, N.Y. 1976) Vol. 34, p.245

10. H. Haken, "Synergetics" (Springer-Verlag, Berlin 1977)
11. R. Graham and H. Haken, Z. Physik 243, 289 (1971)
12. H. Risken, H.D. Vollmer and M. Mörsch, Z. Physik B40, 343 (1981), S.N. Dixit, P. Zoller and P. Lambropoulos, Phys. Rev. A21, 1289 (1980)

STOCHASTIC DIFFERENTIAL EQUATIONS

R. Vasudevan
The Institute of Mathematical Sciences
Madras - 600113, India

Stochastic processes form a basic part of many types of modelling of phenomena in physical and mathematical sciences as well as in engineering. The random elements enter into the dynamics of the systems in very many different ways. Usually the time evolution of the system is governed by a differential equation wherein the parameters like the coefficients of the differential equation may be random. The driving forces may be fluctuating. The initial conditions may be random. Thus we are faced with the problem of integrating a stochastic differential equation which implies the knowledge of the probability density function of the solution. Alternatively we may be satisfied with some of the moments of the solution. The well known equation of Brownian motion with additive noise terms is a famous example (cf. Ramakrishnan [1]).

The breakdown of the exact parallelism between integrals of random and of deterministic functions is the key problem. The physical scientist based most of the conclusions on the Fokker-Planck equation governing the random processes and their integrals. Doob and Itô developed the properties of integrals of certain special random functions [2]. Meyer developed a comprehensive set of theorems on Martingales and stochastic integrals. Usual Riemann-sums of random functions lead to loss of uniqueness. Bartlett's book deals with stochastic integrals and stochastic equations from the view point of mean square convergence. The estimates of the probabilities of sample paths in the context of stochastic integrals were carried out in the work of Ramakrishnan et al [1]. Srinivasan and Vasudevan detailed an account of these in their book [3]. It was the work of Mcshane that provided a basis for a rigorous and unambiguous method of arriving at stochastic integrals. The work of Stratanovich in his book and also the work of Skorohod and Gikman are recommended for a deep study. Weiner's trajectories and the Feynman-Kac path integrals as used in nonrelativistic quantum mechanics are of great interest and specify methods widely in use in integrations of random equations. The list of references relating to these topics are given at the end [1-18].

Stochastic calculus

To discuss stochastic integrals and differentiation the first step is the development of a calculus to define convergence of a sequence of random variables. There are three types of convergence. Usually the mean square convergence is taken to be fundamental apart from convergence in probability and almost sure convergence [19]. With this notion mean square continuity follows.

Definition: A second order stochastic process $x(t)$, $t \in T$, has mean square derivative $\dot{x}(t)$ at t if

$$\lim_{\tau \to 0} \left[\frac{x(t+\tau) - x(t)}{\tau} \right] \longrightarrow \dot{x}(t) \tag{1}$$

exists in the mean square sense.

Higher order m.s. derivatives are defined analogously.

Differentiation in mean square criterion

A second order stochastic process $x(t)$, $t \in T$, is differentiable at t if and only if the second generalized derivative

$$\lim_{\tau, \tau' \to 0} \frac{1}{\tau \tau'} \cdot \left[\Gamma(t+\tau, s+\tau') - \Gamma(t+\tau, s) - \Gamma(t, s+\tau') + \Gamma(t,s) \right] \tag{2}$$

exists at (t,s) and is finite, where

$$\Gamma(t,s) = E\left[x(t)x(s)\right] \tag{3}$$

We remark that the concept of mean square differentiation leads to the notion of mean square Taylor expansion of a stochastic process and the definition of mean square analyticity.

Mean square Riemann integral

Let $x(t)$ be a second order stochastic process defined on $[a,b]$. Let $f(t,u)$ be an ordinary function defined on the same interval for t and Reimann integrable for every $u \in U$. We form the random variable

$$Y_n(u) = \sum_{k'=1}^{n} f(t_{k'},u)\, x(t_{k'})(t_{k'}-t_{k'-1}) \tag{4}$$

If for $u \in U$

$$\lim_{n \to \infty, \Delta t \to 0} Y_n(u) = Y(u) \tag{5}$$

exist for some sequence of subdivisions ρ_n the stochastic process $Y_n(u)$, $u \in U$, is called the mean square Riemann integral of $f(t,u)x(t)$ over the interval $[a,b]$ and it is denoted by

$$Y(u) = \int_a^b f(t,u)x(t)dt \tag{6}$$

Integration in mean square criterion

The stochastic process $Y(u), u \in U$, defined by (6) exists if and only if the ordinary double Riemann integral

$$\int_a^b \int_a^b f(s,u)\Gamma_{xx}(t,s)\, dt\, ds \tag{7}$$

exists and is finite.

Similarly Riemann-Stieltjes integral can be defined and criteria for mean square Riemann-Stieltjes integrals to exist can be analysed.

We will see in the sequal that the definition (4) and the definition of Riemann-Stieltjes integrals will take different forms according to different prescriptions warranted by different types of modelling.

A special class of differential equations which has found important applications in control, filtering and communications theory is one where the vector stochastic process $\underline{Y}(t)$ of equation

$$\dot{\underline{x}}(t) = \underline{f}(\underline{x}(t), \underline{Y}(t), t), \ t \in T, \ \underline{x}(t_o) = \underline{x}_o \tag{8}$$

has only white noise components. More specifically we mean equations of the form

$$\dot{\underline{x}}(t) = \underline{f}(\underline{x}(t), t) + \underline{G}(\underline{x}(t), t)\, \underline{W}(t), \ t \in T, \ \underline{x}(t_o) = \underline{x}_o \tag{9}$$

where $W(t)$ is an m-dimensional stochastic process whose components are Gaussian white noise. $\underline{G}(\underline{x}(t),t)$ is an n x m matrix function and is independent of $\underline{W}(t), t \in T$. The popularity of this model in control and filtering applications is due to two principal reasons. The first is the mathematical simplicity - it is a natural stochastic extension of the powerful state-space approach in classical optimal theory [20-25]. Moreover as we shall see, the solution process generated by (9) is Markovian for which powerful techniques for obtaining its solution exist. The second reason is that, although white noise is a mathematical artifice it approximates closely the behaviour of a number of important noise processes in electrical and electronic systems.

Let $B(t)$, $t \geqslant 0$, be the Brownian motion or the Weiner process. It is Gaussian with mean 0 and covariance

$$\mu_B(t_1, t_2) = 2\, D \min(t_1, t_2) \tag{10}$$

The formal derivative of $B(t)$, $\dot{B}(t)$ is clearly Gaussian with mean zero and its covariance is given by

$$\mu_{\dot{B}}(t_1, t_2) = 2\, D\, \delta(t_1 - t_2) \tag{11}$$

Thus we can formally write

$$\frac{dB(t)}{dt} = W(t), \ t > 0 \tag{12}$$

With this formal representation for the white Gaussian noise $W(t)$ we can interpret (9) as formally equivalent to

$$\underline{x}(t) - \underline{x}(t_o) = \int_{t_o}^{t} \underline{f}(\underline{x}(s), s)\, ds + \int_{t_o}^{t} \underline{G}(\underline{x}(s), s)\, d\underline{B}(s), \ t \in T \tag{13}$$

where $\underline{x}(s)$ is independent of the increment $d\underline{B}(t)$, $t \in T$ (cf [16,20-24] for many applications of stochastic differential equations in various fields).

The Paradox

Here we shall explain a paradox as detailed by [8]. For equations of the type (13) there is the approach of Fokker-Planck partial differential equation.

However a conceptual difficulty arises in writing the Fokker-Planck equation, since the heuristic mathematical idealization of the fluctuating force F(t) as a white noise process can lead to difficulties. The result has been that something of a controversy has appeared in the recent literature concerning the two possible ways of extending ordinary calculus to stochastic functions: The socalled Stratonovich calculus, in which the usual rules continue to apply and the socalled Itô calculus in which the rules are modified [1-17]. As remarked by Mortinsen [8] white noise is much the same kind of mathematical pathology in the theory of random processes as that the Dirac delta function is in the theory of deterministic functions.

Taking the one-dimensional version of (9) one can write

$$x(t) = f(x(t),t) + g(t)\, v(t) \tag{14}$$

where $v(t)$ is a Gaussian noise with

$$E[v(t)] = 0, \quad E[v(t)\, v(s)] = \delta(t-s)$$

or equivalently

$$dx(t) = f(x(t),t)\, dt + g(t) dB(t) \tag{15}$$

or

$$x(t) = x(o) + \int_o^t f(x(t),\tau) d\tau + \int_o^t g(\tau)\, dB(\tau) \tag{16}$$

Since $g(t)$ is a nonrandom function at t, both Stratanovich and Itô interpretations of these integrals agree and we take the Riemann-Stieltjes sum

$$\sum g(t_i)\, [B(t_{i+1}) - B(t_i)] \tag{17}$$

where

$$0 = t_o < t_1 < \dots < t_n = t$$

The probability density $p(x,t|x_o,t_o)$ of the solution Markov process $x(t)$ is obtained by the forward Fokker-Planck equation [7,8,1,2,3].

$$\frac{\partial p(x,t|x_o,t_o)}{\partial t} = -\frac{\partial}{\partial x}[f(x,t)p(x,t|x_o,0)] + \frac{1}{2}g^2(t)\frac{\partial^2 p(x,t|x_o,0)}{\partial x^2}, \quad -\infty < x < \infty,\ t > 0 \tag{18}$$

with the boundary conditions

$$\lim_{t\to 0} P(x,t|x_o,0) = \delta(x-x_o), \quad \lim_{|x|\to\infty} p(x,t|x_o,0) = 0$$

Similarly the backward Fokker-Planck equation is

$$\frac{-\partial p(x,t|x_o,t_o)}{\partial t_o} = f(x_o,t_o)\frac{\partial p(x,t|x_o,t_o)}{\partial x_o} + \frac{1}{2}g^2(t)\frac{\partial^2 (x,t|x_o,t_o)}{\partial x_o^2} \tag{19}$$

If f = 0, g = 1 the solution x(t) is a Wiener process with probability density p (x,t) given by

$$p(x,t) = \frac{1}{\sqrt{2\pi t}} \exp(-x^2/2t) \tag{20}$$

This p(x,t) obeys the Fokker-Planck equation

$$\frac{\partial p(x,t)}{\partial t} = \frac{1}{2} \frac{\partial^2 p(x,t)}{\partial x^2} \tag{21}$$

Let x(t) be passed through a nonlinear device to produce a new process

$$z(t) = \sinh [x(t)] \tag{22}$$

$$\pi(z,t) = p(x,t) \frac{dx}{dz}\Big|_{x = \sinh^{-1}(z)} \tag{23}$$

hence

$$\pi(z,t) = \frac{(1+z^2)^{\frac{1}{2}}}{\sqrt{2\pi t}} \exp[-(\sinh^{-1} z)^2/2t] \tag{24}$$

Now going back to the F.P. equation with the change of variable $x = \sinh^{-1}(z)$ one finds the F-P-equation satisfied by $\pi(z,t)$ as

$$\frac{\partial \pi(z,t)}{\partial t} = -\frac{\partial}{\partial z}[\frac{z}{2} \pi(z,t)] + \frac{1}{2}\frac{\partial^2}{\partial z^2} [(1+z^2) \pi(z,t)] \tag{25}$$

and this equation is the forward equation corresponding to the stochastic differential equation

$$dz(t) = \frac{1}{2}z(t)dt + [1+z^2(t)]^{\frac{1}{2}} dB(t) \tag{26}$$

If we simply compute dz(t) from equation (22) we get

$$dz(t) = (1+z^2)^{\frac{1}{2}}dx(t) \tag{27}$$

which means

$$dz(t) = (1+z^2)^{\frac{1}{2}} dB(t) \tag{28}$$

from equations (13,21)

The stochastic differential equations (26) and (28) differ by $\frac{1}{2}$ z(t)dt. The question is which is the correct s.d.e. for generating the process z(t). Therefore the situation is that one unambiguous way to specify a Markov process mathematically is to specify its transition density or equivalently the F-P-equation obeyed by the transition density. The divergence boils down to two different ways of associating the coefficients in the F-P-equation with the coefficients in the s.d.e. Thus we can have two ways of integrating this stochastic equation.

Using the Itô rules with the equation (26) we will obtain the F-P-equation (25). Hence integrating (26) by Itô rules one should get

$$z(t) = \sinh[x(t)] \tag{29}$$

According to Stratanovich, the equation (28) is perfectly valid and hence integration should also yield

$$z(t) = \sinh[x(t)] \tag{30}$$

Hence starting with a F-P-equation in an unambiguous way there are two possibilities of modelling the process as the solution to a s.d.e. Thus it is open to us to choose the best type of dynamical equation. However the engineer does not start with the transition density, but starts with a differential equation which he has obtained on the basis of known physical laws and then adds a white noise forcing term to get a stochastic model. If the coefficient of the noise itself is random, then there are two possible ways of interpreting the equation leading to two types of differential F-P-equations and hence two different processes. The question is which process does one really get in the physical world? Which kind of calculus one should prefer? In the following sections we will examine the situation in more detail

The Itô - calculus

If we take the Wiener process B(t) and calculate the increment in time

$$\Delta B(t) = B(t+\Delta t) - B(t) \tag{31}$$

it can be shown that

$$P_{\Delta B} d(\Delta B) = \frac{1}{\sqrt{(2\pi\Delta t)}} \exp\left(-\frac{(\Delta B)^2}{2\Delta t}\right) \tag{32}$$

The important point is that the increment ΔB is independent of B(t), the state of the process at time t. Thus B(t), the Wiener process, belongs to a special class of processes with independent increments. We know also

$$\begin{aligned} E[(\Delta B)^k] &= 0 \;; \quad k \text{ odd} \\ &= 1.3.5 \ldots.(k-1)(\Delta t)^{k/2}; \; k \text{ even} \end{aligned} \tag{33}$$

Thus if we take a function $F[B(t+\Delta t)]$ and expand it by Taylor series and calculate the differential we can easily see that

$$\begin{aligned} E\left\{\frac{d}{dt} F[B(t)]\right\} &= \lim_{\Delta t\to 0} \frac{E\{F''[B(t)]\}\Delta t + 0(\Delta t)}{\Delta t} \\ &= \frac{1}{2} E\{F''[B(t)]\} \end{aligned} \tag{34}$$

Since $E[(\Delta B)^2]$ is of order Δt and not $(\Delta t)^2$ using the chain rule of ordinary calculus, one will write

$$dF[B(t)] = F'(B)dB(t) \tag{35}$$

hence

$$E\{dF[B(t)]\} = E[F'(B)]\; E[dB(t)] = 0 \tag{36}$$

Clearly equation (36) is different from equation (34).

Itô showed how the rules of calculus have to modified to handle the above phenomenon. As given by Skorohod and others [cf.12]

$$d_I F[B(t)] = F'[B(t)]\, d\, B(t) + \frac{1}{2} F''[B(t)]\, dt \tag{37}$$

which is the Itô rule for differentiation in the present case. However we obtain

$$d_I \sinh B(t) = \cosh B(t) d\, B(t) + \frac{1}{2} \sinh B(t)\, dt$$
$$\text{if } F[B(t)] = \sinh[B(t)] \tag{38}$$

Hence if

$$z = \sinh[B(t)]$$

$$d_I z(t) = \frac{1}{2} z(t)\, dt + (1+z^2)^{\frac{1}{2}} d\, B(t) \tag{39}$$

which is consistent with the F-P-equation (25).

If we wish to preserve the fundamental property of calculus then we require to have

$$I \int_{t_o}^{t} d_I F[B(t)] = F[B(t)] - F[B(t_o)] \tag{40}$$

Hence we are forced to assume

$$I \int_{t_o}^{t_1} F'[B(t)]\, dB(t) = F[B(t_1)] - F[B(t_o)] - \frac{1}{2}\int_{t_o}^{t_1} F''[B(t)]\, dt \tag{41}$$

Thus

$$I \int g[B(t)]\, dB(t) = \int_{B(t_o)}^{B(t_1)} g(\xi) d\xi - \frac{1}{2}\int_{t_o}^{t_1} g'[B(t)]\, dt \tag{42}$$

Stratanovich calculus treats $\int g(\xi) d\xi$ as an ordinary integral, treating ξ as a deterministic dummy variable of integration. Hence the relation between the Itô and the Stratanovich rules of integration is given by

$$I \int_{t_o}^{t} g[B(t)] d\, B(t) = S\int_{t_o}^{t} g(B) d\, B(t) - \frac{1}{2}\int_{t_o}^{t} g'[B(t)]\, dt \tag{43}$$

where $\quad g'[B(t)] = \left.\frac{dg(\xi)}{d\xi}\right|_{\xi=B(t)}$

Let us take $g[B(t)] = B(t)$. Then we get

$$I\int_{t_o}^{t} B(t) d\, B(t) = \frac{1}{2} B^2(t_1) - \frac{1}{2} B^2(t_o) - \frac{1}{2}(t_1 - t_o) \tag{44}$$

The presence of the term $\frac{1}{2}(t_1-t_o)$ in the equation (44) can be made more plausible by the following considerations

$$E\left[I\int_{t_o}^{t} B(t) d\, B(t)\right] = \int_{t_o}^{t} E[B(t)]\; E[dB(t)] \tag{45}$$

and so

$$E\left[I\int_{t_o}^{t_1} BdB \right] = 0$$

According to (42)

$$E\left[I\int_{t_0}^{t_1} B(t)d\,B(t)\right] = \frac{1}{2}(t_1-t_0) - \frac{1}{2}(t_1-t_0) = 0 \qquad (46)$$

However

$$E\left[S\int_{t_0}^{t_1} B(t)dB(t)\right] = \frac{1}{2}(t_1-t_0) \neq 0 \qquad (47)$$

Thus for Stratanovich integral it cannot be true that dB(t) is independent of B(t).

The Stratanovich increment is given by

$$\Delta_s B(t) = B\left[t+\frac{\Delta t}{2}\right] - B\left[t-\frac{\Delta t}{2}\right] \qquad (48)$$

This increment still has mean zero and variance Δt, but it is not independent of B(t).

For a more rigorous approach to the Itô calculus as the limit of Riemann sums, we define the integral

$$J = I\int_0^T z(t)dB(t)$$

By using n partitions, define

$$I_n = \sum z(t_{k-1})\left[B(t_k) - B(t_{k-1})\right] \qquad (49)$$

which in the limit converges to a limiting random variable J. It is to be noted that $z(t_{k-1})$ is always taken at the beginning of the interval over which the increment $[B(t_k) - B(t_{k-1})]$ is taken. Hence $z(t_{k-1})$ and $\Delta B(t_k)$ are always independent and consequently we have

$$B[I_n] = \sum_{k=1}^{n} E[z(t_{k-1})]\,E[B(t_k) - B(t_{k-1})] = 0 \qquad (50)$$

It can be proved from first principles that the Itô integral is the value given in (44) in mean square sense (cf.[6])

The Stratanovich integral

In this case we take the sequence of sums of the form representing the integral as

$$I_n = -\sum_{k=1}^{n} z\left(\frac{t_{k-1}+t_k}{2}\right)\left[B(t_{k-1}) - B(t_k)\right] \qquad (51)$$

and let us take the generalized form of the s.d.e.

$$dx(t) = f(x(t),t) + g(x,t)dB \qquad (52)$$

Explicitly we have

$$x_I(t) = x_I(0) + \int_0^t f(x,\tau)\,d\tau + I\int_0^t g(x,\tau)\,dB(\tau) \qquad (53)$$

where symbol I denotes the Itô belated integrals. If the symbol S denotes the Stratanovich integration (52) yields

$$x_S(t) = x_S(o) + \int_0^t f(x,\tau)\,d\tau + S\int_0^t g(x_S(\tau),\tau)\,dB(\tau) \tag{54}$$

Because of the relationship explained earlier between the Ito and the Stratanovich integrals we have the equation

$$x_I(t) = x_I(o) + \int_0^t \left[f(x_I,\tau) - \tfrac{1}{2} g(x_I(\tau),\tau)\, g'(x_I(\tau),\tau) \right] d\tau$$
$$+ S\int g(x_I(\tau),\tau)\,dB(\tau) \tag{55}$$

where

$$g'(x,t) = \frac{\partial g(x,t)}{\partial x}$$

and also we have

$$x_S(t) = x_S(o) + \int_0^t \left[f(x_S,\tau) + \tfrac{1}{2} g(x_S,\tau)\, g'(x_S,\tau) \right] d\tau + I\int_0^t g(x_S,\tau)\,dB(\tau) \tag{56}$$

For the Itô equation (53) the F-P-equation is given by

$$\frac{\partial p_I(x,t|x_o,o)}{\partial t} = \frac{\partial}{\partial x}\left[f(x,t)\, p_I(x,t|x_o,o) \right] + \frac{1}{2}\frac{\partial^2}{\partial x^2}\left[g^2(x,t)\, p_I(x,t|x_o,o) \right] \tag{57}$$

For the Stratanovich equation the F-P-equation is given by

$$\frac{\partial p_S(x,t|x_o,o)}{\partial t} = -\frac{\partial}{\partial x}\left[f(x,t)\, p_S(x,t|x_o,o) \right] + \frac{1}{2}\frac{\partial}{\partial x}\, g(x,t)\frac{\partial}{\partial x}\left[g(x,t)\, p_S(x,t|x_o,o) \right] \tag{58}$$

Using the relation between the Stratanovich and the Itô integrals we find that the solutionsof equations (26) and (27) are the same, given by

$$z(t) = \sinh\left[B(t) \right] \tag{59}$$

Also we find that the F-P-equation corresponding to (28) is the same as the F-P-equation corresponding to (26) which is (25) obtained by using the usual Itô calculus.

Additive and multiplicative stochastic processes

In describing the physics of macroscopic systems by the deterministic time evolution of the collective variables we can find globally stable states of the system in which case the statistical nature and the dynamics can be safely neglected. However by changing some parameters large excursions from the deterministically describable values occur. We find that the fluctuations in a system play an important role in their understanding. The equilibrium phase transition is one of such class like liquid-vapour transitions etc. Another class of phenomena consists of phase transition analogous that have been found in nonequilibrium systems such as laser and nonlinear optics, hydrodynamic instabilities etc. In a variety of problems the fluctuations can be taken into account by adding a fluctuating force to the deterministic equations. The most familiar examples of these types of additive fluctuations are the vacuum fluctuations of the electromagnetic field that triggers the spontaneous emission by atoms and in the macroscopic case that of the Brownian motion.

The most important characteristic of the additive fluctuations is that they do not depend on the present state of the collective variables of the system.

In an autocatalytic chemical reaction the production of a molecule of some type is enhanced by the presence of other molecules of the same type that have been produced already. Therefore the only possible reaction channel is the autocatalytic reproduction of the molecules according to the blue print provided by the molecules of the same kind already present. As a result the fluctuations in this case do depend on the state of the system. If this dependence can be described by a function of the macroscopic variables multiplying the 'fluctuating stochastic forces' we call such process a 'multiplicative stochastic process' [18,30].

The Langevin equation for these two types of the system can be generally described by

$$\dot{x}_i(t) = L_i[\{x_j\}] + F_i \tag{60}$$

for the additive system and

$$\dot{x}_i(t) = L_i[\{x_j\}] + G_{ij}[\{x_k\}] F_j \tag{61}$$

for the multiplicative system with properties

$$< F_j^i(t+\tau) F_k^l(t) > = Q_{jk}^{(i,l)} \delta(\tau) \tag{62}$$

where Q is a measure of fluctuations independent of the random variables $\{x_i\}$. $L[\{x_i\}]$ may be a linear or nonlinear function.

In the first case and the case in which $G_{ij}[\{x\}]$ is independent of x_k's the Stratanovich and the Itô calculus give the same F-P-equation. They become different if $G_{ij}[\{x_k\}]$ depend on x_k's. The F-P-equation for the multiplicative noise problem relating to the Langevin equation (61) is

$$\frac{\partial \pi(\{x_k\},t)}{\partial t} = -\frac{\partial}{\partial x_i}[K_i^{(1)}(\{x_k\})\pi] + \frac{1}{2}\frac{\partial^2}{\partial x_i \partial x_j}[K_{ij}^{(2)}(x_k)\pi] \tag{63}$$

with

$$K_i^{(1)}(\{x_k\}) = L_i[\{x_k\}] + \frac{1}{2}\frac{\partial G_{ij}}{\partial x_l} G_{lj}$$

$$K_{ij}^{(2)}(\{x_k\}) = G_{il}[\{x_k\}] G_{jl}[\{x_k\}]$$

assuming the noise has unit variance. This is the Stratanovich description and takes into account the correlations engendered by the multiplicative processes to a great extent [30].

<u>Canonical extension</u>

Consider the differential equation

$$dx(t) = f[x(t)]\,dt + g[x(t)]\,dB(t) \tag{64}$$

McShane [16] has developed the general calculus which includes both the Itô and the Stratanovich cases. Under this calculus the integral equation corresponding to (64) is replaced by a canonical equation which is valid for more general continuous noise processes (including Brownian motion). He has extended this when B(t) is taken to be a point process and hence one can consider the stochastic differential equations driven by jump processes also by a theory analogous to that of Itô differential equations representing Brownian motion.

The differential equation may be of the form

$$dx(t) \quad = \quad f(x,t)dt + \sum_{n=1}^{N(t)} h[x(\tau_n),U_n] \tag{65}$$

where N(t) is the number of incident points during [o,t] regardless of their mark and τ_n and U_n are the times of occurence and the random marks respectively of the nth point (it will be assumed that a counting process N(t) is almost surely left continuous). This can also be written as

$$x(t) \quad = \quad x(o) + \int_o^t f(x(\tau),\tau)d\tau \quad + \int_o^t h[x(\tau),u(\tau)] \; dN(\tau) \tag{66}$$

To arrive at a general canonical extension which can be applied to class of noise processes and also for the Brownian motion carried out by McShane, McKean and others cf. [16,17,25].

Now let us consider the differential equation

$$\dot{x} \quad = \quad g(x(t)), \qquad x(o) \quad = \quad x_o \tag{67}$$

The Lie series solution is given by

$$x(t) \quad = \quad x_o + \sum_{m=1}^{\infty} \frac{t^m}{m!} D^{m-1} [g(x_o)] \tag{68}$$

with

$$D \quad = \quad \sum_{l=1}^{n} g^l(x) \frac{\partial}{\partial x_l} \tag{69}$$

where l denotes the number of components of the process.

In all these f and g are analytic and the solutions are assumed to exist. The integral form of the stochastic differential equation may be obtained for one component system as

$$x(t) \quad = \quad x_o + \int f[x(\tau)]d\tau \; + \; \sum_{m=1}^{\infty} \frac{1}{m!} \int D^{m-1}[g(x(\tau)][dz(t)]^m \tag{70}$$

where the last integral is the Itô belated integral of McShane. If dz satisfies the conditions of the Brownian motion kicks, all the belated integrals in (70) with $m > 2$ vanish. If the s.d.e. is driven by point processes the integral form of (70) will be

$$x(t) = x_o + \int_0^t f[x(\tau)]\, d\tau + \int_0^t [e^{\alpha D} x(\tau) - x(\tau)]\, dN(\tau) \tag{71}$$

if α is the constant size of the jump (cf. [25])

Stationary solutions of Fokker-Planck- equations

Let us take the one-dimensional Fokker-Planck equation for probability density or the transition probability given by

$$\frac{\partial \pi(x,t)}{\partial t} = -\frac{\partial}{\partial x}[k(x)\pi - \frac{1}{2} D \frac{\partial \pi}{\partial x}] = -\frac{\partial}{\partial x} J(x) \tag{72}$$

if we write the current density $J(x,t)$ as

$$J(x,t) = k(x)\pi - \frac{1}{2} D \frac{\partial \pi}{\partial x}$$

This is the result of the Langevin equation with an additive suitable white noise with a deterministic force term given by $k(x)$

$$\dot{x} = k(x) + F(t) \tag{73}$$

For the n-dimensional case the Fokker-Planck equation is

$$\frac{\partial \pi}{\partial t} + \sum_{k=1}^{n} \frac{\partial}{\partial x_k}(K_k \pi - \frac{1}{2} \sum_{l=1}^{n} D_{kl} \frac{\partial \pi}{\partial x_l}) = 0 \tag{74}$$

with the probability current given by $\vec{J} = (J_1, \ldots\ldots, J_n)$ and

$$\frac{\partial \pi}{\partial t} + \nabla . \vec{J} = 0 \tag{75}$$

This is the continuity equation for (π, J). In the stationary case

$\frac{\partial \pi}{\partial t} = 0$ and hence we have for a one dimensional process

$$\frac{1}{2} D \frac{\partial \pi}{\partial x} = K \pi \tag{76}$$

We see easily that the solution is given by stationary π^o

$$\pi^o(x) = N \exp\left(-\frac{2V(x)}{D}\right) \tag{77}$$

with

$$V(x) = -\int_{x_o}^{x} k(x')dx', \qquad \int \pi(x)dx = 1. \tag{78}$$

The normalization condition determines theconstant N. The boundary condition is $\pi(x) = 0$ as $x \rightarrow \pm\infty$. We have $V(x) = \alpha x^2/2$ if $K(x) = -\alpha x$ and $\dot{x} = -\alpha x + F(t)$. The fluctuating force pushes the particle up a slope of the potential well, but it settles down at $x = x_o = 0$ so that $\langle x \rangle = 0$ at $t = \infty$.

If $k(x) = -\alpha x - \beta x^3$

$$V(x) = \frac{\alpha x^2}{2} + \frac{\beta}{4} x^4 \tag{79}$$

This corresponds to the Langevin equation

$$\dot{x} = -\alpha x - \beta x^3 + F(t) \tag{80}$$

When $\alpha > 0$ the minimum of V(x) is at x = 0 and the stationary solution is

$$\pi(x) = N \exp\left[-\frac{2}{D}\left(\frac{\alpha x^2}{2} + \frac{\beta x^4}{4}\right)\right] \tag{81}$$

This is stable at x = 0. However for $\alpha < 0$ and $\beta > 0$ we find the following solution:

x = 0 is unstable and $x_{1,2} = \pm \left(\frac{|\alpha|}{\beta}\right)^{\frac{1}{2}}$ are stable. For the potential $V(x) = \frac{1}{2}\alpha x^2 + \frac{1}{4}\beta x^4$ as α goes from +ve values to -ve values the stability changes and for α negative the potential has two stable points. As α goes through 0,V(x) becomes flat near x = 0 and hence the restoring force becomes smaller and this is the socalled critical slowing down. Also we see from the equation

$$\dot{x} = -\alpha x - \beta x^3$$

that if we change $x \longrightarrow -x$ the equation remains unchanged. If α is -ve, when $x \longrightarrow -x$, $\dot{x} = \alpha x - \beta x^3$. This symmetry breaking occurs by deformation of the potential curve.

Phase transition analogy

We now consider a phase transition analogy of paramagnet becoming a ferromagnet. The ferromagnet has spontaneous magnetisation. The average magnetization $M = (N\uparrow - N\downarrow)m$ can be equated to a quantity q called an order parameter of the system. The free energy F(q,T) as a function of q and T is assumed as

$$F(q,T) = F(0,T) + qF' + \frac{q^2}{2!}F'' + \frac{q^3}{3!}f''' + \ldots$$

with $F' = F''' = 0$ due to symmetry. Hence we take

$$F = F(o) + \frac{\alpha}{2}q^2 + \frac{\beta}{4}q^4 \tag{82}$$

The probability distribution for this thermodynamic state is

$$\pi = N \exp(-F/K_B T) \tag{83}$$

Let $\alpha = a(T-T_c)$. F_{min} occurs at $q_o = 0$ for α being positive and π is maximum. For $T < T_c$, α is negative and $q_o = \pm\left(\frac{|\alpha|}{\beta}\right)^{\frac{1}{2}}$. Entropy is continuous and specific heat is discontinuous. The equation governing q is

$$\dot{q} = -\frac{\partial F}{\partial q}$$

Hence for a model system we have

$$\dot{q} = -\alpha q - \beta q^3 + \text{fluctuating force} \tag{84}$$

The critical slowing down associated with the soft mode happens as $\alpha \longrightarrow 0$. Fluctuations of q become considerable.

For the equilibrium distribution mean square fluctuation is

$$D < q^2 > \;=\; \frac{\int \frac{\partial \bar{V}}{\partial \alpha} \exp(-\bar{V})\, dq}{\int \exp(-\bar{V})\, dq}\,, \qquad \bar{V} = \frac{2V}{D} \tag{85}$$

We expand all the quantities around q_0 far from the transition point

$$D < q^2 > \;=\; -\frac{\partial}{\partial \alpha} \log\left\{ \exp(-\bar{V}(q_0)) \int \exp(-\bar{V}''(q_0)(q-q_0)^2)\, dq \right\}$$

$$= \frac{\partial \bar{V}(q_0)}{\partial \alpha} - \frac{1}{2} \frac{\partial}{\partial \alpha} \log\left(\frac{\pi}{\bar{V}''(q_0)} \right) \tag{86}$$

The second term stems from fluctuations which are large near $\alpha = \alpha_c$. The Ginsberg-Landau functional for the free energy with space-dependent q as q(x) can be written down and values of critical exponents can be arrived at (cf. Kadanoff et al Rev. Mod. Phys. 39, 395 (1967)).

Solution of F-P-equations by the path integral method

Taking the one-dimensional Langevin equation $\dot{x} = K(x) + F(t)$ we see that the Chapman-Kolmogorov equation for $\pi(x,t)$ given by

$$\pi(x, t+\Delta) = N \int_{-\infty}^{\infty} \exp\left\{ -\frac{1}{2D\Delta} (x - x' - K(x')\Delta)^2 \right\} \pi(x', t)\, dx' \tag{87}$$

leads to the F-P-equation for π,

$$\frac{\partial \pi}{\partial t} = \frac{\partial \pi K(x)}{\partial x} + \frac{D}{2} \frac{\partial^2 \pi}{\partial x^2} \tag{88}$$

From the Langevin equation

$$\dot{x} = K(x) + F(t)$$

we see that the jump in x process for each Δ is given by

$$(x - x' - K\Delta) = \Delta F(t) = z.$$

The amount of fluctuation jump should be z, to get x from x' in time Δ. This z is distributed as $\sim N \exp(-\frac{1}{2D\Delta}) z^2$ since it is stationary Gaussian white noise. Hence

$$\pi(x, t+\Delta) = N \int_{-\infty}^{\infty} \exp\left[-\frac{1}{2D\Delta} \left(\xi^2 + 2\Delta \xi K(x+\xi) + \Delta^2 K(x+\xi)^2 \right) \right] \pi(x+\xi, t)\, d\xi \tag{89}$$

where we take $x' = x + \xi$. Expanding K and π around x we have

$$\pi = N \int_{-\infty}^{\infty} d\xi \exp\left(-\frac{\xi^2}{2D\Delta}\right) \left[1 - \frac{K^2 \Delta}{2D} - \frac{\xi^2 K'}{D} - \frac{\xi K}{D} \right] \left[\pi(x,t) + \xi \pi'(x) + \xi^2 \pi''/2 + \ldots \right] \tag{90}$$

Terms odd in ξ are dropped and keeping terms of order Δ only and choosing $N = (1/\sqrt{2D\Delta\pi})$ we obtain the Fokker-Planck equation. We may repeat the process from t_o to t by going from $t \to t+\Delta$, $t+\Delta \longrightarrow t+2\Delta$, ...etc. Hence

$$\Pi(x,t) \sim \iint D(x)\ \exp(-\tfrac{1}{2}\,O\ \Pi(x',o)) \tag{91}$$

with $\Delta \to 0$ and n the number of steps tending to ∞ such that $n\Delta = t$. The symbols D and O in (91) represent

$$D(x) = (2D\Delta\Pi)^{-N/2} dx_o dx_1 \ldots dx_{n-1}$$

$$O = \prod_i \exp\left\{-\frac{1}{2D\Delta}(x_{i+1}-x_i-\Delta K(x_i))^2\right\},\ i = 0,1,2\ldots,n-1 \tag{92}$$

This is the path integral formalism for the time dependent solution of the F-P-equation for given s.d.e. (cf. [28])

Coloured noise induced transitions

In the case of coloured noise we want to demonstrate how the transition density or F-P-equation can be obtained by a method different from the usual techniques adopted by Kitahara et al [27]. Our work (Vasudevan and Parthasarathy [2b]) is based on the method of van Kampen [18]. When the dichotomous noise in the equation goes over to white noise limit it turns out that we obtain the Stratanovich equation rather than the Itô equation. Most models are idealisations to a white noise and not a flat spectral power system. Hence the Stratanovich method leads to more physical results as West et al [28] have shown in many cases. Coloured noise has been used by Kabashima [29] for electrical systems.

Let us take a general nonlinear system of the type $\dot{x} = f(x) + g(x)\ I(t)$ where I(t) is a dichotomous Markov process which takes values $\pm$ d with probabilities P_+ and P_- described by the master equation

$$\frac{d}{dt}\begin{bmatrix} P_+ \\ P_- \end{bmatrix} = \frac{\gamma}{2}\begin{bmatrix} 1 & -1 \\ -1 & 1 \end{bmatrix}\begin{bmatrix} P_+ \\ P_- \end{bmatrix} \tag{93}$$

$\frac{\gamma}{2}$ is the probability per unit time that a transition occurs as a jump process. If $P_o = P_+ = P_- = \frac{1}{2}$ we obtain the solution at any time t

$$\vec{P}(t) = \left(\exp\left\{\frac{\gamma}{2}\,t\,[I-\sigma_x]\right\}\right)\vec{P}(0) \tag{94}$$

where σ_x is the Pauli matrix. The process I(t) has a correlation function

$$\langle I(t)I(t')\rangle = d^2\ \exp(-\gamma|t-t'|) \tag{95}$$

The process I(t) is also called random telegraphic signal process.

Let us now adopt the technique of van Kampen [18] to this problem by going to the solution of the stochastic differential equation and phase space density of such solutions called $\rho(x,t) = \delta(x-x_s)$ where x_s is the solution of this equation. It can be shown that $\langle\rho(x,t)\rangle$ the average over the ensemble of solutions is the probability density of x(t) at any time t, given by

$$p(x,t) = \langle\rho(x,t)\rangle \tag{96}$$

Hence the divergence equation for the Liouville density is

$$\frac{\partial \rho}{\partial t} + \frac{\partial}{\partial x}(\dot{x}\rho) = 0 \quad \text{with} \quad \rho(x,t=0) = \delta(x-x_o) \tag{97}$$

Hence the equation for ρ , using the equation of motion is

$$\frac{\partial \rho}{\partial t} = -\frac{\partial}{\partial x}[\, f(x) + Ig(x)]\, \rho(x,t) \tag{98}$$

and $\rho(x,t)$ is compressible. Following the method of van Kampen [18] we arrive at the F-P-equation as

$$\frac{\partial p}{\partial t} = -\frac{\partial}{\partial x}(fp) + \Delta^2 \frac{\partial}{\partial x}\int_{-\infty}^{t} d\tau \exp\left\{-[\gamma + \frac{\partial f}{\partial x}](t-\tau)\right\}\frac{\partial}{\partial x} gp(x,\tau) \tag{99}$$

Now assume that I is white noise. This can be obtained as the limit of the coloured noise if we take $\Delta^2 \to \infty$, $\gamma \to \infty$ such that Δ^2/γ = finite = $\sigma^2/2$, i.e. $\frac{\Delta^2}{\gamma}\gamma\exp(-\gamma|t-t'|) = \infty$ if $t = t'$ and $= 0$ otherwise if $\gamma \to \infty$, $\Delta^2/\gamma \to$ finite. Hence in this limit the F-P-equation is

$$\frac{\partial p}{\partial t} = -\frac{\partial}{\partial x}[f + \frac{1}{2}\sigma^2 g(x)g'(x)]\, p + \frac{1}{2}\sigma^2\frac{\partial^2}{\partial x^2} g^2 p \tag{100}$$

Going to the limit when $t \to \infty$ we obtain the stationary solution p_s given by

$$p_s = \left(\frac{Ng}{\Delta^2 g - f^2}\right)\exp\left\{\gamma\int dx'\ f(x')/[\Delta^2 g^2 - f^2(x')]\right\} \tag{101}$$

Hence the stable points are given by the solutions of the equation

$$f(x_m) - \left[\frac{\Delta^2}{\gamma}\right] g(x_m)g'(x_m) + \frac{2}{\gamma} f(x_m)f'(x_m) - \frac{1}{\gamma} f^2(x_m)(g'(x)/g(x_m)) = 0 \tag{102}$$

In the white noise limit we have the corresponding equation as

$$f(x_m) - \left[\frac{\Delta^2}{\gamma}\right] g(x_m)g'(x_m) = 0 \tag{103}$$

Hence coloured noise introduces two other terms in the equation for the extremal points due to the finite value of γ. Passing from coloured noise to the white noise limit we obtain the Stratanovich type of F-P-equation which is an interesting point to be noted.

References

1. Alladi Ramakrishnan, Handbuch der Physik Vol.III/2 (1959) (Springer-Verlag) 524-651.
2. J.L. Doob, 'Stochastic Processes', (John Wiley) (1953).
3. S.K. Srinivasan and R. Vasudevan, 'An Introduction to Stochastic Differential Equations' (American Elsivier Pub.) (1971).
4. N. Wiener, J. Math, and Phys. 2 (1923) 131.
5. R.P. Feynman and A.R. Hibbs, 'Quantum Mechanics and Path Integrals' (McGraw Hill, N.Y.)(1965).
6. Jazwinsky, 'Stochastic Processes and Filtering Theory", (Academic Press, N.Y.) (1970).

7. T.T. Soong, 'Random Differential Equations in Science and Engineering'.
8. R.E. Moretenson, J. Stat. Phys. 1, 271 (1969).
9. S. Chandrasekhar, Rev. Mod. Phys. 15, (1943) 1.
10. M.C. Wang and G.E. Uhlenbeck, Rev. Mod. Phys. 16 (1945) 323.
11. R.L. Stratanovich, "Topics in the theory of Random Noise' Vol.I, (Gordon and Breach) (1963)(English translation).
12a. I.I. Gikman and A.V. Skorohod, 'Stochastic Differential Equations' (Springer-Verlag)(1972).
12b. S.K. Srinivasan, 'Stochastic integrals' S.M. Archives, Vol.3 (1978) p 325.
13. R.L. Stratanovich, SIAM Journal of Control 4 (1966) 363.
14. I.E. Wong and M. Zakai, Int. J. Eng. Sci. 3 (1965) 213.
15. K. Itô, Springer Lecture Notes in Physics (1979) p.214.
16. E.J. McShane, 'Stochastic Calculus and Stochastic Models' (Academic Press, N.Y.) (1974).
17. H.P. McKean, Jr. 'Stochastic Integrals', (Academic Press, N.Y.) (1969).
18. N.G. van Kampen, 'Rep. Prog. Phys. 24C (1976) 171.
19. Alladi Ramakrishnan and R. Vasudevan, J. Ind. Math. Soc., (Golden Jubilee Volume) 24 (1961) 457.
20. M. Aoki, 'Optimization of Stochastic Systems' (Academic Press, N.Y.) (1967).
21. Kushmer, 'Stochastic Stability and Control' (Academic Press, N.Y.) (1967).
22. Astrom, 'Introduction to Stochastic Control Theory' (Academic Press, N.Y.)(1970)
23. R.L. Stratanovich, 'Conditional Markov Processes and then Applications to the Theory of Optimal Control' (American Elsivier, N.Y.) (1963).
24. Middleton, 'An Introduction to Statistical Communication Theory' (McGraw Hill, N.Y.) (1960).
25. S.I. Markus, IEEE Trans. Information Theory, II-24 No.2 p.164.
26. R. Vasudevan and K.V. Parthasarathy (to be published).
27. Kitahara et al Phys. Lett. A70, 377.
28. B.J. West et al, Physica A97, (1979) 211, 234.
29. S. Kabashima et al, J. Appl. Phys. 50 (1979) 6296.
30. A. Schenzle and H. Brand, Phys. Rev. A20 (1979) 1628.

ON SOME NEW CONCEPTS IN PROBABILITY THEORY

R. Jagannathan
The Institute of Mathematical Sciences
Madras - 600113, India

I shall first discuss the new concepts 'disparity', 'activity' and 'duality' introduced recently in the study of some stochastic processes by Alladi Ramakrishnan (cf. [1-5] and references therein) and in conclusion advance arguments based on our work [6-8] to indicate the interesting possibility of a fundamental role for classical probability in the quantum theory of subnuclear phenomena.

For a physical system with n possible states let $\pi(j;t)$ denote the probability that it is in the state j at time t and $\pi(j/k;t)$ denote the probability that the system is in the state j at time t given that it is in the stake k at time t = 0. Obviously

$$\sum_j \pi(j;t) = 1, \qquad \sum_j \pi(j/k;t) = 1 \tag{1}$$

Then the Chapman-Kolmogorov equation

$$\pi(j;t_2) = \sum_k \pi(j/k;t_2-t_1)\,\pi(k;t_1) \tag{2}$$

or equivalently the matrix equation

$$\vec{\pi}(t_2) = [\pi(t_2-t_1)]\,\vec{\pi}(t_1) \tag{3}$$

with $\pi(j;t)$ as the j-th element of the probability vector $\vec{\pi}(t)$ and $\pi(j/k;t)$ as the jk-th element of the matrix $[\pi(t)]$ representing the probability transition operator, gives the temporal evolution of the system.

Now if,

$$[\pi(t_2-t_1)] \approx I + R\,(t_2-t_1) \quad \text{for } t_2-t_1 = \Delta \approx 0 \tag{4}$$

where I is the identity matrix and R defines the fundamental transition probability matrix such that

$$\pi(j/k;t) \rightarrow R(j/k)\Delta \quad \text{for } j \neq k \text{ as } t \rightarrow \Delta \approx 0$$

$$R(j/k) \geqslant 0, \quad \text{for } j \neq k, \quad R(k/k) = -\sum_{j \neq k} R(j/k) \tag{5}$$

then the evolution equation for the process, (2) or (3), can be written as

$$\frac{d\vec{\pi}(t)}{dt} = R\vec{\pi}(t) \tag{6}$$

with the formal solution

$$\vec{\pi}(t) = e^{Rt}\vec{\pi}(0) \tag{7}$$

i.e. in (3) $[\pi(t)] = e^{Rt}$. In terms of matrix elements (6) becomes

$$\frac{d\pi(j;t)}{dt} = \sum_{k\neq j} R(j/k)\,\pi(k;t) - \Big[\sum_{k\neq j} R(k/j)\Big]\pi(j;t) \tag{8}$$

When the process evolves according to (8) it is obvious that for given j, t and Δt, $\pi(j;t+\Delta t) - \pi(j;t)$ can assume any value between -1 and 1, <0, >0 or $= 0$. Then, the following question arises. Is there any sense of 'direction' in the evolution of the stochastic process governed by the above equations or in other words does there exist any quantity associated with the above process which strictly decreases (or increases) with time? Recently this question has been answered by Ramakrishnan in [1-3] through the introduction of the new concepts 'disparity' and 'activity' as follows.

Let two identical physical systems evolve according to (8) with the same fundamental R matrix but starting from two different initial distributions $\vec{\pi}_1(0)$ and $\vec{\pi}_2(0)$ at t = 0. According to (7) at time t

$$\vec{\pi}_1(t) = e^{Rt}\vec{\pi}_1(0), \qquad \vec{\pi}_2(t) = e^{Rt}\vec{\pi}_2(0) \tag{9}$$

Now let the 'disparity' between the two distributions at time t be defined by

$$D(t) = \sum_j |\pi_2(j;t) = \pi_1(j;t)| \tag{10}$$

which is always nonnegative with an upper bound 2. Then Ramakrishnan's theorem states that the disparity D(t) as defined by (10) decreases with time t.

Proof of the above theorem can be given easily as follows. Let at a given time t

$$D_+(t) = \sum_{\substack{\text{sum over all}\\ j_+ \text{ such that}\\ \pi_2(j_+;t)\geqslant\pi_1(j_+;t)}} \left[\pi_2(j_+;t) - \pi_1(j_+;t)\right] > 0 \tag{11}$$

and

$$D_-(t) = \sum_{\substack{\text{sum over all}\\ j_+ \text{ such that}\\ \pi_2(j_-;t)<\pi_1(j_-;t)}} \left[\pi_2(j_-;t) - \pi_1(j_-;t)\right] < 0 \tag{12}$$

It is to be noted that for any given t the sets (j_+) and (j_-) do not have any common element. Hence obviously for all t

$$D_+(t) + D_-(t) = 0 \tag{13}$$

so that

$$D(t) = D_+(t) - D_-(t) = 2D_+(t) = 2\,|D_-(t)| \tag{14}$$

and as long as $D(t) > 0$ neither of the sets (j_+) and (j_-) can be empty. Thus it is enough if it is proved that $D_+(t)$ decreases strictly with time. This can be shown as follows. Using (6) and the fact $\sum_{\text{all } j} R(k/k) = 0$ lead to

$$\frac{dD_+(t)}{dt} = \sum_{j_+} \Big\{ \sum_{j'_+} R(j_+/j'_+) [\pi_2(j'_+;t) - \pi_1(j'_+;t)]$$

$$+ \sum_{j'_-} R(j_+/j'_-) [\pi_2(j'_-;t) - \pi_1(j'_-;t)] \Big\}$$

$$= \sum_{j'_+} [\sum_{j_+} R(j_+/j'_+)][\pi_2(j'_+;t) - \pi_1(j'_+;t)]$$

$$+ \sum_{j'_-} [\sum_{j_+} R(j_+/j'_-)][\pi_2(j'_-;t) - \pi_1(j'_-;t)]$$

$$= \sum_{j'_+} [-\sum_{j_-} R(j_-/j'_+)][\pi_2(j'_+;t) - \pi_1(j'_+;t)]$$

$$+ \sum_{j'_-} [\sum_{j_+} R(j_+/j'_-)][\pi_2(j'_-;t) - \pi_1(j'_-;t)] \tag{15}$$

Now using the definition of the elements of R in (5) and the classification of j's into (j_+) and (j_-) it is seen that

$$\frac{dD_+(t)}{dt} < 0 \tag{16}$$

which proves Ramakrishnan's theorem.

When the states of the system are to be labelled by a continuous parameter x instead of the discrete index j, the equations (1), (2),, (5) and (8) become

$$\int dx\, \pi(x;t) = 1, \quad \int dx\, \pi(x/x';t) = 1 \tag{17}$$

$$\pi(x;t_2) = \int dx'\, \pi(x/x';t_2-t_1)\, \pi(x';t_1) \tag{18}$$

$$\pi(x/x';t) \longrightarrow R(x/x')\Delta \qquad \text{for } x \neq x' \quad \text{as } t \longrightarrow \Delta \approx 0 \tag{19}$$

and

$$\frac{d\pi(x;t)}{dt} = \int dx' R(x/x')\, \pi(x';t) - \pi(x;t)\Big(\int dx' R(x'/x)\Big) \tag{20}$$

Consequently the 'disparity' between the distributions of two identical physical systems governed by the same evolution equation (20) becomes

$$D(t) = \int dx\, |\pi_2(x;t) - \pi_1(x;t)| \tag{21}$$

where $\pi_2(t)$ and $\pi_1(t)$ are the distributions reached at time t from the initial distributions $\pi_2(0)$ and $\pi_1(0)$ respectively at $t = 0$ according to

$$\pi_1(x;t) = \int dx'\, \pi(x/x';t)\, \pi_1(x';0)$$

$$\pi_2(x;t) = \int dx'\, \pi(x/x';t)\, \pi_2(x';0) \qquad (22)$$

Now if $\pi_2(x;0)$ is taken as $\pi_1(x;\Delta t)$ then $\pi_2(x;t) = \pi_1(x;t+\Delta t)$. Then the corresponding expression for $D(t) = \int dx\,|\,\pi_1(x;t+\Delta t) - \pi_1(x;t)\,|$ can be associated with system 1 itself meaningfully. Thus Ramakrishnan introduced in [1] the quantity

$$A(t,\Delta t) = \begin{cases} \sum_j |\,\pi(j;t+\Delta t) - \pi(j;t)| & \text{for discrete state-space} \\ \int dx\,|\,\pi(x;t+\Delta t) - \pi(x;t)| & \text{for continuous state-space} \end{cases} \qquad (23)$$

called the 'activity' of the given system in the time interval $(t,t+\Delta t)$. Looked at as a special case of 'disparity' it is clear that the probability 'activity' of the system also decreases with t for any chosen time interval Δt.

Having thus explained the meaning of the 'sense of direction' in the case of the above type of Markovian stochastic processes through the introduction of the concepts 'activity' and 'disparity' in [1-3] Ramakrishnan analyses in [4] the basic problem of inverse probability in such cases which involves the question : Is it possible to make statements about the past given the present state of the system? He shows that a 'duality' is inherent in this respect in the mathematical theory of stochastic processes as follows.

In conventional treatments (2) or (3) is usually written only for $t_2 > t_1$. Ramakrishnan has demonstrated as early as 1955 in [5] that the equation

$$\vec{\pi}(t_2) = [\pi(t_2-t_1)]\,\vec{\pi}(t_1) = e^{R(t_2-t_1)}\,\vec{\pi}(t_1) \qquad (24)$$

is valid even $t_2 < t_1$. In such a case $[\pi(t_2-t_1)]$ can be used meaningfully as an 'operator' to relate $\vec{\pi}(t_2)$ to $\vec{\pi}(t_1)$ for $t_2 < t_1$ though $[\pi(t_2-t_1)]$ will have negative elements without any probability significance. However $\vec{\pi}(t_2)$ can be obtained from $\vec{\pi}(t_1)$ for $t_2 < t_1$ in a different manner also. The joint probability that the system is in state j at time t_2 and in state k at time t_1 for $t_1 > t_2$ can be written in two ways as

$$\pi(j,t_2;k,t_1) = \pi(j;t_2)\,\pi(k/j;t_1-t_2) = \pi(k;t_1)\,P(j/k;t_2-t_1) \qquad (25)$$

where the 'inverse probability'

$$P(j/k;t_2-t_1) = \pi(j;t_2)\;\pi(k/j;t_1-t_2)\;\pi(k;t_1)^{-1} \qquad (26)$$

is nonnegative by definition, of course assuming that for any k $\pi(k;t_1) > 0$. Then $\vec{\pi}(t_2)$ can be obtained from $\vec{\pi}(t_1)$ for $t_2 < t_1$ through the relation

$$\vec{\pi}(t_2) = [P(t_2-t_1)]\,\vec{\pi}(t_1) \qquad (27)$$

also instead of through (24) where the matrix $[P(t_2-t_1)]$ with $P(j/k;t_2-t_1)$ is the typical element can be recognised to be given by

$$[P(t_2-t_1)] = D(t_2)\,[\pi(t_1-t_2)]^T\; D^{-1}(t_1)\;, \quad t_2 < t_1$$

$$[\pi(t_1-t_2)]^T_{jk} = \pi(k/j;t_1-t_2) \quad , \quad D(t)_{jk} = \pi(j;t)\,\delta_{jk} \quad . \tag{28}$$

Thus in the case of $t_2 < t_1$ Ramakrishnan calls $[\pi(t_2-t_1)]$ as the 'analytic inverse' and $P(t_2-t_1)$ as the 'Bayes inverse' and notes that $\vec{\pi}(t_2)$ can be related to $\vec{\pi}(t_1)$ in a 'dual' manner either through $[\pi(t_2-t_1)]$ using (24) or through $[P(t_2-t_1)]$ using (27). This 'duality' in relating $\vec{\pi}(t_2)$ to $\vec{\pi}(t_1)$ is not surprising since more than one matrix can transform one vector to another. However the 'analytic inverse' is fundamental since it is independent of $\vec{\pi}(t_1)$ and $\vec{\pi}(t_2)$ whereas the 'Bayes inverse' transforming $\vec{\pi}(t_1)$ to $\vec{\pi}(t_2)$ depends on $\vec{\pi}(t_1)$ and $\vec{\pi}(t_2)$ (cf. [4] for more details).

I shall now conclude by mentioning an interesting possibility in the quantum theory of subnuclear phenomena. (cf. [6-8] for more details). So far all known particles are described by wavefunctions $\psi(q)$ with $|\psi(q)|^2 dq$ representing the probability of locating the particle somewhere in the position interval (q, q+dq) and it is assumed that q can take all values continuously from $-\infty$ to ∞. This assumption is implicit in prescribing the position operator $\hat{q}$ and the conjugate momentum operator $\hat{p}$ to obey the Heisenberg relation

$$[\hat{q}, \hat{p}] = i\hbar \tag{29}$$

which can have only infinite dimensional realizations as is well known. As pointed out by Weyl in [9] the Heisenberg relation (29) can be written equivalently in the form

$$\exp(i\alpha\hat{p}/\hbar)\exp(i\beta\hat{q}/\hbar) = \exp(i\alpha\beta/\hbar)\exp(i\beta\hat{q}/\hbar)\exp(i\alpha\hat{p}/\hbar) \quad , \; -\infty \leqslant \alpha\, , \; \beta \leqslant \infty \tag{30}$$

and hence $\hat{q}$ and $\hat{p}$ can be thought of as the generators of the continuous group of unitary operators

$$G_{\alpha\beta\gamma} = \exp(i\alpha\hat{p}/\hbar)\; \exp(i\beta\hat{q}/\hbar)\; \exp(i\gamma/\hbar) \; , \qquad -\infty \leqslant \alpha, \beta, \gamma \leqslant \infty$$

$$G_{\alpha\beta\gamma}G_{\alpha'\beta'\gamma'} = G_{\alpha+\alpha',\, \beta+\beta',\, \gamma+\gamma'-\alpha'\beta} \tag{31}$$

Thus Weyl associated the kinematical basis of Schrödinger-Heisenberg-Dirac quantum mechanics with the above continuous unitary group and expressed the prophetic hope : "..... the field of discrete groups offers many possibilities which we have not yet been able to realize in Nature ; perhaps these holes will be filled by applications to nuclear physics".

A fundamental role for discrete analogues of the above unitary group G, as envisaged by Weyl, ofcourse with the natural requirement that the continuous group G underlying the current quantum mechanics of atomic and nuclear phenomena should be the asymptotic limit of such discrete groups (correspondence principle of Bohr), can arise at a deeper, say subnucleonic, level if in that realm the phase-space formed by all eigenvalues of q and p or equivalently α and β has a lattice structure given by, for instance,

$$\alpha = n\epsilon \; , \; \beta = m\eta \; , \; \epsilon\eta = 2\pi\hbar/(2J+1) \, , \; n,m = -J,-J+1,\ldots,-1,0,1,\ldots,J-1,J \tag{32}$$

where J is a positive integer. Then the new position operator Q and the momentum operator P can be looked upon as the 'generators' of the finite group G defined by

$$G_{nmk} = A^n B^m \omega^k, \quad AB = \omega BA, \quad A^{2J+1} = B^{2J+1} = I, \quad \omega^{2J+1} = 1$$

$$G_{nmk} G_{n'm'k'} = G_{(n+n'),(m+m'),(k+k'-n'm)}, \quad (x) = x.\mathrm{mod}(2J+1)$$

$$A = \exp(i\epsilon P/\hbar), \quad B = \exp(i\eta Q/\hbar), \quad \omega = \exp(i\epsilon\eta/\hbar) = \exp(i2\pi/(2J+1)) \qquad (33)$$

and with the faithful and unique irreducible representation

$$\langle n | Q | n' \rangle = n\epsilon\delta_{nn'}$$

$$\langle n | P | n' \rangle = \begin{cases} 0 & \text{for } n = n' \\ (\eta/2i)(-1)^{n-n'} \operatorname{cosec}[(n-n')\pi/(2J+1)] & \text{for } n \neq n' \end{cases}$$

$$n, n' = -J, -J+1, \ldots, -1, 0, 1, \ldots J-1, J \qquad (34)$$

(cf. [6-12] for details)

A formalism of quantum mechanics in which the usual infinite-dimensional Schrodinger-Heisenberg realizations of q and p are replaced by finite-dimensional matrices Q and P respectively should be, in my opinion, an ideal candidate for the quantum mechanics of quark phenomena if it is true that these subnucleonic constituents have permanent confinement or in other words have only a finite spectrum for position as observed in the rest frame of the composite nucleon. Such Weylian Finite-Dimensional Quantum Mechanics (WFDQM) is being developed by us (cf. [6-8]) to describe the quantum mechanics of quarks with reference to the rest frame of the nucleons, based on the early pioneering works of Weyl [9] and Schwinger [10] on the group structure of quantum kinematics and the recent works of Ramakrishnan and collaborators [11,12] on generalized Clifford algebras of the type defined in (33). Gudder and Naroditsky also have analysed in [13] similar forms of 'finite-dimensional quantum mechanics'. A form of quantum mechanics, analogous to our WFDQM, is also implicit in the work of Drell, Weinstein and Yankielowicz [14] on the lattice field theory of fermions wherein the derivative operation on the lattice is taken as the action of the matrix operator $iP/\hbar$ on the concerned vector defined on the lattice with the matrix P defined exactly as in (34) above.

So far in our work [6-8] quarks have been considered as belonging to a formalism of WFDQM based on some particular group G as defined in (33) and (34) with a fixed value for J. For any given J of course the Heisenberg uncertainty principle holds in the sense that any state of the quark can not belong to a single eigenpoint in the phase-space lattice since still the commutator [Q,P] is nonzero. But the question arises whether the phase-space lattice itself can have a sharply defined structure as described above. In other words, can the value of J be taken as a fixed integer for a given type of quark? This may not be so. Consequently J may have characteristic classical-type (in my opinion) of probability distributions so that the actual quantum mechanics of quarks may be a linear superposition of close-lying forms of Weylian quantum mechanics, in the sense described above, with classical probability weights. It should be interesting to explore this possibility further.

References

1. Alladi Ramakrishnan, 'A mathematical feature of evolutionary stochastic processes', Sonderdruck aus Methods of Operations Research 36, 239 (1980)(paper presented at the Oberwolfach Conference in November 1978)
2. Alladi Ramakrishnan, 'Approach to stationarity in stochastic processes', Proc. Tamil Nadu Acad. Sciences, 2, 189 (1979)
3. Alladi Ramakrishnan, 'A new concept in probability theory', J. Math. Anal. Appl. 83, 408 (1981)
4. Alladi Ramakrishnan, 'Duality in stochastic processes', J. Math. Anal. Appl. 84, 483 (1981)
5. Alladi Ramakrishnan, 'Inverse probability and evolutionary Markov stochastic processes', Proc. Indian Acad. Sciences A41, 145 (1955)
6. R. Jagannathan, T.S. Santhanam and R. Vasudevan, 'Finite-dimensional Quantum Mechanics of a particle', International J. Theor. Phys. 20, 755 (1981)
7. R. Jagannathan and T.S. Santhanam, 'Finite-dimensional quantum mechanics of a particle-II', International J. Theor. Phys. 21, 351 (1982)
8. R. Jagannathan, 'Finite-dimensional quantum mechanics of a particle-III:Weylian quantum mechanics of confined quarks' International J. Theor. Phys. 22, (1983) to appear
9. H. Weyl, Theory of Groups and Quantum Mechanics, Dover, New York (1932) Section 4.14
10. J. Schwinger, Quantum Kinematics and Dynamics, Benjamin, New York (1970) Sections 2.12 and 6.18
11. Alladi Ramakrishnan, L-Matrix Theory or Grammar of Dirac Matrices, Tata-Mc Graw Hill, New Delhi, (1972)
12. R. Jagannathan and N.R. Ranganathan, 'Generalized Clifford groups-I and II' Reports on Mathematical Physics, 5, 131 (1974), 7, 229 (1975)
13. S. Gudder and V. Naroditsky, 'Finite-dimensional quantum mechanics', International J. Theor. Phys. 20, 619 (1981)
14. S.D. Drell, M. Weinstein and S. Yankielowicz, 'Strong-coupling field theories II : Fermions and gauge fields on a lattice', Phys. Rev. D14, 1627 (1976)

DECAY OF METASTABLE STATES - KRAMERS, FIRST PASSAGE TIME AND VARIATIONAL APPROACHES

S. Dattagupta and S.R. Shenoy
School of Physics, University of Hyderabad
Hyderabad 500 134, India

1. Introduction

The dynamics of a great variety of problems can be usefully studied on the basis of the Fokker-Planck equation (FPE), that builds in at the outset, a separation of macroscopic and microscopic time scales. The effect of macroscopic drives and dissipations enters through the drift terms, whereas the diffusion terms account for rapidly varying microscopic noise. An important question, conveniently answered in the FPE framework, is : what is the time required for the passage of a prepared initial state to a final stationary state . Physical examples where such questions find relevance include the nucleation rate of a liquid droplet from the vapour phase [1], decay of a supercurrent in a superfluid or superconductor [2] , spinodal decomposition in an alloy [3], optimum sweep rates for hysteresis in first order transitions [4], and so on. It is clear that deterministic forces and random noise will both play a role in determining the transition rates. Quite generally, therefore, one is interested in calculating the decay rate or the relaxation time of a metastable state through the natural tool of the FPE [5] .

Historically, a problem of this kind was first studied in a classic paper by Kramers[6]. He was interested in computing the rate of escape of a particle across a mechanical barrier due to thermal fluctuations. An FPE was used, with the stochastic variable being the position x, the thermal energy $k_B T$ being related to the diffusion constant, and the mechanical force and viscosity determining the drift term. The Kramers treatment is based on an ansatz that should be valid when the barrier height is larger than the thermal energy. A similar high barrier or weak noise assumption is invoked at different stages of other methods as will become clear later on.

An explicit example of a Kramers-like problem would be rotational Brownian motion of a single domain magnetic particle in a highly anisotropic potential[7]. Here the angle θ would replace the variable x. Another example, in a more general context, could be the dynamics of an equilibrium or nonequilibrium phase transition. Then the variable x would be an (single component) order parameter while the potential would be a coarse-grained free energy-like functional[8]. Extensions of the one-dimensional Kramers problem to many variables have also been made by several authors, using variants of the original Kramers ansatz[5]or other techniques[9]. Some comments on this will be made below.

The Kramers method is intuitively appealing and based on sound physical insight into the problem. An alternative approach is to estimate the mean time for a stochastic variable, within a given region, to first reach the boundary of that region. The equation governing the dependence on the initial position of this mean 'First-Passage Time' (FPT), is derived from the FPE. In the high barrier/weak noise limit, the Kramers and FPT estimates for the decay rate of a metastable state are identical, if 'passage' is appropriately defined. The FPT formalism is instructive because it links the problem more directly and systematically to the machinery of stochastic problems [10], it can be formally generalized to the many-variable case and situations where detailed balance does not hold[9]. Physical applications of the FPT ideas include transient phenomena in optically bistable systems [11,12] and analysis of the problem of hysteresis versus jump behaviour in first order equilibrium and nonequilibrium transitions [4] .

Mathematically, the FPE is parabolic : it is a second order partial differential equation in the space derivatives and first order in the time derivative. The time-dependent probability can therefore be written as an eigenfunction expansion, with the eigenvalue$\{\lambda_n\}$ appearing in exponential decay factors associated with each eigenfunction. The zero eigenvalue corresponds to the stationary state, and the nontrivial eigenvalues, therefore, determine the decay rates for metastable states. This allows one to introduce a third, formally elegant approach to the problem, involving variational bounds on these eigenvalues. It turns out that the lowest non-trivial eigenvalue $\lambda_1 \sim \langle T_p \rangle^{-1}$, the inverse of the mean first passage time, in the large barrier limit. A physical application of this approach to the Brownian motion of a magnetic particle is discussed in detail in this volume [13].

The Kramers, FPT and variational approaches are different ways of looking at the same metastable state decay problem. These are all connected in the high barrier/low noise limit, as will be illustrated in the following sections. As mentioned earlier, generalizations of the one dimensional decay problem to many dimensions have been made in the context of Kramers treatment[5] and the FPT approach[9,14] . It is not surprising that similar ideas of saddle-point coordinate systems, small noise expansions, etc., have been independently introduced by mathematicians[9,14] and physicists[5,15] sometimes without knowledge of analogous work done elsewhere.

The rest of the article is organized as follows. In §2, we outline the basic framework of the problem. The Kramers analysis for the one dimensional case is treated in § 3. We describe next the one dimensional FPT approach in § 4. The variational treatment for the one dimensional case is then included in § 5. The generalization to higher dimensions, in the FPT framework, as done by Schuss and Matkowsky, and its relationship to Kramers-like ideas[5], are discussed in §6. Finally, some applications of the FPT ideas are considered in § 7.

2. Basic mathematical picture

For most part of the analysis (see, however, § 6), we shall restrict ourselves to a one-dimensional, multiplicative, but time-homogeneous stochastic process described by the FPE [16]

$$\frac{\partial P}{\partial t}(x,t) + \frac{\partial J(x,t)}{\partial x} = 0 , \tag{2.1}$$

where the 'current'

$$J(x,t) = - A(x)\ P(x,t) - \frac{\partial}{\partial x}(D(x)\ P(x,t)) , \tag{2.2}$$

A(x) and D(x) being the drift and diffusion terms respectively. The stationary state solution (at $t = \infty$) is obtained by setting the current to zero, hence

$$P_o(x) = C \ \exp(-\Phi(x)) , \tag{2.3}$$

where C is a normalization constant, and

$$\Phi(x) = \ln D(x) + \int dx' \frac{A(x')}{D(x')} . \tag{2.4}$$

We shall employ in (2.3) the natural boundary conditions :

$$P_o(\pm\infty) = 0 \tag{2.5}$$

It may be noted here that if one specializes to the case of an additive stochastic process, D(x) is a constant, and

$$P_o(x) = C' \exp(-\int dx' A(x')/D) . \tag{2.6}$$

In addition, if the so-called 'potential condition' is satisfied, i.e. [10]

$$A(x) = \frac{\partial U(x)}{\partial x} , \tag{2.7}$$

then,

$$P_o(x) = \eta \exp(- U(x)/D) , \tag{2.8}$$

where η is yet another normalization constant. In the context of an 'equilibrium' problem, (2.8) has the familiar structure with D being proportional to the thermal energy $k_B T$.

Coming back to the general case of (2.3) and (2.4), we shall direct our attention to a bistable potential indicated schematically in Fig.1. The point X_s, in the one dimensional case, is a maximum of the potential. However, we use the subscript s to indicate that X_s is actually a saddle point in the more general context of a multidimensional process (§ 6). As mentioned in § 1, our analysis is restricted to high barrier/weak noise limit which implies that $A(x_s)/D(x_s)$ is 'suitably' small. The question we want to answer is : starting from an arbitrary initial distribution i.e. $P(x,t=o) = P_{init}(x)$, how long does one have to wait for the probability to evolve into $P_o(x)$, given by (2.3) ? The sequence of time development is expected to

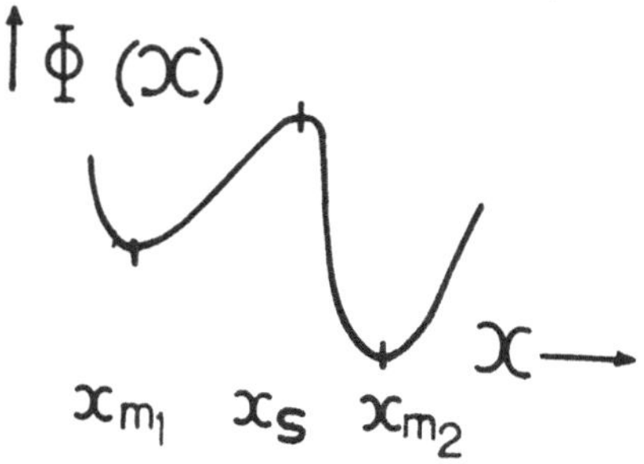

Fig.1. Sketch of $\Phi(x)$. The point X_{m_1} is a metastable minimum, x_{m_2} is a stable minimum, while X_s is an unstable maximum.

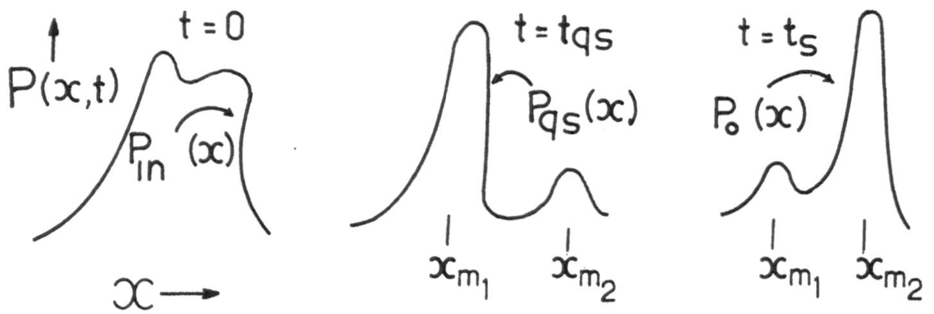

Fig. 2 The development of the probability. The quasistationary distribution $P_{qs}(x)$ is expected to be reached 'very quickly' ($t_s \gg t_{qs}$).

look like the one depicted in Fig.2. It is intuitively expected (and the expectation can be substantiated by WKB-type arguments [17]) that the time t_{qs} to reach the quasi-stationary distribution is much shorter than the time t_s to reach the final stationary distribution, in the large barrier/weak noise limit. We are, of course, interested in studying only the slow evolution of the probability between t_{qs} and t_s.

3. Kramers' argument

In the regime $t_{qs} \leqslant t \leqslant t_s$, the probability $P(x,t)$ is so slowly varying that its time derivative can be neglected. Hence from (2.1) and (2.2), the current can be replaced by J_{qs} which is independent of x and is a slowly varying function of t. Thus, in this domain of time-evolution, population of the state x_{m_1} is depleted while that of 'state' x_{m_2} is increased at an almost steady rate, in view of slow diffusion (or probability leakage) across X_s. Denoting by $P_{qs}(x,t)$ the probability in the quasi-stationary region, we have from (2.2) - (2.4),

$$J_{qs} \exp\left(\bar{\Phi}(x)\right) = -\frac{\partial}{\partial x}\left\{\exp\left(\Phi(x)\right) P_{qs}(x,t)\right\} \tag{3.1}$$

where

$$\bar{\Phi}(x) = \Phi(x) - \ln D(x) \quad . \tag{3.2}$$

Now, consider two 'small' regions $x_{m_1} - \Delta x_1 \leqslant x \leqslant x_{m_1} + \Delta x_1$ and $x_{m_2} - \Delta x_2 \leqslant x \leqslant x_{m_2} + \Delta x_2$, and define the 'density of points' in these regions by

$$n_1 \equiv \int_{x_{m_1}-\Delta x_1}^{x_{m_1}+\Delta x_1} dx \; P_{qs}(x,t) \quad . \tag{3.3}$$

A similar expression defines n_2. It is evident from Fig.2 that $P_{qs}(x,t)$ is sharply peaked around x_{m_1} and x_{m_2}. Accordingly,

$$P_{qs}(x,t) \approx P_{qs}(x_{m_1},t) \exp(-(\Phi(x) - \Phi(x_{m_1})), \; x_{m_1} - \Delta x_1 \leqslant x \leqslant x_{m_1} + \Delta x_1 \tag{3.4}$$

An analogous expression holds in the region around x_{m_2}. Combining (3.3) and (3.4)

$$n_1 \approx I_1 P_{qs}(x_{m_1},t)\exp(\Phi(x_{m_1})), \tag{3.5}$$

where $$I_1 \equiv \int_{x_{m_1}-\Delta x_1}^{x_{m_1}+\Delta x_1} dx \exp(-\Phi(x)). \tag{3.6}$$

The expression for n_2 is obtained from (3.5) upon replacing 1 by 2. Integrating (3.1) from x_{m_1} to x_{m_2}, and using (3.5), we have

$$J_{qs} I_s \approx n_1/I_1 - n_2/I_2, \tag{3.7}$$

where $$I_s \equiv \int_{x_{m_1}}^{x_{m_2}} \exp(\Phi(x))\, dx. \tag{3.8}$$

Finally, we note that at this level of approximation, most of the system points are expected to be concentrated only around x_{m_1} and x_{m_2}. Thus

$$J_{qs} \approx \dot{n}_1 = -\dot{n}_2. \tag{3.9}$$

Combining (3.7) with (3.9), we arrive at the familiar rate equations

$$\dot{n}_1 = -\dot{n}_2 = -\omega_{12} n_1 + \omega_{21} n_2, \tag{3.10}$$

where

$$\omega_{12} \equiv \tau_{12}^{-1} = (I_s I_1)^{-1},\ \omega_{21} = \tau_{21}^{-1} = (I_s I_2)^{-1} \tag{3.11}$$

The times τ_{12} and τ_{21} are the so-called reaction times of Kramers which measure the times of passage from $x_{m_1} \longrightarrow x_{m_2}$ and $x_{m_2} \longrightarrow x_{m_1}$ respectively.

In the high barrier/weak noise limit, the integrals in (3.6) and (3.8) can be evaluated approximately by the method of steepest descents, and we obtain

$$\omega_{12} = \tau_{12}^{-1} \approx (2\pi)^{-1}(\bar{\Phi}''(x_{m_1})\,|\,\bar{\Phi}''(\bar{x}_s)|)^{1/2} \exp(-(\bar{\Phi}(\bar{x}_s)-\bar{\Phi}(x_{m_1})) \tag{3.12}$$

The expression for ω_{21} can be written down from (3.12) by interchanging 1 and 2. Here $\bar{x}_s$ is defined by

$$\bar{\Phi}'(x=\bar{x}_s) = 0,\quad \bar{\Phi}''(x=\bar{x}_s) < 0 \tag{3.13}$$

It may be remarked that the analysis given above is more general than the original Kramers' treatment in that a multiplicative process or x-dependent diffusion has been considered. In the special case of an additive process for which the potential condition is satisfied (cf., (2.6)-(2.8)), (3.12) yields the familiar result [6]

$$\omega_{12} \approx D(2\pi)^{-1}(U''(x_{m_1})\,|U''(x_s)|)^{1/2} \exp(-(U(x_s)-U(x_{m_1}))/D). \tag{3.14}$$

4. First Passage Time (FPT) estimates

In this section we shall outline briefly the FPT method for one dimensional stochastic processes in order to put it at par with the Kramers treatment. Later, in § 6,7, we shall return to a more elaborate discussion of FPT calculations in the context of multi-dimensional processes and mention a few physical applications as well.

Let us focus our attention first to calculating $\mathcal{T}_{12}$ (cf. (3.11)). The system point is assumed to be initially at x_o at time $t = o$, where $-\infty \leqslant x_o \leqslant x_s$. We imagine that an absorbing boundary is erected at x_s such that once the system point reaches x_s it is removed from any further consideration. We denote by $\mathcal{T}(x_o)$ the time taken by the system point to reach x_s for the first time, having started from x_o at $t = o$. This so called first passage time (FPT) is clearly a random variable which varies from realization to realization. We shall later identify the mean FPT $\langle \mathcal{T}(-\infty) \rangle$ with $\mathcal{T}_{12}$ of § 3 provided 'passage' is defined in an appropriate manner.

Let $P(x,t|x_o,o)$ be the conditional probability that the random process is x at time t given that it was x_o at $t=o$. Then, the probability that at time t the system point is still within the interval $-\infty$ to x_s (not having reached x_s even once) is given by

$$G(x_o,t) = \int_{-\infty}^{x_s} P(x,t|x_o,o)\, dx. \tag{4.1}$$

The quantity $G(x_o,t)$ evidently equals $P_r(\mathcal{T}(x_o) \geqslant t)$. Since the conditional probability by definition, is a delta function centred around x_o at $t = o$, it follows from (4.1) that

$$\begin{aligned} G(x_o,o) &= P_r(\mathcal{T}(x_o) \geqslant o) = 1, \quad -\infty \leqslant x_o \leqslant x_s \\ &= o \quad , \text{ elsewhere.} \end{aligned} \tag{4.2}$$

On the other hand, if x_o happens to equal x_s, the absorbing boundary, the process 'dies' immediately, hence

$$G(x_s,t) = P_r(\mathcal{T}(x_s) \geqslant t) = o \tag{4.3}$$

Alternatively

$$G(x_o,t=\infty) = o \tag{4.4}$$

as the system point is expected to reach x_s at least once when $t = \infty$.

Since $G(x_o,t)$ is the probability that passage has not occurred: $\mathcal{T}(x_o) \geqslant t$, and $1-G(x_o,t)$ is the probability that passage has occurred, the mean FPT is given by

$$T_p(x_o) = \langle \mathcal{T}(x_o) \rangle = -\int_o^{\infty} t\, \dot{G}(x_o,t)\,dt = \int_o^{\infty} G(x_o,t)\, dt, \tag{4.5}$$

where the last step follows upon integration by parts and from (4.4). Our next task is to derive an equation for $T_p(x_o)$. This can be done easily if one uses the backward FPE [10] which reads

$$\frac{\partial}{\partial t} P(x,t|x_o,o) = -A(x_o)\frac{\partial}{\partial x_o} P(x,t|x_o,o)+D(x_o)\frac{\partial^2}{\partial x_o^2} P(x,t|x_o,o). \tag{4.6}$$

Therefore, from (4.1) (4.5) and (4.6), we have

$$L^{\dagger}_{x_o} T_p (x_o) = -1 \tag{4.7}$$

where the adjoint operator $L^{+}_{x_o}$ is the one associated with the backward FPE:

$$L^{\dagger}_{x_o} \equiv D(x_o)\frac{\partial^2}{\partial x_o^2} - A(x_o)\frac{\partial}{\partial x_o}. \tag{4.8}$$

The solution of (4.7), consistent with the boundary conditions:

$G(x,t) = o$ at $x = x_s$ (absorbing boundary)

$$\frac{\partial}{\partial x} G(x,t) = o \text{ at } x = -\infty \text{(reflecting boundary)}, \tag{4.9}$$

can be written as

$$T_p(x) = \int_x^{x_s} dx' \exp(\bar{\Phi}(x')) \int_{-\infty}^{x'} dx'' \exp(-\bar{\Phi}(x'')) \quad . \tag{4.10}$$

Equations for higher moments of $\mathcal{T}(x)$ can also be derived from the distribution of FPT. It turns out [12] that in the high barrier/weak noise limit,

$$T_p^{(r)}(x) \equiv \langle \mathcal{T}^r (x) \rangle \simeq r! \langle \mathcal{T}(x) \rangle^r = r! (T_p(x))^r. \tag{4.11}$$

This corresponds to a distribution for $\mathcal{T}$ that for large lifetimes is

$$P(\mathcal{T}) \simeq [T_p(x)]^{-1} e^{-\mathcal{T}(x)/T_p(x)} \tag{4.12}$$

Evaluating the integrals in (4.10) by the method of steepest descent, in the large barrier limit and setting $x = -\infty$, we obtain an answer for $\mathcal{T}_{12} (= T_p(-\infty))$ which is one-half the Kramers estimate (cf.(3.12)). This discrepancy is, however, not serious and can be removed if the absorbing boundary X_s is taken at $+\infty$, which is of course the more appropriate limit, for a meaningful comparison with Kramers result.

5. Variational treatment

We may write the solution of the FPE (cf.(2.1) and (2.2) as

$$P(x,t) = a_o P_o(x) + \sum_{n>o} a_n P_n(x) \exp(-\lambda_n t), \tag{5.1}$$

where the eigenvalues $\lambda_n > o$ for $n>o$. The first term corresponds to zero eigenvalue which yields the stationary state solution (as $t\rightarrow\infty$). Comparing with (2.3), therefore,

$$a_o = c, \quad P_o(x) = \exp(-\Phi(x)) \quad . \tag{5.2}$$

Combining (5.1) with (2.1) leads to the elliptic partial differential equation

$$\frac{\partial}{\partial x} (A(x)\ P_n(x)) + \frac{\partial^2}{\partial x^2} (D(x)\ P_n(x)) = -\lambda_n\ P_n(x) \ . \tag{5.3}$$

Now, the substitution :

$$F_n(x) = P_n(x) \ / \ P_o(x) \ , \tag{5.4}$$

transforms (5.3) into a <u>self-adjoint</u> eigenvalue equation of the Sturm-Liouville form [18] :

$$\frac{\partial}{\partial x} (D(x)\ P_o(x) \frac{\partial}{\partial x})\ F_n(x) = -\lambda_n\ P_o(x)\ F_n(x) \ . \tag{5.5}$$

Equation (5.5) is the Euler-Lagrange equation for the functional (with the density function $P_o(x)$) :

$$I\left[F_n(x)\right] = \int_{-\infty}^{\infty} dx\ P_o(x) \left[D(x) \left(\frac{\partial F_n(x)}{\partial x} \right)^2 - \lambda_n\ F_n^2(x) \right] \ . \tag{5.6}$$

Following standard procedure [18], the eigenvalue λ_n can be shown to obey the Rayleigh-Ritz inequality

$$\lambda_n \leqslant K\left[f_n(x)\right] \ / \ H\left[f_n(x)\right] \ , \tag{5.7}$$

where

$$K\left[f_n(x)\right] = \int_{-\infty}^{\infty} dx\ D(x)\ P_o(x) \left(\frac{\partial f_n(x)}{\partial x} \right)^2 \ , \tag{5.8}$$

$$H\left[f_n(x)\right] = \int_{-\infty}^{\infty} dx\ P_o(x)\ f_n^2(x) \ . \tag{5.9}$$

and $f_n(x)$ is some general trial function.

In order to employ (5.7) for estimating the eigenvalue λ_n one normally assumes a particular form for the trial function $f_n(x)$ in terms of certain variational parameter(s) and minimizes the right hand side of (5.7) with respect to these parameter (s). As mentioned before, one requires only the lowest nontrivial eigen value for describing the passage from the quasistationary to the stationary state. Thus we shall use the inequality (5.7) for n=1 only, and drop the subscript n henceforth for the sake of brevity. The variational choice of the eigenfunction f(x) must be normalized and orthogonal to the lowest eigenfunction $F_o(x)$, i.e.,

$$\int P_o(x)\ f(x)\ dx = 0 \ , \tag{5.10}$$

since $F_o(x)$ is unity in the present case (cf. (5.4)). Equation (5.10) implies that the trial function must change sign as the system point moves from the interval $x_{m_1} \leqslant x \leqslant x_s$ to $x_s \leqslant x \leqslant x_{m_2}$.

The first and only (to the best of our knowledge) attempt to correlate the variational treatment with the Kramers kind of approach was due to Brown[7,13]. While the Brown choice for the trial function is quite adequate for calculating upper bounds to the eigenvalues from the Rayleigh-Ritz principle, it is not suitable for computing certain lower bounds given by Weinstein and Kamke [19]. The Weinstein-Kamke criteria, used in conjunction with the Rayleigh-Ritz principle, can obviously provide better bounds to the eigenvalues [20]. This turns out to be possible only if the trial function has finite first and second derivatives, a condition which the Brown choice does not satisfy. We shall present below an alternate form for the trial function which, in addition to providing Kramers-like result in the large barrier limit, yields also a lower bound to the eigenvalue λ [21].

Guided by the fact that $\frac{\partial f}{\partial x}$ should be concentrated near X_s where $P_o(x)$ has its minimum in order to keep $K[f(x)]$ small (see (5.8)), we choose

$$f(x) = f_1[1+\exp(-a(x-x_s))]^{-1} + f_2[1+\exp(a(x-x_s))]^{-1} , \tag{5.11}$$

where the constants f_1 and f_2 (which will turn out to be of opposite sign in view of (5.10)) can be determined from normalization and orthogonality conditions. Now, in anticipation of the fact that the Kramers - like formula should emerge in the large barrier limit, we take

$$a^2 = -b\ \Phi''(x_s) , \tag{5.12}$$

where b is a variational parameter and $\Phi''(x_s)$ is the curvature of the potential at x_s. Details of the variational calculation will be reported elsewhere [21]. Here, we merely state that substitution of (5.11) into (5.7) and evaluation of the relevant integrals by the method of steepest descent yields

$$\lambda \leqslant \frac{\pi}{8} b\ (1-b)^{-\frac{1}{2}}\ (\omega_{12} + \omega_{21}) , \tag{5.13}$$

where ω's are given by (3.12). Minimization of the b-dependent term in (5.13) leads to b=2, and hence the variational estimate for the lowest upper bound to the eigenvalue is

$$\lambda_v = \frac{\pi}{4} (\omega_{12} + \omega_{21}) , \tag{5.14}$$

which is still lower than the Kramers estimate by a factor of $4/\pi$!

6. First passage time estimates in higher dimensions

We have presented in § 3 - § 5 three distinct approaches to the study of the decay of a metastable state. It has been demonstrated that for a one dimensional stochastic process the Kramers, FPT and variational methods all lead to essentially identical result for the decay rate, in the high barrier/weak noise limit. The equivalence is expected to hold also for higher dimensional stochastic processes which are important in certain phase transitions such as in superfluidity [2] and two-mode lasers [22]. In this section, we outline the extension of the FPT calculation to higher dimension, based on the work of Schuss and Matkowsky [9,14].

We consider an n-component stochastic variable $\vec{x} = (x_1, x_2, \ldots\ldots x_n)$ taking an initial value $\vec{x}_o$ somewhere within an n-dimensional volume Ω. Generalizing the concept introduced in § 4, 'first passage' is now defined by $\vec{x}$, within a given realization, first reaching the boundary $\partial\Omega$, an (n-1) dimensional surface. The mean FPT satisfies an equation analogous to (4.7) :

$$L^+_{\vec{x}_o} \; T_p(\vec{x}_o) = \left(D\,\vec{\nabla}^2_{\vec{x}_o} - \vec{A}(\vec{x}_o)\cdot\vec{\nabla}_{\vec{x}_o} \right) T_p(\vec{x}_o) = -1 \;, \qquad (6.1)$$

where, for simplicity, we consider an $\vec{x}$ - independent (additive noise) diagonal diffusion term [23] and restrict ourselves to the potential case $\vec{A}(\vec{x}) = \vec{\nabla} U(\vec{x})$. As in the one-dimensional case, the boundary condition imposed is (cf.(4.3))

$$T_p(\vec{x}_o \in \partial\Omega) = 0 \qquad (6.2)$$

It may be noted that, unlike in the one dimensional problem, the equation for $T_p(\vec{x}_o)$ is now a partial differential equation in n variables, so obtaining a general solution is difficult. The strategy, therefore, would be to search for a single (or at least only a few) dominant variable and ignore the others. The Schuss-Matkowsky procedure employs just this in terms of a systematic high barrier/low noise expansion.

The first step is to scale (6.1) by ΔU, the smallest barrier height in the problem (defined below more precisely). Thus

$$(\epsilon\,\vec{\nabla}^2_{\vec{x}_o} - \vec{a}(\vec{x}_o)\cdot\vec{\nabla}_{\vec{x}_o})\,\mathcal{T}_p(\vec{x}_o) = -1 \;, \qquad (6.3)$$

where

$$\epsilon \equiv D/\Delta U \;, \quad \vec{a} \equiv \vec{A}/\Delta U \;, \quad \mathcal{T}_p \equiv \Delta U\, T_p \;. \qquad (6.4)$$

In the high barrier/weak noise limit, $\epsilon \ll 1$.

Since in the limit $\epsilon \to o$ the FPT is expected to be infinitely large, the basic idea of Matkowsky and Schuss is to isolate the singular dependence of $\mathcal{T}_p(\vec{x})$ on ϵ and assume the rest to be regular in ϵ, expandable in a power series. Thus

$$\mathcal{T}_p(\vec{x}) = v(\vec{x}) \exp(K/\epsilon) \;, \qquad (6.5)$$

with

$$v(\vec{x}) = v^{(o)}(\vec{x}) + \epsilon\, v^{(1)}(\vec{x}) + \ldots\ldots\ldots\ldots, \qquad (6.6)$$

where K is a constant. Substituting (6.5) and (6.6) into (6.3),

$$(\epsilon\,\vec{\nabla}^2_{\vec{x}} - \vec{a}(\vec{x})\cdot\vec{\nabla}_{\vec{x}})\,(v^{(o)}(\vec{x}) + \epsilon\, v^{(1)}(\vec{x}) + \ldots\ldots\ldots) = -\exp(-K/\epsilon) \qquad (6.7)$$

Now, if we were to retain only the term manifestly independent of ϵ, we would write

$$\vec{a}(\vec{x})\cdot\vec{\nabla}_{\vec{x}}\, v^{(o)}(\vec{x}) = 0 \;, \qquad (6.8)$$

which would imply that $v^{(o)}(\vec{x})$ is a <u>non-zero</u> constant(in regions where $\vec{a}(\vec{x}) \neq o$). This however, would immediately contradict the boundary condition (6.2), keeping in view (6.5). The paradox can be resolved satisfactorily if there is at least one direction along which $v^{(o)}(\vec{x})$ varies rapidly near the surface $\partial\Omega$ on a scale of $\epsilon^{1/2}$ in such a way that its second derivative times ϵ is actually independent of ϵ. This special direction which we call z, is expected to lie along the path of the steepest descent from the saddle point x_s on $\partial\Omega$. It is convenient now to switch to a new set of axes $(\vec{\rho},\vec{z})$, centred on the point $\vec{X}_s$, as indicated schematically in Fig.3. The unit vector $\hat{z}$ points inwards from the boundary and the other n-1 directions $\vec{\rho} = (\rho_2 \ldots\ldots \rho_n)$ are normal to $\hat{z}$. For points near $\vec{x}_s$ i.e. $\hat{z}$ small, a consistent expression to zeroth order in ϵ would include only the steepest descent variation

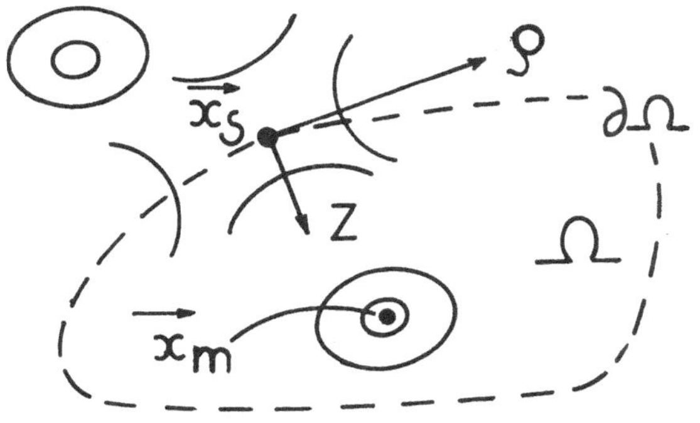

Fig.3. Schematic plot of equipotential (solid) lines and preferred axes $(z,\vec{\rho})$ at saddle point $\vec{x}_s$. $\vec{x}_m$ is the minimum within the boundary $\partial\Omega$ (dashed line).

$$\left(\epsilon \frac{\partial^2}{\partial z^2} - a_z(z,\vec{\rho}) \frac{\partial}{\partial z}\right) v^{(0)}(z,\vec{\rho}) = 0 \quad , \tag{6.9}$$

with terms first order in ϵ neglected :

$$\epsilon \vec{a}(z,\vec{\rho}) \cdot \vec{\nabla}_{\vec{\rho}} v^{(0)}(z,\vec{\rho}) \approx 0 \quad . \tag{6.10}$$

For the sake of simplicity we shall now assume that the potential condition (2.7) holds. From (6.4) then

$$a_z(z,\vec{\rho}) = \frac{1}{\Delta U} \frac{\partial U(z,\vec{\rho})}{\partial z} \quad . \tag{6.11}$$

Expanding around the saddle point $\vec{x}_s$ at z=o, $\vec{\rho}$ =o

$$a_z(z,\vec{\rho}) \approx -\frac{z}{\Delta U} \, |U''(o)| \quad . \tag{6.12}$$

Here the primes denote partial derivatives with respect to z, and the argument is evaluated at the origin of $(z,\vec{\rho})$, i.e. the saddle point.

Substituting (6.12) into (6.9), and using (6.4), we have

$$v^{(0)}(z,\rho) \approx \xi \Delta U \int_0^z \exp\left[-\frac{1}{2}\frac{|U''(o)|}{D} z'^2\right] dz' \tag{6.13}$$

where ξ is a constant independent of $\vec{\rho}$ (cf.(6.10). From (6.4) and (6.5), the mean FPT in the high barrier/weak noise limit is then given by

$$T_p(z) = \xi \exp(K/\epsilon) \int_0^z \exp\left[-\frac{1}{2}\frac{|U''(o)|}{D} z'^2\right] dz' \tag{6.14}$$

The remaining task is to evaluate the constant ξ . To this end, we note from (6.3),

$$\int_{\Omega} d^n x \exp\left(-\frac{U(\vec{x})}{D}\right) = \int_{\Omega} d^n x \exp\left(-\frac{U(\vec{x})}{D}\right) [a(\vec{x}) \cdot \vec{\nabla} - \epsilon \vec{\nabla}^2] T_p(x)$$

$$= -\epsilon \int_{\Omega} d^n x \; \vec{\nabla}_{\vec{x}} \cdot [(\vec{\nabla}_{\vec{x}} T_p(\vec{x})) \exp\left(-\frac{U(\vec{x})}{D}\right)] \quad , \qquad (6.15)$$

where we have used the potential condition (6.11) in writing (6.15) as a divergence in the n-dimensional space. Employing now the Gauss theorem, and noting that the normal to the surface $\partial\Omega$ (at z=o) lies inward along $\hat{z}$, we have

$$\int_{\Omega} d^n x \exp\left(-\frac{U(\vec{x})}{D}\right) = \epsilon \int_{\partial\Omega} d^{n-1}\rho \left[\left(\frac{\partial}{\partial z} T_p(z,\vec{\rho})\right) \exp\left(-\frac{U(z,\vec{\rho})}{D}\right)\right]_{z=o}$$

$$= \epsilon \, \Delta U \, \xi \exp(K/\epsilon) \int_{\partial\Omega} d^{n-1}\rho \; \exp\left[-\frac{U(o,\vec{\rho})}{D}\right] , \qquad (6.16)$$

where the last step follows from (6.14) and the definition of T_p. Equation (6.16) determines ξ and hence the mean FPT from (6.14). Noting that the mean FPT corresponds to the Kramers reaction time (§ 3) when the initial point z lies at infinity, we have from (6.14) and (6.16)

$$T_p \equiv T_p(z=\infty) = \frac{1}{D}\sqrt{\frac{\pi D}{2\,|U''(o)|}} \; \frac{\int_{\Omega} d^n x \exp(-U(\vec{x})/D)}{\int_{\partial\Omega} d^{n-1}\rho \; \exp\left[-\frac{U(o,\vec{\rho})}{D}\right]} \qquad (6.17)$$

Using sharp peaking arguments and assuming a minimum at $\vec{x}_m \in \Omega$, and a saddle point at $\vec{x}_s$, (6.17) can be given a more compact form

$$T_p = \frac{\pi \exp(\Delta U/D)}{[H_n(\vec{x}_n)]^{\frac{1}{2}} \left\{[H_{n-1}(\vec{\rho}=0)]^{-\frac{1}{2}} \, |U''(o)|^{\frac{1}{2}}\right\}} \qquad (6.18)$$

where ΔU, introduced earlier, is defined by

$$\Delta U \equiv U(\vec{x}_s) - U(\vec{x}_m). \qquad (6.19)$$

In (6.18), $H_n(\vec{x})$ is an nxn Hessian with its elements $\partial^2 U/\partial x_i \partial x_j$ evaluated at $\vec{x}_n$. The expression in curly brackers is evaluated at the origin ($\vec{x}_s$) of the coordinate system $(z,\vec{\rho})$, $H_{n-1}(\vec{\rho})$ being a similar (n-1) x (n-1) Hessian. It is easy to check that (6.18) reduces to the one-dimensional result discussed in § 4, upon replacing H_n by a second derivative and H_{n-1} by unity (see also (3.12)). For several saddle points on the boundary that are degenerate, i.e. have the same ΔU, a sum over the curly brackets is taken in the denominator of (6.18).

The result given above in (6.18) is for systems where the potential condition holds. We shall not deal with the 'non potential condition' case here except to remark that it can be done in terms of a series expansion in ϵ [23]. Graham and Schenzle [24] have developed similar ideas for the stationary solution of the FPE describing dispersive optical bistability, in the case where the potential conditions to not hold.

The generalization of the Kramers treatment to n dimensions has also been done [5]. An outline of the argument follows, pointing out the similarity to the FPT treatment above.

The quasistationary distribution, with steady probability flow across a saddle point at $\vec{x}_s$, is taken to be

$$P_{qs}(\vec{x}) = \omega(x)\, e^{-U(x)/D} \tag{6.20}$$

The probability current density describing this flow between wells is then

$$\vec{j}(\vec{x}) = -D(\vec{\nabla}\omega(\vec{x}))e^{-U(x)/D} \tag{6.21}$$

with the Fokker-Planck equation in the quasistationary limit, corresponding to $\vec{\nabla}.\vec{j}=0$. The $\omega(x)$ must have constant values in either well, to enhance/suppress the two peaks of the true stationary distribution $P_o(\vec{x}) = e^{-U(\vec{x})/D}$. By current conservation requirements, $P_o(\vec{x})$ falls off rapidly around $\vec{x}_s$, along the lines of current flow. $\omega(\vec{x})$ must therefore vary rapidly, near $\vec{x}_s$, to compensate. This is clearly reminiscent of the behaviour of $v(x)$ in the FPT treatment.

In fact, the equation for $\omega(x)$ coming from $\vec{\nabla}.\vec{J}(\vec{x}) = 0$ is seen to be similar to the homogeneous version of (6.3) (cf. also (5.4), (5.5), in the variational case). Once again, the arguments of a preferred coordinate system $(z,\vec{\rho}\,)$, retention of only the steepest descent variable z etc., carry through for $\omega(\vec{x})$, just as for $v(\vec{x})$ [5].

With hindsight, there is even some conceptual similarity between the FPT and the Kramers treatments. In the latter, the decay rate is the (integrated) current divided by the initial density of the particles in the metastable well. This is like an initial decay rate, maintained by replenishment of the metastable population. The FPT treatment equates the metastable decay rate to the mean time of first passage of a given realization. Thus the similiarities of both the intermediate arguments and the final results, is not surprising.

7. Applications:

We now briefly discuss some applications of the first-passage time ideas. Applications of the variational approach are discussed elsewhere [13].

The first passage time can be explicitly calculated for models of physical systems. For example a ring laser [22] with two counter-propagating (complex) field modes E_1, E_2, obeys the Langevin equation

$$\frac{\partial E_{1,2}}{\partial t} = (a_{1,2} - |E_{1,2}|^2 - \xi|E_{2,1}|^2)\, E_{1,2} + q_{1,2}(\tau)\,. \tag{7.1}$$

Here a_1 and a_2 are pump parameters for the two modes, ξ is a mode-coupling constant $2 > \xi > 1$, and $q_1(t)$ and $q_2(t)$ are (complex) delta-correlated random forces, of scaled strength 2.

The potential conditions are satisfied, and the potential [22,4] depends only on the intensities $I_{1,2}$ and not on the phases $\theta_{1,2}$, where $E_{1,2} = \sqrt{I_{1,2}}\, e^{i\theta_{1,2}}$. It is given by

$$U(I_1,I_2) \quad = \quad \frac{1}{2}a_1I_1 + \frac{1}{2}a_2I_2 - \frac{1}{4}I_1^2 - \frac{1}{4}I_2^2 - \frac{1}{2}\xi I_1I_2. \tag{7.2}$$

For the homogeneously broadened case, $\xi > 1$, there are wells at $I_1 = 0$, $I_2 = a_2$ and at $I_2 = 0$, $I_1 = a_1$, with a saddle point at $(I_{1s},I_{2s}) = (\xi a_2-a_1)/(\xi^2-1)$, $(\xi a_1-a_2)/(\xi^2-1)$. An estimate of the first passage time has been made [24]. The I_2 dependence of $P_0(I_1,I_2)$ is first integrated out and the one-dimensional T_p formula of (4.10) is used, with an I_1-dependent diffusion constant ansatz to allow for the higher dimensionality of the actual problem. This yields, for large pump parameters,

$$T_p \quad \approx \quad \frac{\pi^{\frac{1}{2}}}{2} \; \frac{e^{\frac{1}{4}(\xi^2-1)\, I_{1s}^2}}{(\xi^2-1)^{3/2}\, I_{1s}^2} \tag{7.3}$$

Similar results can be obtained, using the systematic Schuss-Matkowsky formalism applied to the four-dimensional case [25].

A direct measurement of 'dwell-times' within the wells centred at $I_1 = 0$ (off state) and $I_1 = a_1$ (on state) has been made [24]. The boundary is defined as at I_{1s}. A photodetector measures the intensity of the selected (I_1) laser mode, with $I_1 > I_{1s}$. A limiter changes the photodetector output to a series of rectangular pulses of variable duration τ_{ON}, that is the 'dwell time' in the well. The mean dwell time is equated to the mean first passage time from the 'on' to the 'off' state, $T_p = \langle \tau_{ON} \rangle$. T_p varies from milliseconds to minutes, depending on the parameters, and is in rough agreement with (7.3). The dwell-time statistics are in good agreement with (4.11).

First passage times also enter naturally in estimates of the extent of hysteresis phenomena analogous to superconducting and superheating [4]. Consider a first order phase transition, either dissipative or non-dissipative, described by a (one-component) order parameter x, with a drive parameter μ and a delta correlated random force $f(\tau)$ of strength 2D. The Langevin equation is

$$\dot{x} \quad = \quad -A(x,\mu) \;+\; f(t) \quad = \quad -\frac{\partial U}{\partial x}(x,\mu) \;+\; f(t) \tag{7.4}$$

The stationary states $\bar{x}(\mu)$, defined by

$$A(\bar{x},\mu) \quad = \quad 0 \tag{7.5}$$

have three branches if μ is within the range $\mu_2 < \mu < \mu_{c1}$. $\bar{x}(\mu)$ is an s-shaped curve on an $\bar{x}$- μ plot, with the backward bending branch unstable, corresponding to a maximum of the potential $U(x,\mu)$. The two forward-bending curves correspond to local minima of $U(x,\mu)$, that move relatively, up and down, as μ is varied. The higher, or metastable well disappears at the spinodal points $\mu = \mu_{c1}$ or μ_{c2}, defined by a vanishing relaxation rate $T_r^{-1} \propto \partial U(\bar{x},\mu)/\partial\bar{x}^2 = 0$ at the metastable minimum.

At some $\mu = \mu_X$, between μ_{c1} and μ_{c2}, the two well-depths are equal and thermodynamic 'Maxwell Construction' behaviour says that a system should jump to the absolute minimum of $U(x,\mu)$ as soon as $\mu > \mu_x$. But hysteresis can occur. The question is, what determines when the jump actually takes place?

The answer turns out to be dependent on three relevant time scales. The rates are (i) $\dot{\mu}$ the rate of change of the control parameter; (ii) the 'hop-over' or first passage rate T_p^{-1}; and (iii) the roll-back or relaxation rate T_r^{-1} in the metastable well. The basic idea is that $\dot{\mu}$ must raise the metastable well too fast for a hop-over, but slow enough so the system sits near the moving well minimum (adiabatic following). These two requirements define the brackets of a 'hysteresis window' for $\dot{\mu}$.

For hysteresis to occur, $P_s^{-1} dP_s/dt > T_p^{-1}$ where $P_s \propto e^{-U/D}$ depends on time only through $\mu(t)$. For adiabatic following, the deviation from the minimum $\tilde{x}(t) = x(t) - \bar{x}(\mu(t))$ must be small, $\tilde{x}/\bar{x} \ll 1$, and not increase in time, $\dot{\tilde{x}} = o$. This yields the condition [4].

$$\frac{\bar{x}}{T_r^{\,2}} \left| \frac{\partial A(\bar{x},\mu)}{\partial \mu} \right|^{-1} > \dot{\mu} > \frac{D}{(\partial U(\bar{x},\mu)/\partial \mu)} \quad T_p^{-1} \tag{7.6}$$

The first inequality, involving T_r, is a condition that the hysteresis state has a simple description, in terms of the most probable values, $\bar{x}(\mu)$, alone.

The second inequality, involving T_p, sets an upper bound on the degree of hysteresis that can occur. At the limit of metastability, the first passage time drops to zero, as the intervening barrier disappears [4]:

$$T_p \propto (\mu - \mu_{cl,2})^{\frac{1}{2}} \tag{7.7}$$

The inequality of (7.6) will be violated at some earlier μ, that is closer to $\mu_{cl,2}$ if $\dot{\mu}$ is larger. For $\dot{\mu} = 0$, infinitely slow variation, the hysteresis window shuts and the Maxwell construction obtains.

Numerical estimates of first passage times for condensed matter systems are often extremely large, when one-dimensional formulae corresponding to uniform states are used. The higher dimensional extensions of Section 6 could be used to include spatial variation and droplet formation, leading to improved estimates of T_p and the hysteresis window.

In conclusion, the Kramers, variational, and first-passage time formalism are three complementary and essentially equivalent ways of estimating metastable lifetimes, and can be usefully applied to a variety of systems in quantum optics and condensed matter physics.

<u>References</u>

1. R. Becker and W. Döring, Ann. Physik <u>24</u>, 719 (1935); also J. Frenkel, "Kinetic Theory of Liquids", Chap.VII, Dover Publications, Inc., New York 1955.
2. J.S. Langer and M.E. Fisher, Phys. Rev. Letters <u>19</u>, 560 (1967) and J.S. Langer and V. Ambegaokar, Phys. Rev. <u>164</u>, 498 (1967).
3. J.S. Langer, Ann. Phys. (N.Y) <u>65</u>, 53 (1971).
4. G.S. Agarwal and S.R. Shenoy, Phys. Rev. <u>A23</u>, 2719 (1981) ; See also R. Gilmore, Phys. Rev. <u>A20</u>, 2510 (1979).

5. R. Landauer and J.A. Swanson, Phys. Rev. 121, 1668 (1961) ; see also J.S. Langer, Ann. Phys. (N.Y.) 54, 258 (1969).
6. H.A. Kramers, Physica (utrecht) 7, 284 (1940).
7. W.F. Brown, Jr., Phys. Rev. 130, 1677 (1963).
8. H. Haken, "Synergetics", Chap. 7, Springer-Verlag, Berlin, 1977.
9. Z. Schuss, SIAM J. Review 22, 119 (1980).
10. R.L. Stratonovich, "Topics in the Theory of Random Noise", trans. by R.A. Silverman, Chap. 4, Vol. 1, Gordon and Breach, New York 1963. See also C.W.Gardiner," A Handbook of Stochastic Methods for Physics, Chemistry and Natural Sciences, Chap. 5, Springer- Verlag, Berlin 1982.
11. J.D. Farina, L.M. Narducci, J.M. Yuan and L.A. Lugiato, in "Optical Bistability", eds. C.M. Bowden et al., Plenum 1981, p.337.
12. R. Roy, R. Short, J. Durnin and L. Mandel, Phys. Rev. Lett. 45, 1486 (1980).
13. D. Kumar, this volume, p.
14. Z. Schuss and B.J. Matkowsky, SIAM J. Appl. Math. 35, 604 (1979); also B.J. Matkowsky and Z. Schuss, ibid, 33, 365 (1977).
15. R. Graham and A. Schenzle, Phys. Rev. A23, 1302, (1981).
16. G.S. Agarwal, this volume, p.
17. V. Srinivasan,this volume, p. ; see also the appendix in Ref. [13] .
18. See, for example, J. Mathews and R.L. Walker, "Mathematical Methods of Physics", Chap. 12, Benjamin, Inc., New York 1965.
19. D.H. Weinstein, Proc. Nat. Acad. Sci. U.S.A. 20, 529 (1934) and E. Kamke, Math. Z. 45, 788 (1939).
20. H. Brand, A. Schenzle and G. Schröder, Phys. Rev. A25, 2324 (1982).
21. G.S. Agarwal and S. Dattagupta, unpublished.
22. M.M. Tehrani and L. Mandel, Phys. Rev. A17, 677 (1978) ; 17, 694 (1978) ; F.T.Hioe, S. Singh and L. Mandel, ibid., 19, 2036 (1979).
23. The second of Ref. [14] contains a general treatment with off-diagonal, x-dependent diffusion, and without the 'potential condition', but within the usual $\epsilon \ll 1$ limit.
24. S. Singh and L. Mandel, Phys. Rev. A20, 2459 (1979); L. Mandel, R.Roy and S.Singh p.127 in "Optical Bistability", cited in Ref.[11].
25. G.S. Agarwal and S.R. Shenoy, unpublished.

INSTANTONS IN THE DYNAMICAL EVOLUTION OF FOKKER-PLANCK SYSTEMS

V. Srinivasan
School of Physics, University of Hyderabad
Hyderabad - 500134, India

In this lecture we shall outline the functional integral approach and the W.K.B. method to calculate the distribution P of a single stochastic variable the evolution of which is described by a Fokker-Planck equation. We shall follow closely a series of papers by B. Caroli, C. Caroli and B. Roulet on this subject[1,2]. The spirit of the classic article on instantons pervades [3] through the CCR works. Since we shall be using the concept of instantons we shall illustrate as to what an instanton is and then proceed to develop the subject.

Instantons:

Example I

Consider the following Lagrangian

$$L = \frac{1}{2}\left(\frac{dx}{dt}\right)^2 - \frac{1}{4}(x^2-1)^2 . \quad (1)$$

The corresponding Euclidean action is obtained by setting $t \longrightarrow -i\tau$

$$S_{Eucl} = \int_{-\infty}^{\infty} d\tau \left[\frac{1}{2}\left(\frac{dx}{d\tau}\right)^2 + \frac{1}{4}(x^2-1)^2\right]. \quad (2)$$

The Euler-Lagrange equation is

$$\frac{d^2x}{d\tau^2} = (x^3-x) . \quad (3)$$

This equation has the solution

$$x(\tau) = \pm \tanh\left[\frac{\tau-\tau_0}{\sqrt{2}}\right]. \quad (4)$$

The solution with the plus sign is called the instanton solution and that with the minus sign is called the anti-instanton solution. An instanton is a soliton in the time variable.

Example II

Consider the Lagrangian

$$L(x,\dot{x}) = \frac{1}{2}\dot{x}^2 + V(x) \quad (5)$$

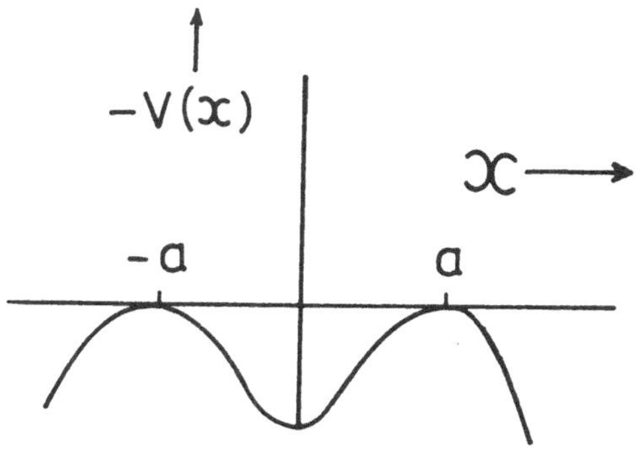

where V(x) has the shape as shown in Figure 1. The Lagrangian (5) represents a particle of mass one unit moving in the potential -V as shown in the Figure 1. The classical equation of motion for the Lagrangian (5) is

$$\ddot{x}_{cl} = \left(\frac{dV}{dx}\right)_{x=x_{cl}} \quad ; \quad x_{cl}(t_i) = x_i \text{ and } x_{cl}(t) = x_1$$

At the top of the hill

$$\frac{\dot{x}^2}{2} = V(x), \text{ (condition for zero energy)} \tag{6}$$

$$t = \int_0^x \frac{dx'}{\sqrt{2V(x')}} + \text{const} . \tag{7}$$

For large t the value of x approaches 'a' and therefore

$$\frac{dx}{dt} = \omega(a-x) \qquad \text{where} \qquad \omega = V''(a)$$

and $x_I(t)$ looks like as shown in Figure 2. This is the instanton solution. If instead the integration is done from +a to -a the solution $x_I(-t)$ is called the antiinstanton solution. The instanton solution looks as shown in Figure 2.

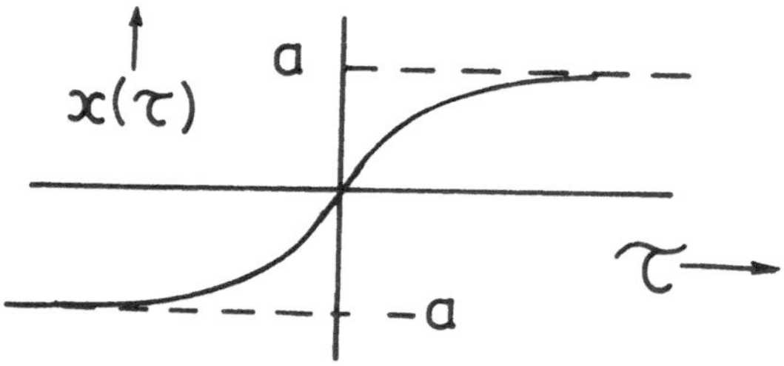

The action corresponding to this solution

$$S_o = \int dt\left[\frac{1}{2}\left(\frac{dx}{dt}\right)^2 + V\right] = \int dt\left(\frac{dx}{dt}\right)^2 = \int_{-a}^{+a} dx\,\sqrt{2V} \tag{8}$$

Such solutions occur when one deals with the Fokker-Planck equation by functional methods.

Fokker-Planck Equation : A functional integral approach

Consider the probability distribution $P(x,t|x_i,t_i)$ of the stochastic variable x whose evolution is described by the Fokker-Planck equation

$$\frac{\partial P}{\partial t} = \frac{\partial}{\partial x}\left[U'(x)P\right] + \theta\frac{\partial^2 P}{\partial x^2} \tag{9}$$

with the initial condition

$$P(x_1,t_i| x_i,t_i) = \delta(x_1-x_i) \tag{10}$$

$U'(x) = \frac{dU}{dx}$ is a nonlinear function of x. If U is a bistable potential and $\theta \leqslant \Delta U$ where ΔU is the height of the barrier separating the wells of U then the solution of (9) satisfying the condition

$x(t_i) = x_i$ and $x(t) = x_1$ is

$$P(x_1,t|x_i,t) = \int_{x_i}^{x_1} Dx(\tau) \exp - \frac{1}{\theta} \int_{t_i}^{t} d\tau \; O(\dot{x}(\tau), x(\tau)) \tag{11}$$

where

$$O(\dot{x},x) = \frac{\dot{x}^2}{4} + V(x) + \frac{\dot{x}}{2} U'(x) \tag{12}$$

and

$$V(x) = \frac{[U'(x)]^2}{4} - \frac{\theta}{2} U''(x) \tag{13}$$

Noticing that $\dot{x}\, U'(x) = \frac{dU}{d\tau}$ we can write

$$P(x_1,t|x_i,t_i) = \exp\left[\frac{U(x_i)-U(x_1)}{2\theta}\right] K(x_1,t|x_i,t_i) \tag{14}$$

where

$$K(x_1,t|x_i,t_i) = \int_{x_i}^{x_1} Dx(\tau) \exp\left[- \frac{1}{\theta} \int_{t_i}^{t} d\tau\, L(\dot{x},x)\right] \tag{15}$$

with

$$L(x,\dot{x}) = \frac{\dot{x}^2}{4} + V(x) . \tag{16}$$

This is the Lagrangian for a particle with mass $\frac{1}{2}$ moving in a potential $-V(x)$. $K(x_1,t|x_i,t_i)$ is the Feyman - Stuckelberg propogator and it has an expansion

$$K(x_1,t|x_i,t_i) = \sum_{n\geqslant 0} \phi_n(x_i)\phi_n(x_1) \exp\left(- \frac{(t-t_i)}{\theta}\lambda_n\right). \tag{17}$$

$\phi_n(x)$ are the eigenfunctions of the equation

$$- \frac{\theta^2 d^2\phi_n}{dx^2} + V\phi_n = \lambda_n\phi_n \tag{18}$$

where V is the potential derived from U.

In the functional integral approach one tries to calculate the Feynman propogator $K(x,t|x_i,t_i)$ while in the W.K.B. approach one calculates λ_n and ϕ_n which are the eigenvalues and eigenfunctions of (18) which is Schrödinger like. The approximations developed depend on the time scales of the problem. We shall illustrate as to how $K(x,t|x_i,t_i)$ is calculated using instanton methods.

Let us now evaluate $K(x,t|x_i,t_i)$ around $x_{cl}(t)$ which is the solution of the Euler-Lagrange equation corresponding to $L(x,\dot{x})$ of equation (16). We find that

$$K(x,t|x_i,t_i) = \exp\left[-\frac{1}{\theta} S_{cl}(x,t\ x_i,t_i)\right].$$

$$\times \int_{y(t_i)=0}^{y(t)=0} D(y(\tau)) \exp\left[-\frac{1}{\theta}\left\{\int d\tau \frac{\dot{y}^2(\tau)}{4} + \frac{y^2(\tau)}{2} V''(x_{cl}(\tau))\right\}\right] \quad (19)$$

where S_{cl} is the solution of the Euler-Lagrange equation of (16). Writing δS as

$$\delta S = \frac{1}{2} \int d\tau \left[y(\tau) - \frac{1}{2}\frac{d^2}{d\tau^2} + V''(x_{cl}(\tau))\right] y(\tau) \quad (20)$$

where

$$y(\tau) = x(\tau) - x_{cl}(\tau) \quad ; \quad y(\tau) = \sum_{n\geqslant 0} c_n y_n(\tau) \quad (21)$$

We can now choose solutions $y_n(\tau)$ which satisfy the equation

$$\left[-\frac{1}{2}\frac{d^2 y_n}{\tau} + V''(x_{cl})\right] y_n(\tau) = \lambda_n y_n \quad (22)$$

where

$$\int y_n(\tau)\, y_m(\tau)\, d\tau = \delta_{mn} \quad (23)$$

This gives immediately

$$K(x,t|x_i,t_i) = \frac{N}{(\pi\lambda_n)^{\frac{1}{2}}} \exp\left[-\frac{1}{\theta} S_{cl}(x,t|x_i,t_i)\right] \quad (24)$$

Here N is a nonmalisation constant which can be determined with help of the normalisation condition on P. This expression is true if λ_n are all non zero.

Now one notices thát the instanton solution $x_I(\tau)$ of figure (2) which is the solution to the potential (1) is a solution to (22) with zero eigenvalue. In such a case this mode must be separated out from the rest. In order to do this we notice

$$\lim S(x_I(\tau-t_1)) = S_o \ .$$

Here we have translated $\tau \to (\tau - t_1)$

$$\int_{-t}^{t/2} S(x_I(\tau-t_1)\, dt_1 = S_o \int_{-t}^{+t/2} dt_1$$

The change in $y(\tau)$ induced by a small change in the location of the centre of the instaton t, is

$$dy = \frac{dy_{cl}}{dt_1} dt_1 \quad (25)$$

Consider y_1 which is the zero eigenvalue solution of (22)

$$y_1 \propto S_o^{-\frac{1}{2}} \frac{dy_{cl}}{dt_1}$$

Therefore

$$dy = a\, S_o^{\frac{1}{2}} y, dt, \tag{26}$$

where a is a constant $\sim \theta^{\frac{1}{2}}$. Integrating over the one instanton, we get

$$K(a,(t/2)|-a,(-t/2)) = \frac{N'}{(\pi'\,\lambda_N)^{-\frac{1}{2}}} \exp - \frac{1}{\theta} S_o . t(S_o)^{-\frac{1}{2}} \tag{27}$$

Similarly if one includes in succession an instanton and an anti-instanton one would have to evaluate integrals of the form $\int_{-t/2}^{t/2} dt_1 \ldots \int_{-t/2}^{t_{n-1}} dt_n$ which gives a factor $\frac{t^n}{n!}(S_o)^{n/2}$ for an instanton and a factor of $(-1)^n\,(t^n/n!)(S_o)^{n/2}$ for an anti instanton. Therefore the net contribution from 'n' instantons gives a power series in t which sums up to an exponential. That is

$$K(a,(t/2)|-a,(-t/2)) \sim \frac{N'}{\pi'(\lambda_n)^{-\frac{1}{2}}} \exp -\left(\frac{1}{\theta} S_o\right) \exp\left(\frac{t}{\theta} S_o^{\frac{1}{2}}\right) \tag{28}$$

A similar contribution will come from 'n' anti-instantons. In this way one computes K and hence $P(x,t|x_i,t_i)$. Knowing $P(x,t|x_i,t_i)$ one can compute the first passage time. Caroli et.al. calculate $K(x,t|x_i,t_i)$ for a potential which is bistable as shown in Figure 3. This give rise to V(x) which is whown in Figure 4.

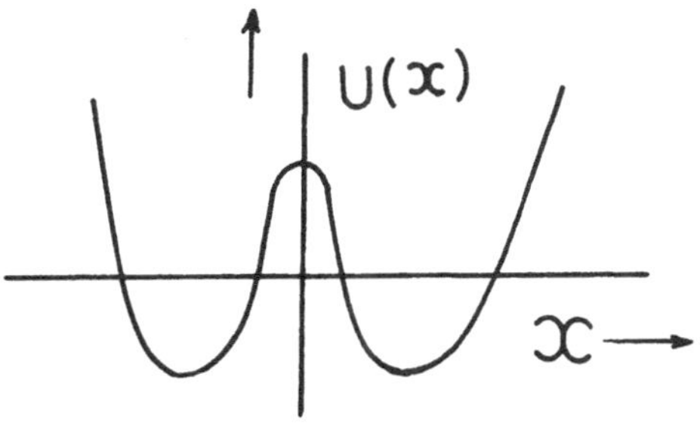

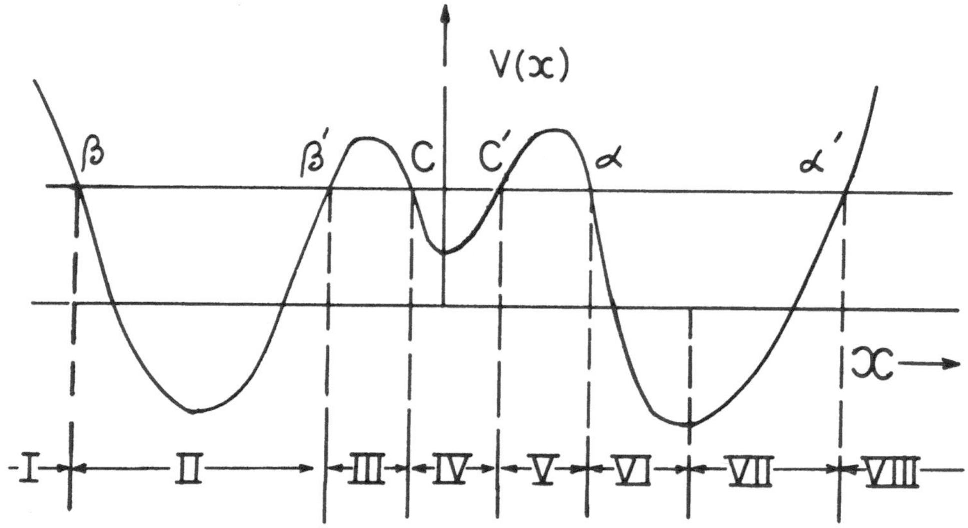

In the case V(x) is bistable has been treated by Coleman [3] by W.K.B. methods and very nicely reviewed by Sakita [6]. The bistable case has also been treated by van Kampen [5].

Caroli et al choose a bistable U(x) for which V(x) is as shown in Figure (4) and solve this by W.K.B. They start from region I and construct the usual exponential W.K.B. solution and match it with the solution in region II which is the harmonic regime. This way continuing to match the solution, and finally demanding that in region VIII the exponentially increasing solution must be zero, they obtain the eigenvalue condition. For this V(x) of figure (4), this condition is an extremely complicated condition. Since usually only the first few eigenvalues of (18) are needed the problem is not too difficult.

It must be pointed out that the functional methods and the W.K.B. methods are complementary to each other and either can be used for such problems.

References

1. B. Caroli, C. Caroli, B. Roulet, J. Stat. Phys. 26, 83 (1981); 28, 757 (1982); 21, 415 (1979).
2. B. Caroli, C. Caroli, B. Roulet and J.F. Gouyet, J. Stat. Phys. 22, 515 (1980).
3. S. Coleman in International School of Subnuclear Physics, Ettore Majorana IX Course, A. Zichichi ed (Academic Press, New York 1975) p.1-76.
4. R. Rajaraman, "Solitons and Instantons", (North Holland Publishing Company, Amsterdam (1982)).
5. N.G. van Kampen, J. Stat. Phys. 17, 71 (1977).
6. B. Sakita, "Tunnelling Phenomena in gauge theories", Kyoto lecture (unpublished).

PROJECTION OPERATOR TECHNIQUES IN STOCHASTIC PROCESSES

G.S. Agarwal
School of Physics, University of Hyderabad
Hyderabad - 500134, INDIA

1. Need for Projection Operator Techniques and Examples of Some Typical Projection Operators

We have already indicated in the lecture on Fokker-Planck equations, that the solution of a multidimensional system is in general not possible. Fortunately for many systems, a complete specification of the probability distribution is not quite needed if one wants to compare with experiments. Often the information about mean and the correlation function suffices. Moreover the system need not exhibit markovian behavior. In such cases we will see that the projection operator techniques are ideally suited[1]. Following examples will show the versality with which projection operator techniques can be used.

(a) Wave propagation in a Random Medium

The propagation of a wave in a random medium can be characterized by

$$[\nabla^2 + \frac{\omega^2}{c^2} n^2] U = 0 \tag{1}$$

where n is the refractive index of the random medium. The amplitude of the wave U becomes stochastic in nature as n is a random function of $\vec{r}$ as is the case of the atmospheric medium. A complete specification of the probability distribution of U is impossible due to (a) nonlinearity of (1) i.e. the random term n appears in the multiplicative form (b) our inadequate knowledge about the fluctuations[2] of n. The quantities that are usually measured correspond to the mean and correlation of U i.e. to $\langle U(\vec{r})\rangle$, $\langle U(\vec{r})U(\vec{r'})\rangle$. Let us introduce the projection operator P that takes the ensemble average over refractive index fluctuations i.e. if

$$U = \sum \int n(\vec{r}_1)\dots n(\vec{r}_m) f(\vec{r}_1)\dots f(\vec{r}_m) d^3r_1 .. d^3r_m \tag{2}$$

then

$$PU = \sum\int \langle n(\vec{r}_1)\dots n(\vec{r}_m)\rangle f(\vec{r}_1)\dots f(\vec{r}_m) d^3r_1 \dots d^3r_m \quad . \tag{3}$$

(b) Adiabatic Elimination of Fast Variables

Our systems may have too many degrees of freedom. In such a case a reduced description is desirable for then, one can hope to obtain some exact solution for example in the lecture on Fokker-Planck equations, we saw that a one dimensional Fokker-Plank equation is always soluble. More specifically consider a system composed of two parts A and B and let τ_A and τ_B be the typical time scales associated with them. If $\tau_A \ll \tau_B$, then any small perturbation on A would get back A quickly in equilibrium. Hence if we are looking for dynamics on the scale $t \gg \tau_A$, then we eliminate the degrees of freedom associated with A. This idea is very extensively used in physical systems. The simplest example being the reduction of the full Fokker-Plank equation for the Brownian motion of a particle in a potential to Smoluchowski equation in the limit of large friction . Even in the context of the laser[3], simpler equations for photon distribution can be obtained by eliminating the stochastic variables for atoms. Let $f(x_A, x_B, t)$ be the joint distribution for the coupled A and B system. The reduced distribution for the system B alone will be $f(x_B, t) = \int dx_A\, f(x_A, x_B, t)$ and hence to obtain the projection of the dynamics, we introduce P defined by

$$P\ldots. = N_A \int dx_A \ldots.., \quad \int N_A dx_A = 1 \ . \tag{4}$$

The choice of N_A depends on the local equilibrium of A. For the derivation of the Smoluchowski equation we can take

$$P\ldots. = e^{-p^2/2mkT} \int dp\ldots. \ . \tag{5}$$

(c) Systems Interacting with Random Perturbations

Usually the behavior of the system is probed by external fields such as electromagnetic fields or magnetic fields which are generally fluctuating as the sources producing such fields are fluctuating [4]. Similarly many times the internal interactions are also modelled as stochastic in nature for example the kinetic Ising Model[5] with random interaction between various sites. The dynamical equation in all such cases can be written as

$$\frac{\partial \rho}{\partial t} = (L_0 + L_1)\rho \ , \tag{6}$$

where L_1 is random either in space or in time. In such a case we have to take the ensemble average of ρ over the randomness contained in L_1. Similarly we need to evaluate the ensemble average of the dipole-dipole correlation function which is important in the line shape problems. The ensemble averaged equations can be derived by using projection operator methods.

2. Ensemble Averaged Equations for One Time Expectation Values

Let us write the basic stochastic equation in the form

$$\frac{d}{dt} A = (C_0 + C_1(t))A + g \tag{7}$$

which can characterize the behavior of both classical and quantum systems e.g. A may represent the expectation values of a complete set of system operators. One can regard A,g as column matrices. Assume that all the randomness is contained in $C_1(t)$. We desire to compute the ensemble average of A i.e. $\langle A \rangle = PA$. The formal derivation of PA is by now standard. One multiplies (7) on the left by both P and (1-P) and then integrates formally the (1-P)A equation and substitutes the result in the PA equation. This procedure leads to the exact relation[1]

$$P\dot{A} = C_oPA+g+PC_1(t)PA+PC_1(t)e^{C_ot}U(t,o)(1-P)A(o)+\int_0^t d\tau\, PC_1(t)e^{C_ot}U(t,\tau)\, e^{-C_o\tau}(1-P)C_1(\tau)PA(\tau), \quad (8)$$

where

$$U(t,\tau) = T\exp\left\{\int_\tau^t dt'(1-P)e^{-C_ot'}C_1(t')e^{C_ot'}(1-P)\right\} \quad (9)$$

The third term on the right hand side of (8) represents the motion due to the coherent part of C_1 whereas the fourth term represents the effect of initial correlations. The last term involves the correlations of C_1 to all orders. For most practical applications Born approximation is sufficient. In such a case (8) reduces to

$$P\dot{A} = C_oPA+g+\int_o^t d\tau\, PC_1(t)e^{C_o(t-\tau)}C_1(\tau)PA(\tau), \quad (10)$$

where we have also assumed for simplicity that $\langle C_1\rangle = 0$, $(1-P)A(o) = 0$. Equation (10) involves the second order correlation of the randomness i.e. if

$$C_1(t) = \sum C_iF_i(t), \quad \langle F_i(t)F_j(\tau)\rangle = D_{ij}(t-\tau) \quad (11)$$

then the Laplace transform of (10) yields:

$$P\hat{A} = [Z-C_o-\sum C_i\hat{D}_{ij}(Z-C_o)C_j]^{-1}(PA(o)+\frac{g}{Z}) \quad (12)$$

where Z denotes the Laplace variable. Note that if the random forces are delta correlated $\langle F_i(t)F_j(t')\rangle = 2D_{ij}\,\delta(t-t')$, then (10) reduces to

$$P\dot{A} = C_o\, PA+g+\sum_{ij} C_iC_jD_{ij}PA, \quad (13)$$

which is an equation local in time. Both (12) and (13) have been extensively used in many physical systems. We like to mention two exact results[6] here — (i) If the random forces $F_i(t)$ are Gaussian and delta correlated, then (13) is <u>exact</u>, though in the derivation we made Born approximation. (ii) If $C_1(t) = C\,F(t)$ where F(t) is a telegraphical signal or a dichotomic markov process, then (12) is <u>exact</u> though <u>not</u> (13). A very important property of dichotomic markov process is

$$\langle x(t)x(t')M(t',t''..)\rangle = \langle x(t)x(t')\rangle\langle M(t',t''...)\rangle, t>t'>t''.., \quad (14)$$

It may be noted that if the random forces have very short correlation time τ_c then for times $t\gg\tau_c$, equation (10) can be made local in time. Applications of (10) - (13) will be discussed in this volume.

3. <u>Ensemble Average of the Multitime Correlations</u>

When a system is interacting with stochastic perturbations, then its average behavior need not be markovian and hence one can not use the fluctuation regression theorem and (8) to obtain the multitime-correlations. However it turns out convenient to obtain directly equations like (8) for the correlation matrix. Let us consider the correlation matrix defined by $R(t+\tau,t) = \langle A_i(t+\tau)B(t)\rangle$, which satisfies equation analogous to (7) :

$$\frac{d}{d\tau} R(t+\tau,t) = (C_o+C_1(t+\tau))\, R(t+\tau,t)+\langle g\, B(t)\rangle . \tag{15}$$

The ensemble average of R over the fluctuations of C_1 can be obtained as before but now the inhomogeneous term $\langle gB(t)\rangle$ will make an additional contribution since $(1-P)\ \langle gB(t)\rangle \neq o$. The final result being [6]

$$\frac{d}{d\tau} PR(t+\tau,t) = C_o PR+Pg\langle B(t)\rangle +PC_1(t+\tau)\, e^{C_o\tau}\, V(\tau,o)(1-P)R(t,t)$$
$$+\int_o^\tau d\alpha\, PC_1(t+\tau)C^{C_o\tau}\, V(\tau,\alpha)\left\{e^{-C_o\alpha}(1-P)g\langle B(t)\rangle\right.$$
$$\left.+(1-P)\, e^{-C_o\alpha}\, C_1(t+\tau)PR(t+\alpha,t)\right\} , \tag{16}$$

$$V(\tau,\alpha) = T\exp\left\{\int_\alpha^\tau dt'(1-P)\, e^{-C_o t'}\, C_1(t+t')\, e^{C_o t'}\, (1-P)\right\}. \tag{17}$$

Note that the initial condition will involve $R(t,t)$ which is a single time expectation value and which can be obtained from the result of Sec II. The result in Born approximation is relatively simple

$$\frac{d}{d\tau} PR(t+\tau,t) = C_o PR+Pg\langle B(t)\rangle +I_1+I_2$$
$$+\int_o^\tau d\alpha\, PC_1(t+\tau)\, e^{C_o(\tau-\alpha)}\, C_1(t+\alpha)PR(t+\alpha,t), \tag{18}$$
$$I_1(t+\tau,t) = \int_o^t d\alpha\, PC_1(t+\tau)\, e^{C_o\tau}\, M\, e^{C_o(t-\alpha)}\, C_1(\alpha)\, PA(\alpha) ,$$
$$I_2(t+\tau,t) = \int_o^\tau d\alpha \int_o^t d\beta\, PC_1(t+\tau)\, e^{C_o(\tau-\alpha)}\, N\, e^{C_o(t-\beta)}\, C_1(\beta)PA(\beta), \tag{19}$$

where the matrices M and N enter through $(1-P)R(t,t) = M(1-P)A$, $(1-P)g\langle B(t)\rangle = N(1-P)A$. Note that the structure of (18) is very similar to (10) except for the new feature that inhomogeneous terms appear. It is because of such terms that the fluctuation-regression theorem is no longer valid and the system's behavior is nonmarkovian. If the random forces are delta correlated, then $I_1 = I_2 = 0$ and the markovian property is recovered. Again it must be noted that the result (18) is exact [6] if $C_1(t)$ has a single stochastic variable of random telegraphical signal type.

We close this chapter with a simple example which will illustrate how the dynamical behavior of the system can change depending on the stochastic nature of the randomness. Consider the stochastic equation

$$\dot{\psi} = i\varsigma X(t)\psi . \tag{20}$$

If $X(t)$ is a delta-correlated Gaussian random process $\langle X(t)X(t')\rangle = X_o^2\, \delta(t-t')$ then (10) and (18) lead to exact results:

$$\langle \dot{\psi} \rangle = \frac{-g^2 x_o^2}{2} \langle \psi \rangle, \frac{d}{d\tau} \langle \psi(t+\tau)\ \psi^*(t) \rangle = \frac{-g^2 x_o^2}{2} \langle \psi(t+\tau)\ \psi^*(t) \rangle \qquad (21)$$

On the other hand for a telegraphic signal with jump rate Γ we get from (10) and (18)

$$\langle \dot{\psi} \rangle = -g^2 x_o^2 \int_o^t d\tau\, e^{-2\Gamma(t-\tau)} \langle \psi(\tau) \rangle \quad ,$$

$$\frac{d}{d\tau} \langle \psi(t+\tau)\, \psi^*(t) \rangle = -g^2 x_o^2 \int_o^\tau dt'\ e^{-2\Gamma(\tau-t')} \langle \psi(t+t') \psi^*(t) \rangle$$

$$+2g^2 x_o^2 \int_o^t dt'\ e^{-2\Gamma(t+\tau-t')} \langle \psi(t') \psi^*(t') \rangle \qquad (22)$$

In this case the memory effects are very important. The differences in the two models are obvious from (21) and (22). It may also be added that (22) allows for the possibility of going from a completely coherent to incoherent situation by changing Γ from a very small value to a very large value.

References

1. G.S. Agarwal in "Progress in Optics" edited by E. Wolf (North-Holland, Amsterdam 1973) Vol.XI, p.1; F. Haake "Springer Tracts in Modern Physics" (Springer-Verlag, Berlin 1973) Vol. 66, p.98.
2. V.I. Tatarskii; "The Effects of the Turbulent Atmosphere on Wave Propagation" (U.S. Department of Commerce, NTIS Springfield, Va 1971); J.W. Strohbehn, "Laser Beam Propagation in the Atmosphere" (Springer-Verlag, Berlin 1978).
3. H. Haken, "Synergetics" (Springer-Verlag, Berlin 1977); Chap.8.
4. J.H. Eberly, in "Laser Spectroscopy IV" edited by H. Walther and K.W. Rothe (Springer-Verlag, Berlin 1979) p.80, G.S. Agarwal, Phys. Rev. A18, 1490 (1978).
5. D. Dhar, this volume.
6. G.S. Agarwal, Z. Physik B33, 111 (1979).

PROJECTION OPERATOR METHODS IN LINEAR STOCHASTIC DIFFERENTIAL EQUATIONS

S. Chaturvedi
School of Physics, University of Hyderabad
Hyderabad 500 134, India

1. Introduction

Consider a linear stochastic differential equation

$$\frac{du}{dt} = \alpha A(t)\, u \tag{1}$$

where u is a vector , A(t) is a stochastic matrix characterised by a correlation time τ_c and α is a measure of the strength of the fluctuations. (In general, the R.H.S. of (1) may contain, in addition to A(t), a non stochastic matrix A_o. This part however can always be transformed away by going over to the interaction picture. Henceforth we shall assume that this has been done). We shall assume that the quantity $\alpha\tau_c$, called the Kubo number, is much less than unity. This serves as an expansion parameter in the perturbation expansions to be developed later.

Before going further, let us see what it means to solve a stochastic differential equation and in what ways do these equations differ from ordinary differential equations. To solve (1) we need to specify the initial conditions. Suppose that at $t=t_o$, $u(t_o)$ has a fixed value a. Now for each realisation of the stochastic process A(t), (1) is simply a deterministic equation. Its solution

$$u(t) = U(t, [A] \mid t_o)a \tag{2}$$

is a functional of A(t) i.e. it depends on all values A(t'), $t_o \leq t' \leq t$. The ensemble of all such realisations corresponding to all possible realisations of A(t) defines the stochastic process u(t) - a solution of (1) characterised by the time t_o at which the initial conditions were specified. If we were to specify the initial conditions at a time $t \neq t_o$ we would obtain a different stochastic process. We may express this alternatively by saying that the characteristic functional for u(t) depends in an <u>essential</u> manner on the time t_o at which the initial conditions are specified. It is in this respect, as has been emphasised by van Kampen [1], that stochastic differential equation differ from ordinary differential equations. Our main aim here is to develop systematic procedures for deriving statistical characteristics of u(t), given those of A(t). We wish to achieve this by introducing a projection operator

$$\mathcal{P}. = \langle \cdot\cdot \rangle \tag{3}$$

and by using two projection operator techniques.

2. Equations for Single Time Averages

2.1 An Integro-differential Equation for $\langle u(t) \rangle$ [2]

Applying $\mathcal{P}$ and $\mathcal{Q} = 1-\mathcal{P}$ to both sides of (2) we can write it, in by now a fairly standard manner, as two coupled equations for $\mathcal{P}u$ and $\mathcal{Q}u$. Solving the equation for $\mathcal{Q}u$

we get

$$\mathcal{Q}u(t) = \mathcal{G}(t,o)\ \mathcal{Q}\,u(o) + \int_o^t d\tau\ \mathcal{G}(t,\tau)\ \mathcal{Q}A(\tau)\ \mathcal{P}u(\tau) \tag{4}$$

where

$$\mathcal{G}(t,\tau) = \overleftarrow{T}\left[\exp \alpha \int_\tau^t ds\ \mathcal{Q}\,A(s)\right] \tag{5}$$

and $\overleftarrow{T}$ prescribes chronological ordering. Substituting (4) in the equation for $\mathcal{Q}\,u$ we obtain an integro-differential equation for $\langle u(t)\rangle$

$$\frac{d}{dt}\langle u(t)\rangle = \int_o^t ds\ \ k(t,s)\ \langle u(s)\rangle + \int_o^t ds\ \ \mathcal{I}(t,s) \tag{6}$$

$k(t,s)$ and $\mathcal{I}(t,s)$ when expanded in powers of α have the following structure

$$k(t,s) = \sum_{m=1}^{\infty} \alpha^m\, k_m(t,s) \tag{7}$$

$$\mathcal{I}(t,s) = \sum_{m=1}^{\infty} \alpha^m\, \mathcal{I}_m(t,s) \tag{8}$$

$$k_1(t,s) = \langle A(t)\rangle \delta_+(t-s) \quad ; \quad \int_o^\varepsilon \delta_+(t)\ dt = 1,\ \varepsilon > o \tag{9}$$

$$k_2(t,s) = \langle A(t)\ \mathcal{Q}\,A(s)\rangle \tag{10}$$

$$k_{m+2}(t,s) = \int_s^t dt_1 \int_s^{t_1} dt_2 \ldots\ldots \int_s^{t_{m-1}} dt_m \langle A(t)\,\mathcal{Q}A(t_1)\ldots \mathcal{Q}A(t_m)\,\mathcal{Q}A(s)\rangle \tag{11}$$

The quantities $\langle A(t)\,\mathcal{Q}\,A(t_1)\ldots\ldots \mathcal{Q}\,A(t_m)\ \mathcal{Q}\,A(s)\rangle$ are known as totally ordered cumulants and often written as $\langle A(t)\ A(t_1)\ldots A(t_m)A(s)\rangle_t$

For $\mathcal{I}_1(t,s)$ we have

$$\mathcal{I}_1(t,s) = \langle A(t)\ \mathcal{Q}\,u(o)\rangle \delta_+(t-s) \tag{12}$$

and the expressions for $\mathcal{I}_m(t,s)$ $m > 1$ can be obtained from those for $k_m(t,s)$ by replacing the final operator $A(s)$ in (10) and (11) by $A(s)\ \mathcal{Q}\,u(o)$.

If we assume that at $t=o$, $\mathcal{Q}\,u(o) = o$ i.e. $u(o)$ is a non-stochastic vector (in the sense of being uncorrelated with $A(t)$) the inhomogeneous term in (6) drops out and we have

$$\frac{d}{dt}\langle u(t)\rangle = \alpha\langle A(t)\rangle + \alpha^2 \int_o^t dt_1 \langle A(t)\ A(t_1)\rangle_t \langle u(t_1)\rangle$$

$$+\alpha^3 \int_o^t dt_1 \int_o^{t_1} dt_2 \langle A(t)\ A(t_1)\ A(t_2)\rangle_t \langle u(t_2)\rangle + \ldots\ldots \tag{13}$$

This equation truncated at second order in α is known as Bourret's integral equation [3]. For the special case in which $A(t)$ involves a single dichotomic stochastic process $\alpha(t)$ i.e. $A(t) = B(t)\ \alpha(t)$ where $B(t)$ is a non stochastic matrix, then all the totally ordered cumulants higher than the second vanish and Bourret's integral equation becomes exact. In general, if $A(t)$ does not have any such special properties, then all one can hope to achieve is to calculate $\langle u(t)\rangle$ from the perturbation expansion in (13). Terwiel [2] has estimated the integrals that appear in (13) by first proving a cluster property for the totally ordered cumulants. It turns out that (13) is not a systematic expansion in powers of $\alpha\tau_c$: the entire integro differential equation has to be rearranged in the form of a differential equation in order to find all the terms of a given order in $\alpha\tau_c$. Thus it will be desirable to derive a differential equation for $\langle u(t)\rangle$ at the very outset.

2.2 A Differential Equation for $\langle u(t)\rangle$: Time Convolutionless Projection Operator Technique [4-6] .

The projection technique above yields an integro differential equation for $\langle u(t)\rangle$ because of the convolution involved in (4). We can formally eliminate this convolution by substituting for $u(\tau)$ in (4) the following expression

$$u(\tau) = G(t,\tau)\, u(t) \tag{14}$$

where

$$G(t,\tau) \stackrel{\wedge}{=} \vec{T}\left[\exp -\alpha\int_{\tau}^{t} ds\, A(s)\right] \tag{15}$$

and $\vec{T}$ prescribes anti-chronological ordering. This gives

$$\mathcal{Q}u(t) = \mathcal{G}(t,o) + \int_{0}^{t} d\tau\, \mathcal{G}(t,\tau)\, \mathcal{Q}A(\tau)\, \mathcal{P}G(t,\tau)\, (\mathcal{P}+\mathcal{Q})\, u(\tau) \tag{16}$$

Solving this for $\mathcal{Q}u(t)$ we get

$$\mathcal{Q}u(t) = [1 - \Sigma(t)]^{-1}\, \mathcal{G}(t,o) + [1 - \Sigma(t)]^{-1}\, \Sigma(t)\, \mathcal{P}u(t) \tag{17}$$

where

$$\Sigma(t) = \int_{0}^{t} d\tau\, \mathcal{G}(t,\tau)\, \mathcal{Q}A(\tau)\, \mathcal{P}G(t,\tau) \tag{18}$$

Equation (17) when substituted in the equation for $\mathcal{P}u(t)$ yields a differential equation for $\mathcal{P}u(t)$

$$\frac{d\langle u(t)\rangle}{dt} = K(t)\, \langle u(t)\rangle + I(t) \tag{19}$$

$K(t)$ and $I(t)$, when expanded in powers of α, are found to have the following structure

$$K(t) = \sum_{m=1}^{\infty} \alpha^m K_m(t) \tag{20}$$

$$I(t) = \sum_{m=1}^{\infty} \alpha^m I_m(t) \tag{21}$$

$$K_1(t) = \langle A(t)\rangle$$

$$K_m(t) = \int_{0}^{t} dt_1 \int_{0}^{t_1} dt_2 \ldots \int_{0}^{t_{m-2}} dt_{m-1}\, \langle A(t)A(t_1)\ldots A(t_{m-1})\rangle_p \tag{22}$$

The quantities $\langle A(t)\ldots A(t_{m-1})\rangle_p$ are known as partially ordered cumulants. These can be written down in terms of the moments using the following rules [7].

1. Write a sequence of m dots.
2. write a zero on the first dot and any permutation of the numerals 1.2....m-1.

3. In every permutation insert a $\mathcal{P}$ between two successive numerals if $i > j$ and a $\mathcal{Q}$ if $i < j$ and place the whole expression between brackets $\langle \ldots\ldots\ldots \rangle$.

4. For each permutation with p-1 operators $\mathcal{P}$ supply a factor of $(-1)^{p-1}$.

5. Replace each numeral m by $A(t_m)$. The numeral o denotes $A(t)$. (This step can be generalised by associating different operators $A_m(t_m)$ with the numeral m and thereby define partially ordered cumulants involving different operators $A_i(t)$).

For $I_m(t)$ we have

$$I_1(t) = \langle A(t) \mathcal{Q} u(o) \rangle \tag{23}$$

$$I_m(t) = \int_0^t dt_1 \ldots \int_0^{t_{m-3}} dt_{m-2} \int_0^{t_{m-2}} dt_{m-1} \langle A(t) \ldots A(t_{m-2}) \; A(t_{m-1}) \mathcal{Q} u(o) \rangle_p \tag{24}$$

These expressions can simply be obtained from those for $K_m(t)$ by replacing $A(t_{m-1})$ in (22) by $A(t_{m-1}) \mathcal{Q} u(o)$ and treating this as a single quantity. If we assume that $\mathcal{Q} u(o) = o$ then $I(t)$ drops out from (19) and we have a differential equation for $\langle u(t_1) \rangle$

$$\frac{d\langle u(t)\rangle}{dt} = [\alpha A(t) + \int_0^t dt_1 \langle A(t) \; A(t_1) \rangle_p + \int_0^t dt_1 \int_0^{t_1} dt_2 \langle A(t) A(t_1) A(t_2) \rangle_p + \ldots\ldots\ldots] \langle u(t) \rangle \tag{25}$$

known as the cumulant expansion [8,9]. For the case in which $A(t) = B\,\alpha(t)$ where B is a constant matrix and $\alpha(t)$ is a Gaussian stochastic process then all the partially ordered cumulants, higher than the second, vanish and the expansion becomes exact in the second order approximation. From the point of view of perturbation expansions, van Kampen [9] and Roerdink [7] have proved a cluster property for the partially ordered cumulants and using this they have estimated the various terms that appear on the R.H.S. of (25). It turns out that (25) yields a systematic expansion in powers of $\alpha \tau_c$ from which $\langle u(t) \rangle$ can be calculated to a desired accuracy.

3. Equations for Higher Moments :

Equations for higher moments can be derived in exactly the same manner as above. For instance, the vector

$$t,t \;) = u(t) \otimes u(t) = \begin{pmatrix} \underline{u}(t) \; u_1(t) \\ \underline{u}(t) \; u_2(t) \\ \vdots \end{pmatrix} \tag{26}$$

consisting of all the second moments of $u(t)$ obeys the equation

$$\frac{d}{dt} \, g(t,t \;) = \alpha C(t) \; g(t,t) \tag{27}$$

where

$$C(t) = A(t) \otimes I + I \otimes A(t). \tag{28}$$

Here $\otimes$ denotes direct product. (27) again has the same structure as (1) and the same methods as above apply here as well.

4. Two Time Correlation Functions :

We now wish to derive equations for the two time correlation functions

$$g(t,t') = u(t) \otimes u(t') \quad , \quad t \geqslant t' \tag{29}$$

assuming that at t=o, u(o) is a non stochastic vector. Again we have two options available

4.1 An integro differential equation for $\langle g(t,t')\rangle$:

The vector $g(t,t')$ obeys the following equation

$$\frac{d}{dt} g(t,t') = B(t)\, g(t,t') \tag{30}$$

$$B(t) = A(t) \otimes I \tag{31}$$

Proceeding as before and keeping in mind that the initial value for g(t,t') i.e. g(t',t') is a stochastic quantity we obtain using (6)

$$\frac{d}{dt}\langle g(t,t')\rangle = \int_{t'}^{t} ds\; k(t,s)\; \langle g(s,t')\rangle + \int_{t'}^{t} ds\; \mathcal{J}(t,s) \tag{32}$$

The structure of the terms on the R.H.S. of (32) is exactly the same as earlier with A(t) replaced by B(t) and $\mathcal{Q}\,u(o)$ by $\mathcal{Q}g(t',t')$. The second term in (32) involves $\mathcal{Q}\,g(t',t')$ and we wish to express it in terms of $\mathcal{P}g(t',t')$. This is easily done using (4) and taking into account the fact that $\mathcal{Q}\,g(o,o) = o$ by assumption. Hence

$$\mathcal{Q}g(t',t') = \int_{o}^{t'} d\tau\; \mathcal{G}(t,\tau)\; \mathcal{Q}\,C(\tau)\; \mathcal{P}g(\tau,\tau) \tag{33}$$

Here

$$\mathcal{G}(t,\tau) = \overleftarrow{T}\left[\exp \alpha \int_{\tau}^{t} ds\; \mathcal{Q}\,C(s)\right] \tag{34}$$

Substituting (33) in the expressions for $\mathcal{J}(t,s)$ we obtain the following integro-differential equation for $\langle g(t,t')\rangle$

$$\frac{d}{dt}\langle g(t,t')\rangle = \int_{t'}^{t} ds\; k(t,s)\; \langle g(s,t')\rangle + h(t,t') \;. \tag{35}$$

k(t,s) has the same structure as for single time averages. The term h(t,t') has the following structure

$$h(t,t') = \sum_{m=2}^{\infty} \alpha^{m}\, h_{m}(t,t') \tag{36}$$

Each h_m (t,t') is a sum of m-1 terms

$$h_2(t,t') = \int_{o}^{t'} dt_1 \langle B(t) \quad C(t_1)\rangle_{t} \langle g(t_1,t_1)\rangle \tag{37}$$

$$h_m(t,t') = \int_{t'}^{t} dt_1 \int_{t'}^{t_1} dt_2 .. \int_{t'}^{t_{m-3}} dt_{m-2} \int_{o}^{t'} ds_1 \langle B(t)B(t_1)..B(t_{m-2})C(s_1)\rangle_{t} \langle g(s_1,s_1)\rangle$$

$$+ \int_{t'}^{t} dt_1 \cdots \int_{0}^{t'} ds_1 \int_{0}^{s_1} ds_2 \; <B(t) \cdots B(t_{m-3}) \; C(s_1) \; C(s_2) >_t <g(s_2,s_2)>$$

$$+ \cdots \cdots \cdots \cdots \cdots$$

$$+ \int_{0}^{t'} ds_1 \int_{0}^{s_1} \cdots \cdots \int_{0}^{s_{m-2}} ds_{m-1} <B(t) \; C(t_1) \cdots \cdots C(s_{m-1})>_t <g(s_{m-1},s_{m-1})> \qquad (38)$$

Equation (35) has been derived by Agarwal [10] in the context of quantum optics. The same equation trunctated at the second order has been previously derived by Morrison and Mckenna [11] .

As is clear the above equation for $<g(t,t')>$ is coupled to the equation for the second moments $<g(t,t)>$ and therefore becomes somewhat complicated. However simplification occurs in the case in which $A(t) = B(t) \; \alpha(t)$ where $\alpha(t)$ is a D.M.P. In this case the R.H.S. of (35) in the second order approximation becomes exact.

4.2 A Differential Equation for $<g(t,t')>$:

If we proceed exactly as in 4.1 using the time convolutionless projection technique instead, we obtain an equation for $<g(t,t')>$ having the following form

$$\frac{d}{dt} <g(t,t')> = K(t,t') <g(t,t')> + \; f(t,t') \; <g(t',t')> \qquad (39)$$

where $K(t,t')$ has exactly the same form as (20) with $A(t)$ replaced by $B(t)$. This equation is not very useful. Among other things it does not even lead to exact results for the scalar Gaussian case i.e. the case in which (1) is a single variable equation with $A(t)$ Gaussian. This equation should therefore be abandoned.

An alternative differential equation for $<g(t,t')>$ using methods not based on projection operator techniques has recently been derived somewhat heuristically in [12] and in a more systematic manner in [13] . This equation has the following form

$$\frac{d}{dt} <g(t,t')> = \; M(t,t') \; <g(t,t')> \qquad (40)$$

where $M(t,t')$ has the following structure

$$M(t,t') \; = \; \sum_{m=1}^{\infty} \alpha^m M_m(t,t') \qquad (41)$$

$$M_1(t,t') \; = \; <B(t)> \qquad (42)$$

$$M_m(t,t') = \int_{t'}^{t} dt_1 \int_{t'}^{t_1} dt_2 \cdots \int_{t'}^{t_{m-2}} dt_{m-1} <B(t) \; B(t_1) \cdots B(t_{m-1})>_p$$

$$+ \int_{t'}^{t} dt_1 \int_{t'}^{t_1} dt_2 \cdots \int_{t'}^{t_{m-3}} dt_{m-2} \int_{0}^{t'} <ds_1 \; B(t) \; B(t_1) \cdots B(t_{m-2}) \; C(s_1)>_p$$

$$+ \cdots \cdots \cdots \cdots \cdots \cdots$$

$$+ \int_{t'}^{t} dt_1 \int_{0}^{t'} ds_1 \ldots\ldots \int_{0}^{s_{m-3}} ds_{m-2} \langle B(t)\, B(t_1)\, C(s_1) \ldots C(s_{m-2}) \rangle_p$$

$$+ \int_{0}^{t'} ds_1 \int_{0}^{s_1} dt_2 \ldots\ldots \int_{0}^{s_{m-2}} ds_{m-1} \langle B(t)\, C(s_1) \ldots\ldots C(s_{m-1}) \rangle_p \qquad (43)$$

Equation (40) has an advantage over (35) in having a simpler structure and in not being coupled to the equations for the second moments. It becomes exact in the second order approximation in the scalar Gaussian case. Also note that the passage from the integro differential equation (35) to (40) simply consists in pulling out $g(s,t)$ and $g(s,s)$ in (35) and (38) from the integrals and replacing them by $g(t,t')$ and subsequently replacing the totally ordered cumulants that appear in the integrals,by partially ordered cumulants.

In the foregoing discussion we have developed general methods for obtaining statistical properties of solutions of linear stochastic differential equations. We have used, wherever possible, projection operator techniques to achieve this. These techniques, although by no means indespensible, prove to be convenient book keeping methods for perturbation expansions. We have seen that in some cases viz., scalar Gaussian case and the case in which $A(t)$ is a D.M.P. we obtain exact results. Mention must be made of other special cases in which exact solutions are possible. These include linear stochastic differential equations in which $A(t)$ is a (i) Kubo-Anderson process [14] (ii) Kangaroo process [14] (iii) a compound Poisson process [16,17] . Although so far we have been concerned with linear stochastic differential equations only, these methods apply to nonlinear stochastic differential equations as well. This is done by first writing a continuity equation - a linear equation corresponding to the given non linear stochastic differential equation and then making use of van Kampen's lemma [1] . In the above we have not included stochastic or non stochastic inhomogeneous terms in our starting equation (1). These can be incorporated within the framework of the formalisms presented above. For details and an excellent account of the topics discussed here we refer the reader to the works by Roerdink [7,13] .

The methods developed above have many applications in a number of areas such as resonance phenomena in quantum optics, nuclear magnetic resonance Mössbauer line shapes, wave propagation in a random medium, diffusion in a turbulent fluid etc. Some of these have been discussed at length in these proceedings [17].

References

1. N.G. van Kampen "Stochastic Processes in Physics and Chemistry", North Holland, Amsterdam, New York, Oxford (1981).
2. R.H. Terwiel, Physica 74, 248 (1974).
3. R.C. Bourret, Can. J. Phys. 40, 782 (1962).
4. F. Shibata, Y. Takahashi and N. Hashitsume, J. Stat. Phys. 17, 171 (1977).
5. M. Tokuyama and H. Mori, Prog. Theor. Phys. 55, 411 (1976).
6. S. Chaturvedi, and F. Shibata, Z. Physik B35, 297 (1979).
7. J.B.T.M. Roerdink, Physica 109A, 23 (1981).

8. R. Kubo, J. Math. Phys. 4, 174 (1963).
9. N.G. van Kampen, Physica 74, 215,239 (1974).
10. G.S.Agarwal, Z. Physik B33, 111 (1979).
11. J.A. Morrison and J. Mckanna, Siam-AMS Proceedings (Stochastic Differential Equations) Vol. VI, Providence R.I., 97 (1973).
12. S. Chaturvedi, Phys. Letts. 90A, 444 (1982).
13. J.B.T.M. Roerdink, Physica 112A, 557 (1982).
14. A. Brisaud, U. Frisch, J. Math. Phys. 15, 524 (1974).
15. B.J. West, K. Lindenberg and V. Seshadri, Physica 102A, 470 (1980).
16. N.G. van Kampen, Physica 102A, 489 (1980).
17. See lectures by G.S. Agarwal in these proceedings.

CONTINUOUS-TIME RANDOM WALK THEORY AND NON-EXPONENTIAL DECAYS OF CORRELATION FUNCTIONS

V. Balakrishnan
Department of Physics
Indian Institute of Technology
Madras - 600036

1. Introduction

Relaxation, or approach to equilibrium, is customarily associated with a damped exponential time dependence. However, there are notable and interesting exceptions. In spin glasses, for example, the remanent magnetization M(t) exhibits a logarithmic fall-off of the form A-B log (λt). A similar behaviour has long been known in the case of rock magnetism, where it can be understood as originating from the occurrence of randomly dispersed magnetic clusters of varying sizes. The magnetic moments of these clusters face a whole spectrum of activation barriers to their re-orientation. This leads to a continuous superposition of conventional exponential relaxation functions. Thus M(t) takes the form $\int dE\ \rho(E) \exp[-\lambda(E)t]$, where $\rho(E)$ is some weight factor. If the standard Arrhenius form $\lambda_0 \exp(-E/kT)$ is assumed for the relaxation rate $\lambda(E)$, it is easily shown (e.g., by a change of integration variables from E to $\lambda(E)$) that M(t) has a logarithmic decay of the form referred to above.

The emergence of non-exponential relaxation (including power-law decays) from a continuous spectrum of exponentials is a rather general feature, common to numerous physical situations. With the help of the relaxation-response relationship underlying linear response theory, one may trace this back to a similar time-dependence for certain associated two-time correlation functions. The latter are frequently determined most conveniently by stochastic techniques, as this permits modelling closely guided by the physics of the problem (such as the identification of the relevant time scales, etc.) It is of interest, therefore, to identify the circumstances and modifications under which the random processes commonly used in physical applications can display a superposition of 'relaxation' times in their two-time correlation functions. For the sake of clarity, attention will be restricted to a single, scalar, stationary random process x(t), and its autocorrelation function.

This task may be approached in (at least) two different ways. In the first (described in Sec.2), one stays within the framework of Markov processes. After some brief remarks on the situation with respect to continuous Markov processes, we consider the simplest and most commonly employed jump processes (the dichotomic and the Kubo-Anderson process), and lead up to a generalization (the "kangaroo process") that has the desired property. The second approach is in many respects a more powerful and more generally applicable one : continuous-time random walk (CTRW) theory[1]. In Secs. 3 and 4, we describe the essentials of this theory, and show how time dependences of the kind referred to in the foregoing arise very naturally. We also exhibit a step-wise constant random process that is a non-Markovian generalization of the well known Kubo-Anderson process.

It must be mentioned that CTRW theory is a particularly appropriate technique for the investigation of diffusion and related phenomena in disordered materials. An extensive literature exists on this application [2]. A number of interesting questions of a technical nature arises in this connection, including the much-debated one on the correctness or otherwise of incorporating a first-waiting-time distribution that is distinct from the holding time distribution for the CTRW. We shall not be concerned here with any of these aspects. Our discussion should, however, be of some help also as an introduction to the analyses carried out in some of the papers listed under [2].

2. Some special Markov jump processes

It is common practice, on physical grounds, to model physical random variables in terms of Markov processes. In many applications, it is convenient to treat $x(t)$ as a continuous Markov process (a "diffusion" process), specified by the stationary probability density $W(x)$ and the conditional density $P(x,t|x_0)$. The latter obeys a generalized Fokker-Planck equation of the form

$$\frac{\partial}{\partial t} P(x,t|x_0) = -\frac{\partial}{\partial x}[A(x)P] + \frac{\partial^2}{\partial x^2}[B(x)P] , \tag{1}$$

where $A(x)$ and $B(x)$ are the 'drift' and 'diffusion' coefficients, respectively. An extensive literature exists on the properties of such generalized Fokker-Planck equations. For our purpose, it suffices to note that the autocorrelation function

$$\begin{aligned} C(t) &= \langle x(o)\ x(t)\rangle \\ &= \int dx_0 \int dx\ W(x_0)\ P(x,t\mid x_0)\ xx_o \end{aligned} \tag{2}$$

is a single exponential function provided certain conditions are satisfied [3]; namely, if

$$\lim_{t \downarrow o} P(x,t\mid x_0) = W(x) \tag{3}$$

(which is true in the context of systems in equilibrium); and if $B(x)$, $A(x)$ are respectively second and first order polynomials in x, together with the requirement that $B(x)\ W(x)$ vanish at the end points of the range of variation in x, and the condition that

$$\langle x\rangle = \int dx\ W(x)x\ ,\quad \langle x^2\rangle = \int dx\ W(x)x^2 \tag{4}$$

be finite. In other cases, $C(t)$ may be a discrete or partially continuous superposition of exponentials.

While continuous Markov processes are mathematically quite complicated to handle in general, in many physical applications it may be quite appropriate to treat $x(t)$ as a jump process that changes discontinuously from one value to another under the triggering action of a random pulse sequence. Let us consider some specific examples [4]. The simplest such process is the telegraph process or the dichotomic Markov process

(DMP). Here the variable $x(t)$ is step-wise constant, and can assume just two values, say $+\xi$ and $-\xi$. It is triggered from one value to another by a stationary Poisson pulse sequence with a (constant) mean pulse rate λ, i.e., the probability that exactly n pulses occur in a time interval t is given by

$$P(n,t) = \frac{(\lambda t)^n}{n!} \exp(-\lambda t) . \tag{5}$$

It is easy to see that the conditional density for the process $x(t)$ is given by

$$P(x,t|x_o) = [\delta(x-x_o) \cosh \lambda t + \delta(x+x_o) \sinh \lambda t] \exp (-\lambda t) , \tag{6}$$

where the initial value x_o can be either $+\xi$ or $-\xi$. The autocorrelation of x is given by

$$C(t) = \xi^2 \exp (-2\lambda t) , \tag{7}$$

a single exponential. A minor generalization of the above DMP allows $x(t)$ to take on the values ξ_1 and ξ_2, with different transition rates λ_{12} and λ_{21}. The condition

$$\lambda_{12} \xi_2 + \lambda_{21} \xi_1 = o \tag{8}$$

ensures that the mean $\langle x \rangle$ vanishes.

A non-trivial generalization of the DMP leads to the Kubo-Anderson process (KAP). This is again a stepwise constant Markov process, the jumps being triggered by a Poisson pulse sequence as before. However, the variable $x(t)$ is completely 'randomized' at each jump, and can take on any value characterized by a stationary probability density $W(x)$. Since the probability of zero pulses occurring in an interval t is $\exp(-\lambda t)$, it is easily seen that

$$P(x,t|x_o) = \delta(x-x_o) \exp (-\lambda t) + W(x) \quad 1-\exp (-\lambda t) \tag{9}$$

in this case. Hence, if $\langle x \rangle = o$, we again find the exponential behaviour

$$C(t) = \langle x^2 \rangle \exp (-\lambda t) . \tag{10}$$

To obtain a superposition of exponentials for $C(t)$, a further generalization of the KAP (still within the framework of stationary Markov processes) is needed. The transition-causing pulse rate λ may itself be a function of the current value of the random variable. If x happens to be far out in the tail of the distribution $W(x)$, for instance, one may intuitively expect the pulse rate $\lambda(x)$ to be considerably larger than its mean value, so as to bring x back towards more probable values more rapidly. The transition rate equation for such a Markov process reads

$$P_{tr}(x, \Delta t|x_o, o) = \delta(x-x_o) [1-\lambda(x_o) \Delta t] + \tilde{W}(x) \quad \lambda(x_o) \Delta t + 0 [(\Delta t)^2]. \tag{11}$$

Here $\tilde{W}(x)$ is a probability distribution that is distinct from, but related to, the distribution $W(x)$, as we shall see below. Equation (11) yields the master equation

$$\frac{\partial}{\partial t} P(x,t | x_o) = \lim_{\Delta t \downarrow o} \int dx_1 \{ P_{tr} (x,t + \Delta t | x_1, t) P(x_1, t | x_o)$$

$$- P_{tr}(x_{1,} t + \Delta t | x,t) \; P(x,t|x_o)\Big\}$$

$$= - \lambda(x) \; P(x,t|x_o) + \widetilde{W}(x) \int dx_1 \; P(x_{1,} t|x_o) \; \lambda(x_1) \; . \qquad (12)$$

Using the limit (3) in Eq.(12), we obtain

$$\widetilde{W}(x) = \lambda(x) \; W(x) \; / \langle \lambda \rangle \; , \qquad (13)$$

where the mean rate $\langle \lambda \rangle$ is defined as

$$\langle \lambda \rangle = \int dx \; W(x) \; \lambda(x) \; . \qquad (14)$$

It is also clear from Eq.(12) why one could not have simply written W(x) instead of $\widetilde{W}(x)$ in Eq.(11). The process described by Eq.(11) or Eq.(12) is called a "kangaroo process" (KP). It reduces trivially to the KAP if $\lambda(x) = \lambda$, a constant. The solution of Eq.(12) has the Laplace transform

$$\widetilde{P}(x,s|x_o) = \frac{\delta(x-x_o)}{s+ \lambda(x)} + \frac{\lambda(x_o) \; \lambda(x) \; W(x)}{s(s+ \lambda(x_o))(s+ \lambda(x)) \; \langle \lambda(s+\lambda)^{-1} \rangle} \qquad (15)$$

where

$$\langle \lambda(s+\lambda)^{-1} \rangle = \int dx \; W(x) \; \lambda(x) / [s+ \lambda(x)] \; . \qquad (16)$$

For simplicity, let us assume that the range of x is $(-\infty, \infty)$, and that $\lambda(x)$, W(x) are even functions of x. Then Eq.(15) leads to

$$C(t) = \int dx \; W(x) \; x^2 \exp \; - \lambda(x) t \; . \qquad (17)$$

This is to be compared with Eq.(10) for a KAP. A change of integration variables from x to $\lambda(x)$ in Eq.(17), shows that a KP has a continuous superposition of 'relaxation times' in general. Among the physical applications of KP's, we may mention in particular the case of wave propagation in random media.

In most cases of practical interest, one works at the level of the first two moments, i.e., the mean and the autocorrelation function. The Markov assumption is therefore not very crucial. It can be sacrificed in favour of a tractable generalization (such as CTRW) that provides greater flexibility in modelling in order to accommodate the underlying physics.

3. The continuous-time random walk method : construction of the pulse sequence

CTRW theory is useful whenever the random process of interest can be regarded as a stochastic sequence that is an ongoing 'renewal' process. Let us first obtain the statistics of the underlying pulse sequence, i.e., the probability P(n,t) of n transition-causing pulses occurring in time t. In general, of course, this is not given by Eq.(5).

The holding time distribution P(t) may be regarded as the fundamental quantity in a CTRW. This is the following conditional probability : given that a transition (jump) has just occurred at some epoch t_o, p(t) is the probability that no further jumps have occurred till epoch t_o+t. Evidently, $p(o)\equiv 1$, $p(t_2) \leqslant p(t_1)$ if $t_2>t_1$, and $p(\infty)\to o$. The corresponding (conditional) transition probability that a jump does occur between (t_o+t) and (t_o+t+dt) is easily seen to be equal to $-\dot{p}(t)dt$. Further, the mean life-time of a 'state' of the random variable, i.e., the mean time between pulses or jumps, is therefore given by

$$\tau = \int_o^\infty dt\ t\,[-\dot{p}(t)] \Big/ \int_o^\infty dt\,[-\dot{p}(t)] \;=\; \int_o^\infty dt\ p(t) \tag{18}$$

Before the required distribution $P(n,t)$ can be constructed in terms of $p(t)$, it is necessary to take care of the first-waiting-time complication. The origin of time in the quantity $P(n,t)$ is arbitrary. Hence it would be incorrect to suppose that the probability for the first pulse (after the clock is started at this arbitrary origin, labelled $t=o$) to occur at epoch t is given by $-\dot{p}(t)dt$, for the latter is a <u>conditional</u> probability. It pre-supposes that the preceding pulse occurred exactly at $t = o$, whereas we have no way of knowing the precise epoch $t_o (\leqslant o)$ at which it did occur. As far as the first pulse in $P(n,t)$ is concerned, therefore, we require the <u>unconditional</u> counterpart of $-\dot{p}(t)$. Let us denote this transition probability by $-\dot{p}_o(t)$, in anticipation of the fact that its integral $p_o(t)$ is just the first-waiting-time distribution-the unconditional counterpart of $p(t)$. One can find $-\dot{p}_o(t)$ as follows. Let a pulse have occurred at epoch $t_o(\;\;o)$, and let the next pulse occur at epoch $t(\;\;o)$. The co-rresponding transition probability is $-\dot{p}(t-t_o)dt$. On the other hand, one may view the event as a two-step process, with an associated probability $p(-t_o)\;\;-\dot{p}_o(t)dt$. The two quantities may be equated, provided we sum over all possible values of the epoch t_o. Thus

$$\int_{-\infty}^{o} dt_o\ [-\dot{p}(t-t_o)] = \int_{-\infty}^{o} dt_o\ p(-t_o)\ [-\dot{p}_o(t)] \quad , \tag{19}$$

which yields the result

$$-\dot{p}_o(t) = (1/\tau)\ p(t) \quad . \tag{20}$$

The associated first-waiting-time distribution is therefore

$$p_o(t) = (1/\tau) \int_t^\infty dt'\ p(t') \ . \tag{21}$$

We are now in a position to write down the distribution $P(n,t)$ for the pulse sequence. Obviously,

$$P(o,t) \equiv p_o(t) \ . \tag{22}$$

Further (and this clarifies the meaning of the term 'renewal' process), for $n \geqslant 1$,

$$P(n,t) = \int_o^t dt_n \ldots \int_o^{t_2} dt_1 (-1)^n\ p(t-t_n)\ \dot{p}(t_n-t_{n-1}) \ldots \dot{p}(t_2-t_1)\dot{p}_o(t_1) \ . \tag{23}$$

It is immediately evident that if $p(t) = \exp(-\lambda t)$, we have $p_o(t) \equiv p(t)$. The first-waiting-time distribution is identical with the holding time distribution in this case. Further, Eq.(23) simplifies to yield the Poisson sequence of Eq.(5). While such an explicit reduction of $P(n,t)$ is not possible in general, it is easy to see that a compact expression obtains for the Laplace transform of $P(n,t)$, and for its generating function

$$H(z,t) = \sum_{n=0}^{\infty} P(n,t)\, z^{n} . \tag{24}$$

We find

$$\tilde{H}(z,s) = \frac{1}{s} + \frac{(z-1)\,\tilde{p}(s)}{s\,\tau[1 - z\{1 - s\tilde{p}(s)\}]} , \tag{25}$$

where $\tilde{p}(s)$ is the Laplace transform of $p(t)$. This completes the specification of the pulse sequence.

4. Jump processes in the CTRW formalism

With Eq.(25) at hand, an entire class of stepwise constant random processes can be investigated. It is convenient to use an operator notation : let $\mathcal{J}$ be the "collision" operator that changes the value of the random variable at each pulse. The matrix element $\mathcal{J}_{x \to x'}$ represents the transition probability for the variable to jump from the value x to the value x' under the action of a pulse. Similarly, let $\mathcal{P}(t)$ denote the effective time development operator whose matrix element $[\mathcal{P}(t)]_{x_0 \to x}$ is simply the conditional density $P(x,t|x_0)$. Inserting the operator $\mathcal{J}$ at each pulse in the multiple integral of Eq.(23), one can construct $\mathcal{P}(t)$. It is evident that the formal operator solution for $\mathcal{P}(t)$ is just the inverse Laplace transform of

$$\tilde{\mathcal{P}}(s) = \tilde{H}(\mathcal{J}, s) , \tag{26}$$

i.e. one replaces z by the operator $\mathcal{J}$ in the expression quoted in Eq.(25). This solution is indeed one of considerable generality, given the flexibility in the choice of both $p(t)$ and $\mathcal{J}$.

A plausible, simple model for $\mathcal{J}$ is as follows. Instead of assuming that each pulse completely randomizes the variable $x(t)$, and throws it from its pre-pulse value x_0 to an arbitrary, x_0-independent value in the stationary distribution $W(x)$, let us suppose that

$$\mathcal{J}_{x_0 \to x} = \gamma W(x) + (1-\gamma)\;\delta(x-x_0), (0 \leqslant \gamma \leqslant 1) . \tag{27}$$

In other words, this model interpolates [5,6] between a strong collision limit ($\gamma = 1$) that is in the spirit of the KAP and KP considered in the foregoing, and a weak collision limit ($\gamma = 0$) that does not alter the pre-pulse value. The required operator inversion can be carried out in this model for $\mathcal{J}$, and one obtains finally [6]

$$\tilde{P}(x,s|x_0) = \delta(x-x_0)\,\frac{[1-\tilde{F}(s)]}{s} + W(x)\,\frac{\tilde{F}(s)}{s} , \tag{28}$$

where

$$\tilde{F}(s) = \gamma\tilde{p}(s)/\,\tau[\gamma + (1-\gamma)\;s\tilde{p}(s)] . \tag{29}$$

Equation (28) is to be compared with Eq.(15) for a KP (and its special case, Eq.(9), for a KAP). Of course, Eq.(28) does not refer to a Markov process in general. Assuming as before that $\langle x \rangle = o$, we now find for the autocorrelation function

$$C(t) = \langle x^2 \rangle [1 - \int_o^t F(t')\, dt'] = \langle x^2 \rangle \int_t^\infty dt'\, F(t') \quad , \tag{30}$$

where F(t) is the inverse Laplace transform of $\tilde{F}(s)$. This compact expression encompasses a very wide range of possible decays of C(t).

Comparison with the results for the KAP and the KP is facilitated by going over to the strong collision limit, $\gamma = 1$, in the above. As stated earlier, this is in keeping with the assumption inherent in the KAP and the KP. Equations (28) and (29) simplify in this limit, to yield

$$P(x,t|x_o) = \delta(x-x_o)\, p_o(t) + W(x)\,[1-p_o(t)] \quad . \tag{31}$$

If $p(t) = \exp(-\lambda t)$, i.e., if the pulse sequence is Poissonian, we recover the KAP. In other cases, one may regard Eq. (31) as a non-Markovian generalization of the KAP. The autocorrelation function becomes

$$C(t) = \langle x^2 \rangle p_o(t) = \langle x^2 \rangle (1/\tau) \int_t^\infty p(t')\, dt' \quad . \tag{32}$$

In particular, suppose the holding time distribution is a (continuous) superposition of exponentials :

$$p(t) = \int_o^\infty d\nu\, \sigma(\nu) \exp(-\nu t) \quad . \tag{33}$$

While this may arise in some cases from a distribution of activation barriers (as already mentioned), it is worth bearing in mind that Eq.(33) is in fact a fairly general representation, valid for a wide class of functions p(t). Using this representation in Eq.(32), we obtain

$$C(t) = \langle x^2 \rangle \int_o^\infty \frac{d\nu}{\nu}\, \sigma(\nu) \exp(-\nu t) \Big/ \int_o^\infty \frac{d\nu}{\nu}\, \sigma(\nu) \quad . \tag{34}$$

This is a (continuous) superposition of exponentials, as promised earlier. The actual t-dependence of C(t) is governed by the distribution $\sigma(\nu)$. The latter is directly amenable to physical modelling, perhaps more so than the rate $\lambda(x)$ in Eq. (17), with which the above result is to be compared. Indeed, at the level of the autocorrelation function, the non-Markov process described by Eq.(31) and the KP may be regarded as equivalent in some sense; and the distribution $\sigma(\nu)$ of Eq.(34) can be related to the functional $x^2\, W[x] / \lambda'[x]$, where $\lambda(x) = \nu$.

It remains to observe that the non-analytic (e.g., power law) behaviour that is frequently associated with a continuous range of relaxation times as in Eq. (34) is actually a general feature, independent of the model for the transition operator $\mathcal{J}$. In fact, the random process need not even be stepwise constant. To see this heuristically, consider Eq.(25) for the generating function of the pulse sequence. If the Laplace transform of the representation (33) is inserted for $\tilde{p}(s)$ in that equation, one obtains

$$\tilde{H}(z,s) = \frac{1}{s} + \frac{(z-1)\int_0^\infty \frac{d\nu\sigma(\nu)}{(s+\nu)}}{s\tau\left[1-z\int_0^\infty d\nu\frac{\sigma(\nu)}{(s+\nu)}\right]} , \tag{35}$$

where

$$\tau = \int_0^\infty d\nu\sigma(\nu)/\nu . \tag{36}$$

The integrals in Eq.(35) immediately suggest that $\tilde{H}(z,s)$ is singular at s=o . This is a consequence of the possible non-analyticity of $\tilde{p}(s)$ at the same point. More detailed remarks may be made if the properties of the distribution $\sigma(\nu)$ are specified. For instance, if $\int d\nu\,\sigma(\nu)/\nu^3$ diverges, then the small s expansion of $\tilde{p}(s)$ reads

$$\tilde{p}(s) \sim \tau + c_1 s + c_2 s^\alpha + \ldots\ldots\ldots, \tag{37}$$

where $1 < \alpha < 2$. Among other consequences, this sort of behaviour leads to a $1/f^\alpha$ current noise due to the random hopping of carriers in a disordered medium [7].

The simple formalism described above can be generalized in various directions. Some of these are : randomly interrupted deterministic evolution; multi-state renewal processes; quantum mechanical (operator) complications; etc. We conclude with the following comment. The emergence of "non-analytic behaviour" (e.g., power-law decays of correlations, critical exponents, cusps, singularities, and so on) from an underlying continuous spectrum in the problem is a rather universal phenomenon. Examples may be cited from Regge poles to rock magnetism. This "search for non-analyticity", in one form or another, may indeed be regarded as one of the central themes of contemporary theoretical physics.

References

1. E.W. Montroll and G.H. Weiss, J. Math. Phys. 6, 167 (1965); E.W. Montroll and H. Scher, J. Stat. Phys. 9, 101 (1973); see also W. Feller, An Introduction to Probability Theory and its Applications, Vol.II (Wiley, New York, 1966).
2. H. Scher and M. Lax, J. Non-cryst. Solids, 8-10, 497 (1972); Phys. Rev. B7, 4491, 4502 (1973); H. Scher and E.W. Montroll, Phys. Rev. B12, 2455 (1975); J.K.E. Tunaley, Phys. Rev. Letters 33, 1037 (1974); J. Stat. Phys. 11, 397 (1974); ibid., 14 461 (1976); ibid., 15, 167 (1976); M. Lax and H. Scher, Phys. Rev. Letters 39,781 (1977); K.W. Kehr and J.W. Haus, Physica 93A, 412 (1978); J. Klafter and R.Silbey, Phys. Rev. Letters 44, 55 (1980).
3. E. Wong, Proc. Amer. Math. Soc. Symp. Appl. Math. 16, 264 (1963).
4. See, for example, A.T. Barucha-Reid, Elements of the Theory of Markov Processes and their Applications (McGraw-Hill, New York, 1960); A. Brissaud and U. Frisch, J. Math. Phys. 15, 524 (1974),
5. M. Fixman and K. Rider, J. Chem. Phys. 51, 2425 (1969); S. Dattagupta and A.K. Sood, Pramana 13, 423 (1979).
6. V. Balakrishnan, Pramana 13, 337 (1979).
7. J.K.E. Tunaley, J. Stat. Phys. 15, 149 (1976).

ON THE APPROXIMATE SOLUTIONS OF THE NONLINEAR LANGEVIN EQUATIONS

G. Ananthakrishna
Reactor Research Centre, Kalpakkam 603 102, India

1. Introduction

Langevin approach in study of fluctuations in nonlinear systems consists in adding a fluctuating term $\beta(x)\,\eta(t)$ to the deterministic equation $\dot{x} = c_1(x)$ to obtain

$$\dot{x} = c_1(x) + \beta(x)\,\eta(t) \qquad (1)$$

It is conventional to take the noise $\eta(t)$ to be a Gaussian white noise with zero mean and

$$\langle \eta(t)\ \eta(t') \rangle = 2\epsilon\,\delta(t-t') . \qquad (2)$$

When $\beta(x)$ is a non-constant function, (1) models fluctuations arising due to the coupling of the original subsystem to external surroundings. In this case the magnitude of fluctuations in the driving noise $\eta(t)$ is amplified by an amount proportional to the strength of coupling $\beta(x)$ depending on the state variable x. When $\beta(x)$ is constant, it refers to internal fluctuations [1] (often, however, the distinction between the subsystem of interest and external surroundings is fuzzy). In the following, we will consider the case when $\beta(x) = 1$. We will discuss a few approximate methods of solutions which serve to represent fluctuations adequately in different situations.

In the following when we refer to a nonlinear Langevin equation (NLE), we will also use the equivalent Fokker-Planck equation.

The first method we consider is the system size expansion method [2] which corresponds to calculating fluctuations around the deterministic solution of (1) . In this case, fluctuations are expected to be small so that one can expand

$$c_1(x) = c_1(\langle x \rangle) + (x - \langle x \rangle\, c_1'(\langle x \rangle) \qquad (3)$$

and

$$\frac{dz}{dt} = z\, c_1'(\langle x \rangle) + \eta(t) \quad ; \quad z = x - \langle x \rangle$$

where z is the fluctuating part. This method holds good as long as the extensivity ansatz is not violated [3] . When the system evolves from an intrinsically unstable state, this property has to be replaced by a more general one namely, $P(X,t) = \exp[\Omega^{-1}\phi_0(x,t) + \phi_1(x,t) + \ldots]$ where X is the extensive random variable and Ω is the size of the system, and $x = \frac{X}{\Omega}$. For this case, the scaling theory [4-7] demonstrates that for large times the second term becomes as dominant as the first.

The plan of the lecture is as follows. We first outline the Ω - expansion. Drawing on results of a specific example, the limitations of the method are discussed. An alternate scheme which has essentially the same features as the Ω-expansion is discussed [8,9]. The method is shown to be useful even when multiple steady states are allowed. Finally scaling theory for decay from unstable states due to Suzuki is discussed with a specific example.

2. Ω - Expansion

Following the Kramers-Moyal expansion it is straight forward to derive the Fokker-Planck equation [10]

$$\frac{\partial P(x,t)}{\partial t} = -\frac{\partial}{\partial x} C_1(x)\ P(x,t) + \frac{\epsilon}{2}\frac{\partial^2}{\partial x^2} C_2(x)\ P(x,t) \tag{4}$$

where $C_1(x)$ and $C_2(x)$ are the first and the second jump moments. The system size expansion depends on the fact that the fluctuating part is of the order of $\epsilon^{1/2}$ compared to the mean. This can be easily seen by considering the linear Langevin equation [7]. This means that we could express x and P(x,t) as follows :

$$x = \phi + \epsilon^{1/2}\xi \ ; \quad \pi(\xi,t)d\xi = P(x,t)dx \ ; \quad \epsilon = \Omega^{-1} \ . \tag{5}$$

where $\pi(\xi,t)$ corresponds to the distribution for ξ. The corresponding linearized equation can be obtained by using (5) in (4). By equating terms of ϵ^{0} and $\epsilon^{1/2}$ we get

$$\dot{\phi}(t) = -C_1(\phi(t)) \tag{6}$$

and

$$\frac{\partial\pi}{\partial t} = -C_1'(\phi(t))\frac{\partial\xi\pi}{\partial\xi} + \frac{1}{2}C_2(\phi(t))\frac{\partial^2\pi}{\partial\xi^2} \tag{7}$$

The solution of (7) with the initial condition $\lim_{t\to o}\pi(\xi,t)\to\delta(\xi)$ is

$$\pi(\xi,t) = \frac{1}{\sqrt{2\pi\sigma^2(t)}}\exp -\frac{\xi^2}{2\sigma^2(t)} \tag{8}$$

where σ^2 is determined by

$$\frac{d\sigma^2}{dt} = -2C_1'(\phi(t))\,\sigma^2(t) + C_2(\phi(t)) \ . \tag{9}$$

It can be ascertained that $\langle\xi\rangle$ satisfies the variational equation of the deterministic solution (6). This means that the dependence of $\langle\xi\rangle$ and σ^2 are controlled by $C_1'(\phi(t))$. If the system is allowed to relax from an arbitrary initial unstable point (i.e. $C'(0))>0$), then fluctuations are initially amplified. The maximum amplification can be shown to be

$$\sigma_m^2 \simeq \left(\sigma_o^2 + \frac{C_2(\phi^*)}{2\gamma}\right)\frac{C_1^2(\phi_m)}{\delta^2\gamma^2} \tag{10}$$

where $\phi^* = \phi(0)$, $C_1(\phi) = \gamma > 0$, $C_1'(\phi_m) = 0$ and $\delta = \phi^* - \phi_u$. Here ϕ_u is the unstable point of $C(\phi) = 0$. Thus the fluctuation enhancement is a $O(\delta^{-2})$ and the fluctuations are thus anomalously large [3].

Ω-expansion holds good as long as Gaussian representation is adequate. This however is not the case when multiple steady states are accessible from an initial unstable state. In this case the extensitivity property [3] which is implicit in Ω-expansion is violated.

To see the limitations of Ω-expansion consider

$$\frac{dn}{dt} = (\beta-\alpha)n - \frac{\gamma}{\Omega}n^2 + \eta(t) \tag{11}$$

where Ω is the size of the system. This model represents Malthus-Verhulst equation where the range of n is $[0,\infty]$. The steady states are $n_1 = 0$ and $n_2 = \Omega(\beta-\alpha)/\gamma$ of which n_1=o is unstable and the other is stable. We shall consider a special case when the states are very close to each other namely n_2 is close to n_1 i.e., $\beta - \alpha \simeq 0$. Choose $\beta - \alpha = \Omega^{-1/2}\Delta$, $\beta + \alpha = 1$ and $\gamma = 1$. In this case, one can show that the equation satisfied by $\langle\xi\rangle$ is no longer the variational equation of the deterministic solution of n. ($n = \Omega\phi + \Omega^{1/2}\xi$ is the choice here). For $t \to \infty$, $\langle\xi\rangle \to \frac{\Delta}{2}$ and $\langle\xi^2\rangle - \langle\xi\rangle^2 \to \frac{1}{4}$. This means that the position of the peak (which in the present case is determined by $\langle\xi\rangle$) is of the same order as the variance. (See ref.3 for details). Since negative values of n are unphysical, the Ω-expansion is valid only for

$$\phi(t) = \frac{\phi(0)}{1+t\phi(0)} \gg \Omega^{\frac{1}{2}} [\langle\xi^{\frac{1}{2}}\rangle]^{\frac{1}{2}}$$

This means that the Ω-expansion is valid only for $t \ll \epsilon^{1/2}$. Now suppose we allow negative values of n also and pose the question slightly differently, namely for what values of $\phi(0)$ is the Ω-expansion valid for all t ? Then, we see that

$$\phi(0) \gg \Omega^{1/2} = \epsilon^{1/2} \tag{12}$$

Although, this result has been derived when the two steady states are close, the central point that has emerged namely, the mean square of the position should be much larger than the variance for the Ω-expansion to be valid, is a general result. Equivalently if $\delta \ll \epsilon^{1/2}$, the approximation breaks down since the spread becomes large to allow for the two possible steady states. In such a situation the initial time development is slow and the effect of the random force is important. For the intermediate time regime nonlinearity plays a crucial role (leading to strong non-Gaussian features) and the effect of fluctuations can be ignored. In the final regime, again the effect of fluctuations cannot be ignored. Thus, we see that there are qualitatively two different kinds of time dependence for the two cases $\delta \gg \epsilon^{1/2}$ and $\delta \ll \epsilon^{1/2}$.

3. Generalized Statistical Linearization Scheme

The basic idea of this method is to replace the original NLE by an equivalent linear equation [8,11,12]

$$\dot{x} = \tilde{\gamma}(t)x + \tilde{C}(t) + \eta(t) \tag{13}$$

With this replacement we should not expect that the non-Gaussian features of (1) to be retained. However, we can obtain the first and the second moments of $x(t)$ in some optimal sense by making use of the arbitrary functions $\tilde{\gamma}(t)$ and $\tilde{C}(t)$. This is done by demanding that the ensemble average of the error due to the replacement of (1) by (13) be minimum [11,12]. This leads to an optimal choice of $\gamma(t)$ and $C(t)$. For the example $C_1(x) = \gamma x - gx^3$, after substituting for $\gamma(t)$ and $C(t)$, we get

$$\frac{d}{dt}\langle x\rangle = \gamma\langle x\rangle - g\langle x^3\rangle , \tag{14}$$

and

$$\frac{d}{dt}\langle x^2\rangle = 2[\gamma\langle x^2\rangle - g\langle x^4\rangle + \epsilon]. \tag{15}$$

[See ref.8, for details.] These equations are identical to the equations obtained starting from the original NLE. Since (13) is linear, the distribution is Gaussian. Hence a good approximation is obtained if we use a Gaussian decoupling of the higher moments (than two). Obviously, this approximation is meaningful only when $\delta \gg \epsilon^{1/2}$ (i.e. the extensive regime where Ω-expansion is valid).

Since both the steady states become accessible for ($\delta \ll \epsilon^{1/2}$, the intrinsically unstable regime) the method fails. However, for $\phi(0) = 0$, both the steady states are equally probable and $\phi(t)=0$ for all t, it is possible to get a reasonable representation of fluctuations by choosing a bimodal Gaussian distribution for decoupling higher order moments. The distribution we choose is [8]

$$P(x,t) = K\left[H(-x)\exp - \frac{(x+x_1)^2}{2\sigma_1^2} + H(x) \exp - \frac{(x-x_1)^2}{2\sigma_1^2}\right] \tag{16}$$

where

$$K = 1/\sqrt{2\pi}\,\sigma_1\left[1 + \operatorname{erf}(x_1/\sqrt{2}\,\sigma_1)\right]$$

Here, $H(x)$ is the step function, x_1 is the position of the peak and σ_1^2 is the variance as defined for one part of the distribution. Using (16) we get

$$\frac{d}{dt}\langle x^2\rangle = 2\Big[\gamma\langle x^2\rangle - g\langle x^2\rangle^2 + \epsilon - 2g\,\sigma_1^2\,\Big\{2\langle x^2\rangle - \sigma_1^2 - Kx_1\langle x^2\rangle\exp\Big(-\frac{x_1^2}{2\sigma_1^2}\Big)\Big\}\Big] \tag{17}$$

since $\sigma_1^2 = \langle x^2\rangle_s - \langle x\rangle_s^2$ (where s refers to one segment), we expect it to be small. Hence to the first approximation we can retain the first three terms, which is Suzuki's self consistent linearization scheme. (He uses this scheme to approximately demonstrate the scaling behaviour). One can do better by writing a similar equation for the fourth moment and determine x_1 and σ_1 self consistently.

The Gaussian decoupling scheme can be effectively used in higher dimensions whenever only the first two moments are of interest. As an example we consider a model which has a basis in plastic flow [13]. The coupled equations are

$$\dot{x} = xy - x^2 + \eta_1 \,, \tag{18}$$

$$\dot{y} = b_o - b_1x^2y + b_2x^3 + \eta_2 \,, \tag{19}$$

where η_1 and η_2 are taken to be Gaussian white noise with zero mean and

$$\langle \eta_i(t)\ \eta_j(t')\rangle = 2\,\epsilon_i\ \delta_{ij}\ \delta(t-t') \,. \tag{20}$$

In the physical problem x is a dimensionless variable related to the square root of dislocation density and y is related to dimensionless stress. There is only one steady state given by $x = y = [b_o/(b_1-b_2)]^{1/3}$ and the constants b_i's are material parameters.

The moment equations can be easily obtained to be

$$\frac{d}{dt}\langle x\rangle = \langle xy\rangle - \langle x^2\rangle \tag{21}$$

$$\frac{d}{dt} < y > = b_o - b_1 < x^2 y > + b_2 < x^3 > \tag{22}$$

$$\frac{d}{dt} < x^2 > = 2[< x^2 y > - < x^3 > + \epsilon_1] \tag{23}$$

$$\frac{d}{dt} < y^2 > = 2[b_o < y > - b_1 < x^2 y^2 > + b_2 < x^3 y > + \epsilon_2] \tag{24}$$

$$\frac{d}{dt} < xy > = < xy^2 > - < x^2 y > + b_o < x > - b_1 < x^3 y > + b_2 < x^4 > \tag{25}$$

where we have used $< x \eta_1 > = \epsilon_1$, $< y \eta_2 > = \epsilon_2$, and $< x \eta_2 > = < y \eta_1 > = 0$, in the linear approximation. Using a bivariate Gaussian distribution, we express higher order moments occuring in (21-25) in terms of the first two moments. These coupled equations can be solved on a computer. The results obtained for the variance agrees very well with the Monte Carlo results for 4900 tracks as can be seen from Fig.1. (See for details ref.8).

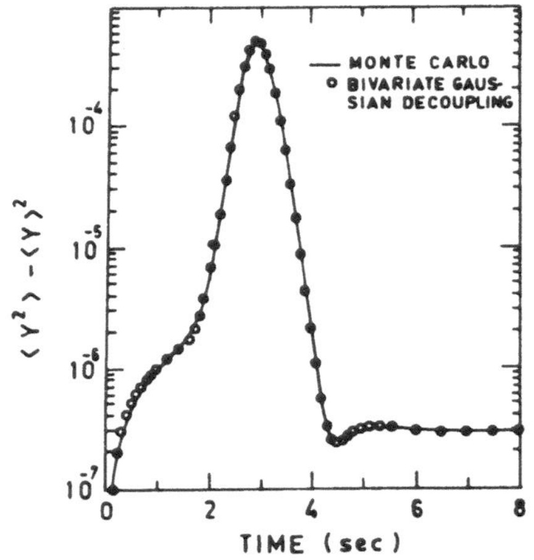

Fig.1. σ_y^2 as a function of time as obtained by Gaussian decoupling and Monte Carlo.

4. Scaling Theory

We shall briefly return to the self consistent linearization scheme, since it brings out the salient feature of the scaling theory. Suzuki linearizes the equation by writing $x^3 = < x^2 > x$ and then determines $< x^2 >$ self consistently. The solution can be expressed in terms a scaled variable τ given by

$$< x^2(t) > = < x^2 >_{st} \frac{\tau}{1+\tau} \tag{26}$$

where

$$\tau = \frac{g}{\gamma}[< x^2(o) > + \frac{\epsilon}{\gamma}] \exp(2\gamma t) \tag{27}$$

and $< x^2 >_{st} = \gamma/g + O(\epsilon)$. In the initial regime since $\tau \sim \epsilon$, $< x^2 > \sim \epsilon$, whereas for the intermediate time $< x^2(t) > \sim 1$. This means fluctuation enhancement of the order ϵ^{-1}. (This should be contrasted with the fluctuation enhancement of δ^{-2} for the extensive regime $\delta^2 \gg \epsilon$). This fluctuation enhancement can be looked upon as formation of order. This time is

$$t = \frac{1}{2\gamma} \ln \frac{g}{\gamma}[\langle x^2(o)\rangle + \frac{\epsilon}{\gamma}]^{-1}$$

Even within the framework of the simple decoupling scheme we see the utility of the scaling variable. Although the decoupling scheme appears ad hoc the analysis presented in the previous section offers some justification [8].

We shall now outline the scaling theory via a nonlinear time independent transformation which essentially transforms the NLE into a linear Langevin equation in terms of the transformed variable.

Consider

$$\frac{dx}{dt} = C_1(x) + \eta(t) \tag{28}$$

For an unstable state, $\gamma = C_1'(0) > 0$, where we have assumed $x = 0$ is the unstable point of $C_1(x)$.

Define

$$\xi = F(x) = \exp \int_{a_0}^{x} \frac{\gamma}{C_1(y)} dy \tag{29}$$

where a_0 is taken such that $F'(o) = 1$. For the special case $C_1(x) = \gamma x - gx^3$ we have

$$\xi = F(x) = x[1 - \frac{g}{\gamma} x^2]^{-1/2} \tag{30}$$

and

$$x = F^{-1}(\xi) = \xi[1 + \frac{g}{\gamma} \xi^2]^{-1/2} \tag{31}$$

Then

$$\frac{d\xi}{dt} = \gamma\xi + \frac{\gamma\xi\,\eta(t)}{C_1(F^{-1}(\xi))} \tag{32}$$

Define

$$\frac{\gamma\xi}{C_1(F^{-1}(\xi))} = 1 + f(\xi) \tag{33}$$

For small ξ, $f(\xi) \rightarrow \xi$.

Then

$$\frac{d\xi}{dt} = \gamma\xi + \eta(t)\,(1 + f(\xi)) \tag{34}$$

Dropping $f(\xi)$ (in the scaling approximation), we get

$$\frac{d\xi}{dt} = \gamma\xi + \eta(t) \tag{35}$$

This has a Gaussian structure in ξ. Therefore we have

$$\xi_{sc}(t) = e^{\gamma t} \int_0^t e^{-\gamma t'} \eta(t')dt' + e^{\gamma t} \xi(o) \tag{36}$$

and

$$\langle \xi_{sc}^2 \rangle = [\langle \xi^2(o)\rangle + \frac{\epsilon}{\gamma}] e^{2\gamma t} - \frac{\epsilon}{\gamma} \tag{37}$$

From this we can calculate $\langle x_{sc}^2 \rangle$. We shall do it for the particular example considered.

$$\langle x^2(t)\rangle_{sc} = \langle \xi_{sc}^2(t)\,[1 + (\tfrac{g}{\gamma})\;\xi_{sc}^2(t)]^{-1}\rangle = \sum_{n=1}^{\infty} (-\tfrac{g}{\gamma})^{n-1} \langle \xi_{sc}^{2n}(t)\rangle$$

$$= \sum_{n=1}^{\infty} (-\tfrac{g}{\gamma})^{n-1}\, e^{2n\gamma t} \int_0^t dt_1\, e^{-\gamma t_1} \,..\, \int_0^t dt_{2n}\, e^{-\gamma t_{2n}} \langle \eta(t_1)\,..\,\eta(t_{2n})\rangle$$

$$= \sum_{n=1}^{\infty} (-\tfrac{g}{\gamma})^{n-1}\,(2n-1)!!\; e^{2n\gamma t}\Big[\int_0^t dt_1 \int_0^t dt_2\; e^{-\gamma(t_1+t_2)} \langle \eta(t_1)\;\eta(t_2)\rangle\Big]^n$$

$$\langle x^2(t)\rangle_{sc} = \sum_{n=1}^{\infty} (-\tfrac{g}{\gamma})^{n-1}\,(2n-1)!!\;\Big[\frac{\epsilon}{\gamma}\,(e^{2\gamma t} - 1)\Big]^n \tag{38}$$

Using

$$\frac{1}{\sqrt{2\pi}} \int_{-\infty}^{\infty} e^{-\frac{1}{2}\xi^2}\; \xi^{2n}\, d\xi = (2n-1)!! \tag{39}$$

in (38), and switching order of summation and integration, we get

$$\langle x^2(t)\rangle_{sc} = \langle x^2\rangle_{st}\, \frac{1}{\sqrt{2\pi}} \int_{-\infty}^{\infty} e^{-\xi^2/2}\; \frac{\xi^2\tau}{1 + \xi^2\tau}\; d\xi\;, \tag{40}$$

where

$$\tau = \frac{g\epsilon}{\gamma^2}\,(e^{2\gamma t} - 1) \tag{40a}$$

If $\langle x^2(0)\rangle \neq 0$, then

$$\tau = \frac{g}{\gamma}\Big[\frac{\epsilon}{\gamma}\,(e^{2\gamma t} - 1) + \langle x^2(0)\rangle\; e^{2\gamma t}\Big] \tag{40b}$$

To see how good the Gaussian decoupling scheme (where the first two non-vanishing moments are considered) and the self consistent linearization scheme perform, we have plotted in Fig.2, the variance as calculated by these methods along with the scaling result. Also shown is the variance calculated by Monte Carlo method (for 4900 tracks), and numerical solution of Fokker-Planck equation [9]. As expected the scaling theory performs the best (compare with Monte Carlo result which is numerically exact). In spite of the fact that the bimodal Gaussian decoupling is a simple representation of the actual non-Gaussian process it does perform reasonably well.

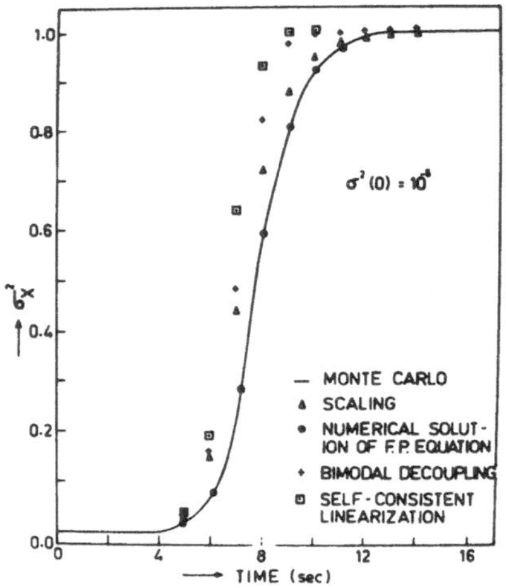

Fig. 2 σ_x^2 as a function of time as obtained by various methods.

The results of scaling theory hold good only for intermediate time regime where nonlinearities play an important role. If one wishes to consider the evolution of the system for $t \rightarrow \infty$ limit, we have to include the effect of fluctuations. Such an extention has been carried out by Suzuki [6].

The problem of relaxation from an intrinsically unstable state has been studied by a number of other authors [14-20]. For example de Pasquale and co-workers [16-18] introduce a time dependent nonlinear transformation given by

$$\xi = F^{-1}(e^{-\gamma t} F(x))$$

It represents the characteristic curve of the stochastic process and a constant along the deterministic path $\frac{dx}{dt} = C_1(x)$. The results via this transformation are not significantly different. The relation of some of these apparently different methods to Suzuki's work has been analysed in detail by Suzuki in his recent reviews on the subject[6,7]. For an interested reader we refer to these reviews.

It is my pleasure to thank M.C. Valsakumar, R. Indira and K.P.N.Murthy.

References

1. N.G. van Kampen, J. Stat. Phys. 25, 431 (1981).
2. N.G. van Kampen, in Advances in Chemical Physics, Vol.34, I. Prigogine and S.Rice eds. (Wiley, New York, 1976) p.245.
3. R. Kubo, K. Matsuo and K. Kitahara, J. Stat. Phys. 9, 51 (1973).
4. M. Suzuki, Prog. Theor. Phys. 56, 77, 477 (1976).
5. M. Suzuki, J. Stat. Phys. 16, 11 (1977).
6. M. Suzuki, in Advances in Chemical Physics, Vol. 41, I. Prigogine and S. Rice, eds. (Wiley, New York, 1981) p.195 and references therein.
7. M. Suzuki, in Proceedings of the XVIIth Solvay Conference, Brussels (John Wiley, New York, 1978).
8. M.C. Valsakumar, K.P.N.Murthy and G. Ananthakrishna, J. Stat. Phys. 31, 637 (1983).
9. R. Indira, M.C. Valsakumar, K.P.N. Murthy and G. Ananthakrishna, to be submitted.
10. G.S. Agarwal, this volume.
11. A. Budgor, J. Stat. Phys. 15, 375 (1976).
12. J.O. Eaves and W.P. Reinhardt, J. Stat. Phys. 25, 127 (1981).
13. G. Ananthakrishna, J. Phys. D15, 77 (1982)
14. F. Haake, Phys. Rev. Lett. 41, 1685 (1978).
15. B. Caroli, C. Caroli and B. Roulet, J. Stat. Phys. 21, 415 (1979); 22, 515 (1980).
16. F. de Pasquale, P. Tartaglia and P. Tombesi, Physica 99A, 587 (1979).
17. F. de Pasquale, P. Tartaglia and P. Tombesi, Z. Phys.B43, 353 (1981) and references therein.
18. F. de Pasquale, P. Tartaglia and P. Tombesi, Phys. Rev. A25, 466 (1982).
19. F.T. Arecchi and A. Politi, Phys. Rev. Lett. 45, 1219 (1980).
20. M.C. Valsakumar, this volume.

SOLUTION OF FOKKER-PLANCK EQUATIONS USING TROTTER'S FORMULA

M.C. Valsakumar
Reactor Research Centre, Kalpakkam 603 102, India

1. Introduction

Fokker-Planck equations (FPE) with nonlinear drift terms are rarely amenable to closed form solutions [1,2]. The noteworthy methods of exact solution of FPE are the eigenfunction method [2,3] and the path integral formulation [4,5]. Both these methods, for their application, demand the FPE be first set in its self adjoint form. If the complete eigen spectrum of the corresponding Hermitian operator is obtainable, then the eigenfunction method can be used. The latter method, on the other hand, provides a formal Feynman path integral representation of the propagator of the above operator. In the present paper we show that the path integral solution to the FPE can be obtained without setting it in its self adjoint form. The method makes explicit use of the Trotter's product formula [6] widely used in perturbation theory [7]. The integral representation of the solution process is easily amenable to approximations. First order approximation gives the scaling solution [8], which has been demonstrated to be of remarkable success in the treatment of diffusion from intrinsically unstable states in a bistable potential [8-10].

2. The Method

We consider solving the FPE given by

$$\frac{\partial}{\partial t} P(x,t) = L\, P(x,t); \quad L = -\gamma \frac{\partial}{\partial x} C(x) + \epsilon \frac{\partial^2}{\partial x^2} \tag{1}$$

for the initial condition

$$P(x,o) = \delta(x-y) = (2\pi)^{-1} \int_{-\infty}^{\infty} dk \; e^{-ik(x-y)} \tag{2}$$

Since the operator L is a sum of two noncommuting operators, the formal solution exp (tL) P(x,o) of eq.(1) cannot be of any immediate use. However, by exploiting the Trotter's formula [6], which reads as

$$e^{A+B} = \lim_{n\to\infty} \left[e^{A/n} \; e^{B/n} \right]^n , \tag{3}$$

we can represent the formal solution of eq.(1) in a more convenient form as

$$P(x,t) = \lim_{n\to\infty} \frac{\theta^n}{2\pi} \int_{-\infty}^{\infty} dk \; e^{-ik(x-y)} \; ; \quad \theta = e^{-b \frac{d}{dx} C(x)} \; e^{a \frac{d^2}{dx^2}} \tag{4}$$

In the above expression $a = \epsilon t/n$ and $b = \gamma t/n$.

We now illustrate the method with the Ornstein-Uhlenbeck (O-U) process for which $C(x) = -x$. Using the relations

$$e^{a\frac{d^2}{dx^2}} e^{-ikx} = e^{-[ak^2 + ikx]} \;, \quad e^{b\frac{d}{dx}x} e^{-ikx} = e^{(b-ike^b x)} \;, \tag{5}$$

eq. (4) for this stochastic process reduces to

$$P(x,t) = \lim_{n\to\infty} \frac{1}{2\pi} \int_{-\infty}^{\infty} dk \; e^{-ak^2 \sum_{r=1}^{n} e^{2(r-1)b}} \; e^{nb-ik(xe^{nb}-y)} \tag{6}$$

Summing the series in the integrand, and recognising that

$\lim_{n\to\infty} a/(e^{2b}-1) = \epsilon/(2\gamma)$, we get the O-U solution

$$P(x,t) = [2\pi \frac{\epsilon}{\gamma} (1-e^{-2\gamma t})]^{-\frac{1}{2}} \; e^{-(x-ye^{-\gamma t})^2 / \frac{2\epsilon}{\gamma}(1-e^{-2\gamma t})} \tag{7}$$

3. Solution for General Nonlinear Drift Term

The solution for a general $C(x)$ is not so simple as in the case of the O-U process. The reason for this is that the quantity Θe^{-ikx} does not go to the form $\alpha e^{-\beta x}$ where α and β are independent of x. In fact we have

$$e^{-b\frac{d}{dx}C(x)} f(x) = \frac{C[G(x)]}{C[x]} f[G(x)] \tag{8}$$

where

$$G(x) = \xi = F^{-1}(F(x)\, e^{-b}); \; F(x) = \exp [\int^x dx'/C(x')] \tag{9}$$

Therefore we get

$$\Theta e^{-ikx} = \frac{C[G(x)]}{C[x]} e^{-[ak^2 + ikG(x)]} = H(k,x) \;, \tag{10}$$

which in the Fourier representation reads

$$\Theta e^{-ikx} = (2\pi)^{-1} \int_{-\infty}^{\infty} dk_1 e^{-ik_1 x} \int dx_1 e^{ik_1 x_1} H(k,x_1) \tag{11}$$

In eq.(11), the limits of x_1 integral is such that $G(x_1)$ is real. Repeating the process n times yields

$$P(x,t) = \lim_{n\to\infty} (2\pi)^{-n} \int \cdots \int dk \prod_{i=1}^{n-1} dk_i dx_i \; e^{ik_i x_i} \; e^{iky} H(k_{n-1},x) \ldots H(k,x_1) \tag{12}$$

Using the fact that

$$\frac{dG(x)}{dx} = \frac{\partial \xi}{\partial x} = \frac{C[G(x)]}{C[x]} \quad \text{and} \quad P(\xi,t)d\xi = P(x,t)dx \;, \tag{13}$$

$$P(\xi,t) = \lim_{n\to\infty} (2\pi)^{-n} \int_{-\infty}^{\infty} \cdots \int_{-\infty}^{\infty} dk \prod_{i=1}^{n-1} dk_i d\xi_i \; e^{ik_i G^{-1}(\xi_i) - ak_i^2}$$
$$e^{iky-ak^2} \; e^{-i[k_{n-1}\xi + \ldots\ldots + k\xi_1]} \tag{14}$$

Performing the $\{k\}$ integrations gives the equivalent result

$$P(\xi,t) = \lim_{n\to\infty} (4\pi a)^{-n/2} \int_{-\infty}^{\infty} .. \int_{-\infty}^{\infty} \prod_{i=1}^{n-1} d\xi_i$$

$$e^{-\frac{1}{4a}\left\{[\xi - G^{-1}(\xi_{n-1})]^2 + \ldots + [\xi_1 - y]^2\right\}} \tag{15}$$

These results [eqs.(14) and (15)] are equivalent to the path integral result [4,5].

4. Approximate Closed Form Solution

We repeat that the results given by eqs.(14) and (15) are only formal and they do not lead to simple closed form solutions for arbitrary drift terms. So approximations are imperative in the useage of these. An expansion in b (which is necessarily small) may be a choice. But one can easily satisfy himself, with explicit application of this procedure on the O-U process, that this will lead to erroneous results. In what follows, we show that an expansion of $G^{-1}(\xi)$ in powers of ξ gives very good results. We illustrate this with the well known model• for diffusion in a bistable potential for which

$$C(x) = x - \frac{g}{\gamma} x^3 \tag{16}$$

When the diffusion is from the extensive regime (initial point sufficiently far from the unstable steady state), the probability distribution function is practically unimodal. Hence the approximation schemes based on 'linearisation' works well [9-11]. On the other hand, when the evolution is from the intrinsically unstable states (initial point close to the unstable steady state), the distribution function is necessarily bimodal and hence the linearisation approximations fail. Only the scaling theory (and its equivalent ones) gives the best results [8,9].

For this problem, the transformation $\xi = G(x)$ and its inverse are

$$\xi = x e^{-b} [1 - \frac{g}{\gamma} x^2 (1 - e^{-2b})]^{-\frac{1}{2}} \tag{17}$$

$$G^{-1}(\xi) = \xi e^{b} [1 + \frac{g}{\gamma} \xi^2 (e^{2b} - 1)]^{-\frac{1}{2}} \tag{18}$$

To the first order in ξ

$$G^{-1}(\xi) = \xi e^{b} \tag{19}$$

Suzuki [8] has demonstrated this approximation to be good in the scaling limit (i.e. $\epsilon \to 0$ and b fixed). On the substitution of this in eq.(14), the $\{\xi\}$ integrals yield delta function in $\{k\}$. Performing the $\{k\}$ integrals and changing the argument of ξ to the actual time variable of interest (γt), we get

$$P(\xi(\gamma t),t) = [2\pi \frac{\epsilon}{\gamma}(1-e^{-2\gamma t})]^{-\frac{1}{2}} e^{-(\xi-y)^2/\frac{2\epsilon}{\gamma}(1-e^{-2\gamma t})} \tag{20}$$

This is the same as the scaling solution of Suzuki [8]. Suzuki has shown that the approximation given by eq.(19) is valid for any C(x) with x as the leading term. Secondly it is the presence of the small diffusion constant ϵ that allows this approximation to be good. Hence under the same circumstances and because of the same reason, our formal results also lead to the scaling solution.

References

1. M.O. Hongler, Physica D2, 353 (1981) and references therein.
2. M.C. Valsakumar, submitted to Physica D.
3. N.G. van Kampen, J. Stat. Phys. 17, 71 (1977).
4. W. Horsthemke and A. Bach, Z. Phys. B22, 189 (1975).
5. R. Graham, Z. Phys. B26, 281 (1977).
6. H.F. Trotter, Proc. Amer. Math. Soc. 10, 545 (1959).
7. W.G. Faris, Bull. Amer. Math. Soc. 73, 211 (1967).
8. M. Suzuki, Adv. Chem. Phys. 46, 195 (1981) and references therein.
9. G. Ananthakrishna, in the same volume.
10. R. Indira, M.C. Valsakumar, K.P.N.Murthy and G. Ananthakrishna, to be published.
11. M.C. Valsakumar, K.P.N.Murthy and G.Ananthakrishna, J. Stat. Phys. 30, 637 (1983) and references therein.

MONTE CARLO METHODS : AN INTRODUCTION

K.P.N. Murthy
School of Physics, University of Hyderabad
Hyderabad - 500134, India

and

Reactor Research Centre, Kalpakkam 603 102
Tamil Nadu, India

1. Introduction

Monte Carlo can be defined as a numerical method that uses random numbers to solve a problem. The methods of Monte Carlo have become quite popular in the recent times, thanks to the availability of high speed computers, and are being applied to a growing variety of problems. Deterministic as well as stochastic problems can be handled. For solving a deterministic problem, we construct first a stochastic model. Then we carry out a Monte Carlo simulation of this artificial stochastic model on a computer. Multi-dimensional integration, matrix inversion, solution of Dirichlets problem etc. are a few examples of the deterministic problems that have been successfully solved using Monte Carlo.

For stochastic problems, however, we can resort to direct simulation. A notable example of this is the study of neutron transport in nuclear reactors. Indeed, the identification and systematic exploitation of Monte Carlo as a powerful tool for solving practical problems started with nuclear applications and the first investigators were von Neumann, Ulam, Fermi, Harris, Herman Kahn, Metropolis and others. A brief account of the interesting history of Monte Carlo and reference to some of the early works can be found in the monograph of Hammersley and Handscomb [1].

If the stochastic problem on hand is complex (or if we are not interested in the microscopic details of the solution process), we can construct a simpler stochastic model of the problem and then simulate this model on a computer. Monte Carlo simulation of Brownian motion modelled by the Langevin equation, study of kinetic Ising models, percolation problems etc. come under this category.

In this lecture I shall try to present briefly the basic features of Monte Carlo and illustrate its application to the study of nonlinear Brownian motion.

2. Elements of Monte Carlo

The important features of Monte Carlo can be easily understood by considering the following simple problem.

Let x be a random variable and f(x) its probability density function. We set to determine the average of a function h(x) given by

$$\mu = \int_{-\infty}^{\infty} h(x) f(x) dx \qquad (1)$$

Monte Carlo method of evaluating μ consists of selecting a set of N independent random values x_i, i = 1, N from the density f(x) and carrying out the following summation

$$\bar{h}_N = \frac{1}{N} \sum_{i=1}^{N} h(x_i) \qquad (2)$$

$\bar{h}_N$ given above is a Monte Carlo estimate of the desired answer μ. Sampling of x from the density f(x) is accomplished most conveniently as follows. First generate a set of random numbers between 0 and 1 from a uniform density. (We shall refer to these as random numbers in all subsequent discussions, and denote them by ξ). Having generated a random number ξ, we shall perform some appropriate transformation to obtain x for use in (2). Such techniques are called random sampling techniques. For example in the inversion technique, we set $X = F^{-1}(\xi)$ where F is the probability distribution (or cumulative density) function of X. There are numerous random sampling techniques like rejection technique, equiprobability table method, composition and decomposition methods etc. for different kinds of density functions. For a review of these techniques see [2]. There are many algorithms to generate random numbers on a computer. These numbers should be strictly called pseudo random numbers since there is nothing random about their source; for, they are produced by deterministic procedures. We are justified in using these numbers in our problem since our concern lies not in the source of these numbers but on whether they are 'correctly' distributed. A sequence of numbers that pass through the required set of statistical tests for randomness should suffice for our purpose. Examples of (pseudo) random number generators are the early mid square method of Metropolis and von Neumann (see [3]), the multiplicative congruential method of Lehmer [4] etc. We shall not go into the details of these techniques. Those interested can see [1-5] and the references therein. See also Chaitin [6] for an interesting discussion on randomness and mathematical proof.

Returning to (2) it is desirable to quantify the confidence we repose on $\bar{h}_N$ as an approximation to μ and this is afforded by the central limit theorem. It states that as the sample size N tends to ∞ the probability density of the random variable $\bar{h}_N$ tends to be Gaussian with mean μ and variance σ^2/N where σ^2 is the variance of the problem. Thus $\pm \sigma/\sqrt{N}$ gives one-sigma confidence interval (also called statistical error). Since σ^2 is unknown we use its Monte Carlo estimate S_N^2 (the sample variance) to obtain the statistical error. The expression for S_N^2 is

$$S_N^2 = \frac{1}{N-1} \sum_{i=1}^{N} h^2(x_i) - \frac{N}{N-1} (\bar{h}_N)^2 \qquad (3)$$

3. Variance Reduction Techniques

It is clear that the statistical error in Monte Carlo estimates can be decreased by increasing N, the sample size or decreasing σ. The latter implies that we distort the problem on hand so that the altered problem has a smaller variance but the same mean. Such variance reduction techniques have become imperative to Monte Carlo simulation and considerable work is being carried out in this area, see for e.g. [7-9]. Let me illustrate the technique of variance reduction by considering importance sampling. This essentially consists of using an importance density function $g(\alpha, x)$ instead of the actual density $f(x)$ for evaluating μ (see (1)). α here is an adjustable parameter, significance of which will become clear soon. Now, for purpose of preserving mean, we consider the function $H(x)$ in place of $h(x)$. Clearly if $H(x)$ is chosen as $h(x)f(x)/g(\alpha, x)$ the mean of $H(x)$ in the distorted space of $g(\alpha, x)$ is the same as μ but the variance is

$$\sigma_1^2(\alpha) = \int_{-\infty}^{\infty} \left\{ \frac{h(x)f(x)}{g(\alpha, x)} \right\}^2 g(\alpha, x)\, dx \tag{4}$$

A judicious choice of $g(\alpha, x)$ would lead to considerable variance reduction. Indeed one can do better by optimizing the importance function during the process of simulation itself, by splitting it into various stages and improving the parameter from stage to stage. This can be accomplished as follows. We rewrite (4) as

$$\sigma_1^2(\alpha) = \int_{-\infty}^{\infty} \frac{h(x)f(x)}{g(\hat{\alpha}, x)} \frac{h(x)f(x)}{g(\alpha, x)} g(\hat{\alpha}, x)\, dx \tag{5}$$

where $\hat{\alpha}$ is an initial guess of the parameter α. The procedure consists of first generating a small sample of x from $g(\hat{\alpha}, x)$. From this sample, using (5) evaluate the sample variance at different values of α. Determine the value of α at which σ_1^2 is minimum. Use this value in the next stage and proceed as before. After a few stages, α converges to its optimum value and this can be used in the final stage where a comparatively large sample is chosen to estimate the mean. This powerful self learning technique due to Spanier has been successfully used in many problems [7-9,11].

Besides importance sampling there are many variance reduction techniques like systematic sampling, stratified sampling, correlated sampling, control variates, antithetic variates, regression methods etc. Ref.[10] provides a review of the numerous variance reduction techniques.

4. Monte Carlo Simulation of Non-Linear Brownian Motion

In this part we shall consider an application of Monte Carlo to study nonlinear Brownian motion, modelled by the Langevin equation,

$$\frac{dx}{dt} = c(x) + \eta(t) \tag{6}$$

where $x(t)$ is the driven stochastic process, $c(x)$ is the nonlinear drift term and $\eta(t)$ is the driving Gaussian white noise satisfying the following properties

$$\langle \eta(t) \rangle = 0 \; ; \quad \langle \eta(t)\, \eta(t') \rangle = 2\, \epsilon\, \delta(t-t') \tag{7}$$

ϵ in the above is the diffusion constant. To simulate the stochastic process $x(t)$ we first discretise the time variable. Correspondingly the random force $\eta(t)$ is also replaced by a set $\{\eta_i\}$ at discrete times $\{t_i\}$. The property (7) now becomes

$$<\eta_i> = 0 \quad ; \quad <\eta_i \eta_j> = \sigma^2 \delta_{ij} \tag{8}$$

We consider finite difference approximation to (6) given by

$$x_j = x_{j-1} + \Delta t.\ c(x_{j-1}) + \Delta t \eta_{j-1} \tag{9}$$

Monte Carlo simulation of x(t) is carried out as follows. A sample track $\{x_0, x_1, x_2\ x_3 \ldots\ldots\}$ is generated by first sampling x_0 from the given initial distribution of the random process. Then we add successively a deterministic increment, $c\,\Delta t$ and a random increment $\eta \Delta t$ where η is sampled from a Gaussian of mean 0 and variance σ^2. The evolution of x(t) is considered upto the desired time. The whole process is repeated and a sample of N tracks is obtained. The required time dependent statistics of the driven stochastic process like mean, variance, correlations etc. are calculated by explicit averaging over the finite ensemble of N tracks.

In the above, random force η is to be sampled from a Gaussian of variance σ^2. The relation between σ^2 and the diffusion constant ϵ is obtained as follows. It is clear that

$$\int_0^t \eta(t)dt = \underset{\Delta t \to o}{\mathrm{Lt}} \sum_i \eta_i \Delta t \tag{10}$$

Then (7) yields

$$<\int_0^t \eta(t')\ \eta(t'')\ dt''> = 2\epsilon \tag{11}$$

from which it follows

$$\sum_i <\eta_i \eta_j> \Delta t = 2\epsilon \tag{12}$$

Thus we get

$$\sigma^2 = 2\epsilon/\Delta t \tag{13}$$

For generating the random forces η_j, j = 1,2 ......we use central limit theorem. This consists of selecting M random numbers and taking their sum $S = \sum \xi_i$. The desired Gaussian variate (mean zero and variance σ^2) is given by

$$\eta = \sigma\left\{\sqrt{12/M} - S - \sqrt{3.M}\right\} \tag{14}$$

For large M, η has the required Gaussian density. In practice M = 12 is sufficient [5].

For computing deterministic evolution of the process over the time interval Δt, one can use sophisticated algorithms instead of the finite difference scheme shown in (9). This would help improve the simulation considerably. Recently we studied the problem of diffusion in bistable potential [12,13] where the nonlinear drift term of (6) has the form $c(x) = \gamma x - gx^3$. Here γ and g are positive real constants. We have used Runge-Kutta Hill method for the deterministic evolution. In Fig.1 we

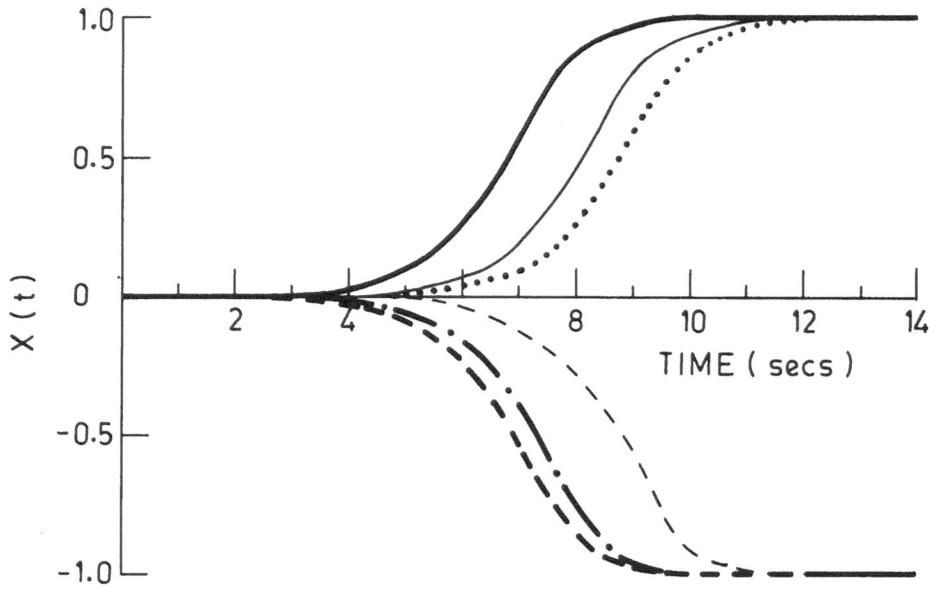

Fig.1 Sample tracks of Nonlinear Brownian Motion from an initial unstable steady state

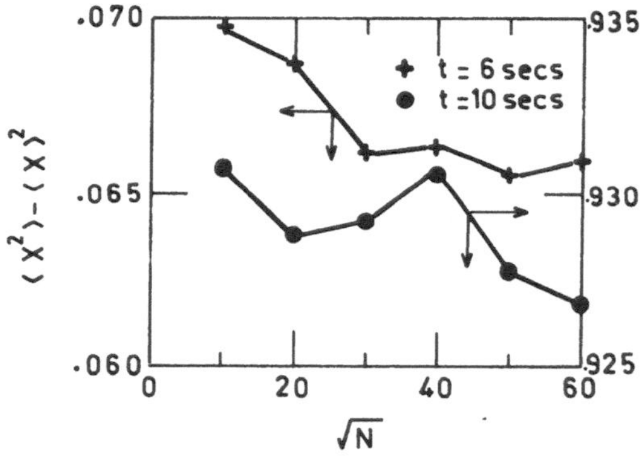

Fig.2 Variation of ensemble average of fluctuations with sample size at two typical times

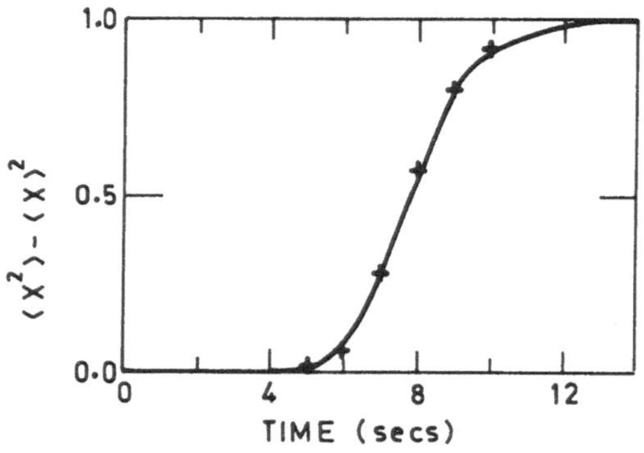

Fig.3 Fluctuations as a function of time. Ensemble average over 3600 tracks.

show a few of the tracks that were generated with the parameters $\epsilon = 0.5 \times 10^{-6}$, $\gamma = g = 1$ and the deterministic initial condition $x_o = 0$. Figure 2 depicts Monte Carlo estimates of the variance at a two typical times as a function of the sample size N. This illustrates the nature of convergence of the sample estimates. We show the fluctuation averaged over an ensemble of 3600 tracks, as a function of time in Fig.3.

5. Conclusions

I have tried to present the elementary notions of the method of Monte Carlo. This presentation undoubtedly suffers from my personal bias. Unlike other numerical techniques, Monte Carlo has to be developed each time taking into account the special requirements of the problem. There is indeed a great scope for ingenuity both in modelling of the problem and in developing special methods of simulation.

References

1. Hammersley J.M. and Handscomb D.C., Monte Carlo Methods, Chapman and Hall (1979).
2. McGrath E.J., and Irving D.C., Random Number Generation for Selected Probability Distributions, Vol.II of Techniques of Efficient Monte Carlo Simulation, Report ORNL-RSIC-38, ORNL (1975)
3. Forsythe G.E., NBS Applied Mathematics Series 12, 13 (1951)
4. Lehmer D.H., Ann. Comp. Lab. Harvard Univ. 26, 141 (1951)
5. James F., Monte Carlo Theory and Practice, Report DD/80/6, CERN Data Handling Division (1980)
6. Chaitin G.J., Scientific American 232, 47 (1975)
7. Spanier J., SIAM J. Appl. Maths., 18, 172 (1970)
8. Murthy K.P.N., Ann. Nucl. Ener. 7, 389 (1980)
9. Ragheb M.M.H et al, Atomkernenergie, Kertechnik 37, 188 (1981)
10. McGrath E.J. and Irving D.C., Variance Reduction Vol.III of Techniques of Efficient Monte Carlo Simulation, Report ORNL-RSIC-38, ORNL (1975)
11. Murthy K.P.N., Atomkernenergie, Kettechnik 34, 125 (1979)
12. Suzuki M., Adv. Chem. Phys. 46, 195 (1981) and the references therein
13. Valsakumar M.C., Murthy K.P.N. and Ananthakrishna G., J. Stat. Phys. 30, 637 (1983)

NUMERICAL SOLUTION FOR THE NONLINEAR FOKKER-PLANCK EQUATION

R. Indira
Reactor Research Centre
Kalpakkam 603102, India

1. Introduction

Nonlinear Fokker-Planck (F.P.) equations are not amenable to closed form solutions in most of the cases. Some approximate methods have been developed, which are valid for certain cases [1,2]. It is possible to obtain exact numerical solution of the nonlinear Fokker-Planck equation and this will be the content of this talk.

The finite difference methods [3] employed to obtain the numerical solution, and a modified alogorithm, which reduces the computational time drastically, are discussed. The method is illustrated with a sample problem of diffusion in bistable potential [4].

Method of Solution

Let us consider the nonlinear F.P. equation of the form

$$\frac{\partial P(x,t)}{\partial t} = -\frac{\partial}{\partial x}\left[f(x)P(x,t)\right] + \epsilon\frac{\partial^2 P(x,t)}{\partial x^2} \tag{1}$$

where $f(x)$ is a nonlinear function of x. This is a second order partial differential equation, with boundary conditions specified by $P(x,t) \longrightarrow 0$ as $x \longrightarrow \pm\infty$. The initial conditions are given by specifying $P(x,t)$ at $t = 0$. The finite difference method for solution of this equation proceeds as follows. A mesh structure is imposed on the variable x and t, the mesh widths being Δx and Δt respectively. Let j and k represent the x-mesh and t-mesh indices. Then the finite difference equations for $P(x,t)$ at $t = t_{k+1}$ are given by

$$a_1P_{j,k+1}+a_2P_{j-1,k+1}+a_3P_{j+1,k+1}-P_{j,k} = 0\,, \quad j = 2,\dots(N-1) \tag{2}$$

Here N represents the number of meshes in the x-domain, k denotes the time step, previous to the time step under consideration. a_1, a_2 and a_3 are given by

$$a_1 = \left\{-f'(x_j) - \frac{2\epsilon}{\Delta x^2}\right\}\Delta t - 1.0 \tag{3}$$

$$a_2 = \left\{\frac{2\epsilon}{\Delta x^2} + \frac{f(x_j)}{2\Delta x}\right\}\Delta t \tag{4}$$

$$a_3 = \left\{\frac{2\epsilon}{\Delta x^2} + \frac{f(x_j)}{2\Delta x}\right\}\Delta t \tag{5}$$

$f'(x_j)$ denotes the first derivative of $f(x)$ at $x = x_j$. Starting from the given initial distribution of $P(x,t)$ at $t = 0$, the equations reduce to a tridiagonal set of equations involving $P_{j,k+1}, P_{j-1,k+1}$ and $P_{j+1,k+1}$. These equations are solved to obtain $P_{j,k+1}$ for $j = 2,\dots N-1$. Now the entire solution can be obtained in steps of Δt.

It can be easily seen that the other physical restrictions (1) P(x,t) is nonnegative for all x at all times and (2) P(x,t) is normalized to unity at time t, are automatically satisfied, if one chooses an initial distribution, which is nonnegative and normalized to unity, (provided proper choice of Δx and Δt are made). Caution should be exercised in the choice of Δx and Δt, since finite difference approximations are valid only for small Δx and Δt. The choice of Δx is dictated by the accuracy with which P(x,t) is computed at time t. It is essential that the function P(x,t) does not vary much over the meshwidth Δx, since the finite difference approximation assumes P(x,t) to be constant over the interval Δx. This in turn implies a small Δt, since the numerical procedure becomes unstable otherwise. It can be shown that, the numerical solution obtained with a large meshwidth Δx, depicts unphysical behaviour of being negative over some region of x. The normalization of P(x,t) is also not preserved.

The computational time required for the numerical solution goes as N^2. As Δx is necessarily small, N is a large number in most cases and an appreciable amount of computational time is called for, to obtain accurate solutions. Hence it is advantageous to have algorithms which reduce the computer time requirements. The algorithm that has been developed by us, is based on continuous modification of Δx and the range in which P(x,t) is computed. The method becomes transparent in the context of a particular example, which we will now discuss.

Consider the nonlinear F.P. equation given by

$$\frac{\partial P(x,t)}{\partial t} = -\left[(\gamma x-gx^3)\, P(x,t)\right] + \epsilon \frac{\partial^2 P(x,t)}{\partial x^2} \tag{6}$$

(For relevance of this equation to diffusion in bistable potential, see [4]). We choose $\gamma = g = 1$ and $\epsilon = 0.5 \times 10^{-6}$. Consider the case where the initial condition is specified to be a Gaussian distribution of mean = .005 and variance = 1.0×10^{-6}. This is a sharply peaked function around the mean value. Hence, though the actual boundaries are at $\pm\infty$, one can set the boundaries at values close to the peak position (say x_{b1} and x_{b2}) to evaluate the initial evolution of P(x,t). The number of meshes required to represent the function accurately is now determined as follows. Choose an arbitrary value of Δx and represent the initial distribution by the corresponding histogram. Compare the moments of the histogram, with the actual moments of the initial distribution. If the comparison is not good, decrease Δx until good comparison is achieved. Now with this value of Δx and with the boundary condition of $P(x,t+\Delta t)$ equal to zero at $x = x_{b1}$ and x_{b2}, solve for $P(x,t+\Delta t)$. This is continued for, say, ten steps of Δt (i.e. upto time $t = t_1$). Now the boundaries and Δx are modified based on the distribution of P(x,t) at the time $t = t_1$. The evolution of P(x,t) in the next few steps of Δt is obtained with the boundary conditions applied at the new boundaries and with the modified meshwidth. This is continued till P(x,t) is obtained in the entire range of time.

2. Discussion

The numerical solution for the sample problem, using usual finite difference methods took one hour with Δx set equal to .0001. The modified algorithm required only ten minutes resulting in a reduction of computer time by a factor of ten. P(x,t) obtained using this algorithm has been shown in figure 1, at three typical times. It is seen that the expected nongaussian feature of P(x,t) is well represented. The first two moments of P(x,t) computed from this solution, were found to compare extremely well with the results of the Monte Carlo simulation of the corresponding nonlinear Langevin equation [4].

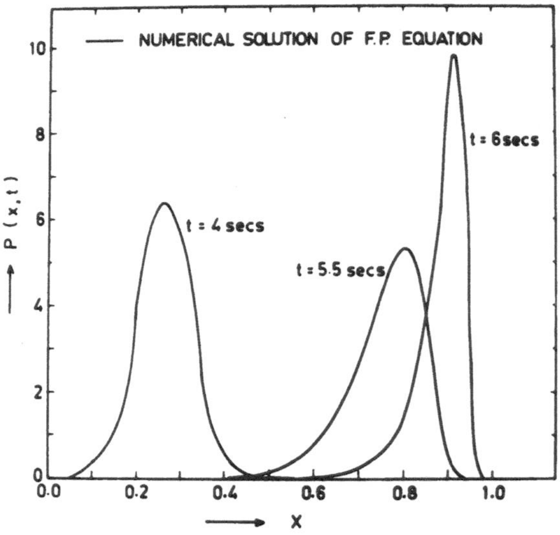

Fig.1 Numerical solution of (5) for γ = g=1; $\epsilon = 0.5 \times 10^{-6}$ and an initial Gaussian distribution of mean = .005 and variance = 1.0×10^{-6}

The numerical solution to (5) was also obtained, in the case of a initial Gaussian distribution with zero mean and an initial variance of 1.0×10^{-6}. In this case, the steady state solution has two peaks at $x = \pm 1$. The numerical solution obtained for this case has been reported in (4). Since more than one peak is found, the modified algorithm discussed above could not be applied, asymptotically. This is because, in the above algorithm, uniform Δx is assumed in the entire x-domain, at any time t. A superior algorithm applicable in any general case, is possible, with variable mesh widths in the x-domain. The meshwidth in any region of x-domain, can be fixed based on the value of the first derivative of the function P(x,t) in that region. This algorithm is presently being developed by us.

References

1. M. Suzuki, Adv. Chem. Phys. 46, 195 (1981) and references cited therein.
2. N.G. van Kampen, Adv. Chem. Phys. 34, 245 (1976).
3. L. Collatz, The numerical treatment of differential equations (Springer-Verlag, 1966).
4. R. Indira, M.C. Valsakumar, K.P.N. Murthy and G. Ananthakrishna (To be published).

STABILITY OF STOCHASTIC SYSTEMS

G.V. Anand
Department of Electrical Communication Engineering
Indian Institute of Science
Bangalore - 560012, India

1. Introduction

The analysis of stability plays a crucial role in the study of parametrically excited linear systems. Similar problems also appear during the investigation of the local stability of the equilibrium states of nonlinear systems under non-parametric excitation. The concepts of stability and asymptotic stability of deterministic systems were rigorously defined by Lyapunov. But there is no unique way of extending these concepts of stability to systems with stochastic inputs.

This article presents a brief survey of the literature on stochastic stability. Various definitions of stability are given in Section 2. A sufficient condition for almost sure asymptotic stability is derived in Section 3. Necessary and sufficient conditions for almost sure asymptotic stability are discussed in Section 4. A new result concerning the necessary and sufficient condition for almost sure asymptotic stability of a second order system with narrowband excitation is derived in Section 5.

2. Definitions of stability

Consider the linear system represented by the equation

$$\frac{dx}{dt} = \left[A + F(t)\right] x \quad , \tag{1}$$

where $x = x(t)$ is an n-dimensional vector

$$x^T = \left\{x_1, x_2 \ldots\ldots x_n\right\} \quad , \tag{2}$$

and A and F(t) and n x n matrices. We are interested in the stability of the equilibrium solution $x(t) = 0$.

Stability in the Lyapunov sense is uniform convergence of the solution with respect to the initial conditions. Let $x(t;x_o,t_o)$ denote the solution at time t corresponding to an initial state x_o at time t_o, and let $\|x\|$ denote a suitable norm, e.g.,

$$\|x\| = (x_1^2 + x_2^2 + \ldots\ldots\ldots + x_n^2)^{\frac{1}{2}} \quad . \tag{3}$$

For deterministic systems, i.e., when all elements for F(t) are deterministic functions of t, definitions of Lyapunov stability are stated as follows [1] :

1. Lyapunov stability

The equilibrium solution is stable if, given $\varepsilon > 0$, there exists $\delta(\varepsilon, t_o) > 0$ such that

$$\|x_o\| < \delta \text{ implies } \sup_{t \geq t_o} \|x(t;x_o,t_o)\| < \varepsilon \;.$$

2. Asymptotic Lyapunov stability

The equilibrium solution is asymptotically stable if (i) it is stable and (ii) there exists $\delta > 0$ such that

$$\| x_o \| < \delta \text{ implies } \lim_{t \to \infty} \| x(t;x_o,t_o) \| = 0 \; .$$

In the case of stochastic systems, one or more elements of F(t) are random functions of t, and the definition of stability varies depending on the chosen mode of convergence. Application of the three commonly employed modes of stochastic convergence [2], viz. (i) convergence in probability, (ii) convergence in the mean, and (iii) almost sure convergence (or convergence with probability 1), leads to the following definitions of stochastic stability [3]:

I_p. Lyapunov stability in probability

The equilibrium solution possesses stability in probability if, given $\varepsilon, \varepsilon' > 0$, there exists $\delta(\varepsilon, \varepsilon', t_o) > 0$ such that

$$\| x_o \| < \delta \text{ implies } P\left[\sup_{t \geq t_o} \| x(t;x_o,t_o) \| > \varepsilon'\right] < \varepsilon .$$

I_m. Lyapunov stability in the mean

The equilibrium solution possesses stability in the mean if, given $\varepsilon > 0$, there exists $\delta(\varepsilon, t_o) > 0$ such that

$$\| x_o \| < \delta \text{ implies } E\left[\sup_{t \geq t_o} \| x(t;x_o,t_o) \| \right] < \varepsilon .$$

$I_{a.s.}$ Almost sure Lyapunov stability

The equilibrium solution possesses almost sure stability if

$$P\left[\lim_{\|x_o\| \to 0} \; \sup_{t \geq t_o} \| x(t;x_o,t_o) \| = 0\right] = 1 .$$

II_p. Asymptotic Lyapunov stability in probability

The equilibrium solution is asymptotically stable in probability if I_p holds and there exists $\delta > 0$ such that

$$\| x_o \| < \delta \text{ implies } \lim_{T \to \infty} \left[\sup_{t \geq T} \| x(t;x_o,t_o) \| > \varepsilon\right] = 0$$

for any $\varepsilon > 0$.

II_m. Asymptotic Lyapunov stability in the mean

The equilibrium solution is asymptotically stable in the mean if I_m holds and there exists $\delta > 0$ such that

$$\| x_o \| < \delta \text{ implies } \lim_{T \to \infty} E\left[\sup_{t \geq T} \| x(t;x_o,t_o) \|\right] = 0 .$$

$II_{a.s}$ Almost sure asymptotic Lyapunov stability

The equilibrium solution has almost sure asymptotic stability if $I_{a.s}$ holds and if

$$\lim_{T\to\infty} P\left[\sup_{t\geq T} \|x(t;x_o,t_o)\| = 0\right] = 1 \quad .$$

In all these definitions of stochastic stability, it is the random variable $\sup_{t\geq t_o} \|x(t;x_o,t_o)\|$ whose convergence is tested relative to the parameter x_o, where $x(t;x_o,t_o)$ now represents a sample solution. This random variable depends on the behaviour of the sample solution over the entire interval(t_o,∞).

In the early stages of development of the subject, most studies were concerned with the stability of various moments of the solution. For instance, the following definition of stability was used extensively.

III. Stability of the mean

The equilibrium solution possesses stability of the mean if the mean exists and if

$$\lim_{\|x_o\|\to 0} E\left[\|x(t;x_o,t_o)\|\right] = 0 \quad \text{for all } t \geq t_o .$$

Similar definitions can be formulated for stability of moments of higher order. Investigation of this type of stability involves the study of statistical behaviour which is simpler than the study of sample behaviour. But when we test a real system subjected to random excitation, we observe a sample solution. Hence, it is more appropriate to study sample stability rather than moment stability. Most of the work on stochastic stability during the last twenty years is concerned with almost sure sample stability as defined by $I_{a.s}$ or $II_{a.s.}$

3. Sufficient conditions for almost sure Asymptotic Stability

We consider, once again, a system governed by equation (1). in which all non-identically zero elements of the matrix F(t) are assumed to be sample functions of ergodic random processes. Choosing the norm

$$\|x\|_p = x^T Px \quad , \tag{4}$$

where P is a positive definite constant matrix, it can be shown that

$$\|x(t)\|_p = \|x(0)\|_p \exp\left(\int_0^t g(s)\, ds\right) \quad , \tag{5}$$

where

$$g(t) = \frac{x^T\left\{\left[A + F(t)\right]^T P + P\left[A + F(t)\right]\right\} x}{x^T P x} \tag{6}$$

It can also be shown that

$$\max_x g(t) = \lambda(t) \quad , \tag{7}$$

where $\lambda(t)$ is the maximum eigenvalue of the matrix $\left\{\left(A+F(t)\right)^T P+P\left(A+F(t)\right)\right\} P^{-1}$.

From equations (5) and (7) we get

$$\|x(t)\|_P < \|x(0)\|_P \exp\left(\int_0^t \lambda(s)\, ds\right) . \tag{8}$$

Since F(t) is a matrix of ergodic processes, we have

$$\lim_{t\to\infty} \frac{1}{t} \int_0^t \lambda(s) = E\left[\lambda(t)\right] . \tag{9}$$

Hence, a sufficient condition for almost sure asymptotic stability is

$$E\left[\lambda(t)\right] < 0 . \tag{10}$$

The stability condition (10) was first derived by Kozin [4]. A lot of effort [5-7] has gone into sharpening the sufficient condition for stability through a proper choice of the matrix P. Kozin and Wu [8] have exploited the first order probability density functions of the co-efficient processes to further sharpen the sufficient condition for stability. Sufficient conditions for almost sure stability employing the spectral density of the parametric excitation have been determined by Gray [9] and Ariaratnam [10].

4. Necessary and Sufficient Conditions for Almost Sure Asymptotic Stability

It is evident from equation (5) that $\lim_{t\to\infty} \|x(t)\| = 0$ with probability 1 if

$$P\left(\lim_{t\to\infty} \frac{1}{t}\int_0^t g(s)\, ds < 0\right) = 1. \tag{11}$$

Hence, if equation (11) is satisfied, we have almost sure asymptotic stability of the equilibrium solution. On the other hand, if the limit in equation (11) is positive with probability 1, we have instability. Hence, equation (11) is a necessary and sufficient condition for almost sure asymptotic stability. It is therefore necessary to establish the existence of the limit in equation (11) with probability 1, and then to evaluate this limit. Khasminskii [11] has shown that such limits can be evaluated for linear systems with Gaussian white noise coefficients. Kozin and Prodromou [12] have applied this technique to a second order linear system with Gaussian white noise coefficients. But the problem of determining the necessary and sufficient stability conditions in the case of 'coloured noise' coefficients remains unsolved.

5. Stability of Linear Second Order Systems with Narrowband Coefficients

A method of determining the regions of stability and instability of a linear second order system with ergodic narrowband coefficients is presented in this Section. The system considered is governed by the equation

$$\frac{d^2x}{dt^2} + 2k\frac{dx}{dt} + \left(1 + F(t)\right) x = 0 , \tag{12}$$

where F(t) is a sample function of an ergodic zero-mean narrowband random process. Equations of this type arise during the investigation of stability of nonlinear systems of the type

$$\frac{d^2y}{dt^2} + 2k\frac{dy}{dt} + y + \varepsilon g(y) = f(t) \; , \tag{13}$$

where f(t) is a sample function of an ergodic random process. If $y_o(t)$ is a solution of equation (13) and x(t) represents a perturbation of this solution, the evolution of x(t) with time is governed by an equation of the form (12). Local stability of the solution $y_o(t)$ of equation (13) depends upon the stability of the trivial solution x(t) = 0 of equation (12).

Since F(t) is a sample function of a narrow band process, it can be written as

$$F(t) = f_1(t)\cos 2\omega t + f_2(t)\sin 2\omega t \quad , \tag{14}$$

where 2ω is the centre frequency of F(t). The functions $f_1(t)$ and $f_2(t)$ vary slowly with time as compared to the trigonometric functions in equation (14).

Equation (12) can be converted into a system of first order equations

$$\dot{x}_1 = x_2 \quad , \tag{15}$$

$$\dot{x}_2 = -\Big(1 + f_1(t)\cos 2\omega t + f_2(t)\sin 2\omega t\Big)\, x_1 - 2kx_2 \; , \tag{16}$$

where the overhead dot denotes differentiation with respect to t. The solution of equations (15) and (16) can be written as

$$x_1 = a_1(t)\cos\omega t + a_2(t)\sin\omega t \quad , \tag{17}$$

$$x_2 = \omega\Big(a_2(t)\cos\omega t - a_1(t)\sin\omega t\Big) \quad , \tag{18}$$

where $a_1(t)$ and $a_2(t)$ are also slowly varying functions of t. On substituting equations (17) and (18) into equations (15) and (16) and solving for $\dot{a}_1$ and $\dot{a}_2$, we get

$$\omega\dot{a}_1 = g\sin\omega t, \qquad \dot{a}_2 = -g\cos\omega t \quad , \tag{19}$$

$$g = (1 - \omega^2 + f_1\cos 2\omega t + f_2\sin 2\omega t)\,(a_1\cos\omega t + a_2\sin\omega t)$$
$$+ 2k\omega(a_2\cos\omega t - a_1\sin\omega t) \; . \tag{20}$$

We solve equations (19) by the method of averaging [13]. The expression for g contains the stochastic terms $a_1(t)$ and $a_2(t)$ and also the oscillatory terms $\cos\omega t$, $\sin\omega t$, etc. Since $a_1(t)$ and $a_2(t)$ are slowly varying functions of time, the change in their values during a period $(2\pi/\omega)$ is very small. Hence, equations (19) may be replaced by their time averages over a period $(2\pi/\omega)$, assuming a_1 and a_2 to be constant. If this is done, the resulting equations are

$$\omega\dot{a}_1 = \Big(\frac{1}{4}f_2(t) - k\omega\Big)\,a_1 - \frac{1}{2}\Big(\omega^2 - 1 + \frac{1}{2}f_1(t)\Big)\,a_2 \quad , \tag{21}$$

$$\omega\dot{a}_2 = \frac{1}{2}\Big(\omega^2 - 1 - \frac{1}{2}f_1(t)\Big)\,a_1 - \Big(\frac{1}{4}f_2(t) + k\omega\Big)\,a_2 \quad . \tag{22}$$

If the co-efficients appearing in equations (21) and (22) were constant, the solutions would be of the form

$$a_i(t) = c_i \, e^{\lambda t}, \quad i = 1,2 . \tag{23}$$

In the present case, the co-efficients are slowly varying functions of time. So, we seek solutions of the form

$$a_i(t) = c_i \exp\left(\int_0^t \lambda(s) \, ds \right) . \tag{24}$$

On substituting equation (24) into equations (21) and (22), we get the following characteristic equation

$$\omega^2 \lambda^2 + 2k\,\omega^2 \lambda - \left(\frac{1}{16} (f_1^2 + f_2^2) - \frac{1}{4} (\omega^2 - 1)^2 - k^2 \omega^2 \right) = 0 , \tag{25}$$

whose roots are

$$\lambda_{1,2} = -k \pm \frac{1}{4\omega} \left(f_1^2 + f_2^2 - 4(\omega^2 - 1)^2 \right)^{\frac{1}{2}} . \tag{26}$$

Since $F(t)$ is an ergodic process, the processes $\lambda_1(t)$ and $\lambda_2(t)$ are also ergodic. Hence, we can write

$$\lim_{t \to \infty} a_i(t) = \lim_{t \to \infty} \sum_{j=1}^{2} c_{ij} \exp\left(\int_0^t \lambda_j(s) \, ds \right)$$

$$= \sum_{j=1}^{2} c_{ij} \exp\left(E(\lambda_j) t \right), \quad i = 1,2 , \tag{27}$$

where $E(\cdot\cdot)$ denotes expectation. Hence, the solutions represented by equations (17) and (18) are stable if and only if both $E(\lambda_1)$ and $E(\lambda_2)$ have negative real parts. Let

$$(f_1^2 + f_2^2)^{\frac{1}{2}} = R , \tag{28}$$

and let $P_R(r)$ be the probability density function of R. Then, we have

$$\text{Re } E\left[\left(f_1^2 + f_2^2 - 4\,(\omega^2-1)^2 \right)^{\frac{1}{2}} \right]$$

$$= \int_{|2(\omega^2-1)|}^{\infty} \left(r^2 - 4\,(\;\,^2-1)^2 \right)^{\frac{1}{2}} P_R(r) \, dr . \tag{29}$$

Hence, the necessary and sufficient condition for stability is

$$\int_{|2(\omega^2-1)|}^{\infty} \left(r^2 - 4(\omega^2-1)^2 \right)^{\frac{1}{2}} P_R(r) \, dr < 4k\omega . \tag{30}$$

If $F(t)$ is a Gaussian process with mean 0 and variance σ^2, R has a Rayleigh density, i.e.

$$P_R(r) = \frac{r}{\sigma^2} \exp\left(- \frac{r^2}{2\sigma^2} \right) . \tag{31}$$

On substituting (31) and (30) and evaluating the integral, the condition for stability is obtained as

$$\sigma \exp\left(- \frac{2(\omega^2-1)^2}{\sigma^2} \right) < 4(2/\pi)^{\frac{1}{2}} \, k\omega . \tag{32}$$

References

1. L. Casari : Asymptotic Behaviour and Stability Problems in Ordinary Differential Equations, Springer-Verlag, Berlin (1959).
2. A. Papoulis : Probability, Random Variables and Stochastic Processes, McGraw-Hill, New York (1965), Chap. 8, pp.260-261.
3. F. Kozin : A Survey of Stability of Stochastic Systems, Automatica, 5, 95-112 (1969).
4. F. Kozin : On Almost Sure Stability of Linear Systems with Random Coefficients, J. Math. and Phys. 42, 59-67 (1963).
5. T.K. Caughey and A.H. Gray, Jr. : On Almost Sure Stability of Linear Dynamic Systems with Stochastic Coefficients, J. Appl.Mech., 32, 365-372 (1965).
6. E.F. Infante : On the Stability of Some Linear Non-autonomous Random Systems, J. Appl. Mech. 35, 7-12 (1968).
7. F.T. Man : On the Almost Sure Stability of Linear Stochastic Systems, J. Appl. Mech. 37, 541-543 (1970).
8. F. Kozin and C.M. Wu : On the Stability of Linear Stochastic Differential Equations, J. Appl. Mech. 40, 87-92 (1973).
9. A.H. Gray, Jr. : Frequency-Dependent Almost Sure Stability Conditions for a Parametrically Excited Random Vibrational System, J.Appl.Mech.34,1017-1019 (1967).
10. S.T. Ariaratnam : Stability of Mechanical Systems under Stochastic Parametric Excitation, Proceedings of International Symposium on Stability of Stochastic Dynamical Systems, Lecture Notes in Mathematics No.294, R.F. Curtain, ed., Springer-Verlag, Berlin (1972), pp. 291-302.
11. R.Z. Khasminakii : Necessary and Sufficient Conditions for the Asymptotic Stability of Linear Stochastic Systems, Theory. Prob. Appl. 12, 124-157 (1967).
12. F. Kozin and S. Prodromou, Necessary and Sufficient Conditions for Almost Sure Sample Stability of Linear Its equations, SIAM J. Appl.Maths. 21, 413-424 (1971).
13. R.L. Stratonovich, Topics in the Theory of Random Noise, Vol. II, Gordon and Breach, New York (1967), Chap. 4, pp. 97-100.

APPLICATION OF STOCHASTIC PROCESSES TO LINE SHAPES

OPTICAL RESONANCE IN PARTIALLY COHERENT FIELDS

G.S. Agarwal
School of Physics, University of Hyderabad
Hyderabad - 500 134, India

1. Introduction :

In this lecture, I will deal with a class of problems in optical resonance and nonlinear optics in situations where the exciting sources are not necessarily monochromatic and deterministic functions of space and time. In such a case the exciting source can be modelled by a stochastic process, so that both statistics and the temporal coherence properties of the source will be important. We will see that the dynamical characteristics of the source will be quite crucial if the time scales of the fluctuations of the exciting source are of the same order as other time scales which determine the dynamical evolution of the system. The effects treated in this lecture are not only important as fundamental physical effects but also for the interpretation of the experimental results in the high resolution spectroscopy.

In the usual class room problems, we know that the interaction of radiation and matter can be described by the perturbation theory, for example, the transition rate R , say for stimulated emission or absorption in a monochromatic field of frequency ω_ℓ is

$$R = R(\omega_\ell)\, I \quad , \tag{1}$$

where I in the intensity of the field. If the exciting field is not strictly monochromatic but characterized by some spectrum $\Gamma(\omega_\ell)$, then one can get the transition rates by averaging (1)

$$R = \int d\omega_\ell\, R(\omega_\ell)\; \Gamma(\omega_\ell) \tag{2}$$

and thus the bandwidth and the shape of the spectrum will determine the transition rates. An important consequence of (2) can be seen in the resonance fluorescence studies. The spectrum of the scattered radiation by a two level atom with energy separation ω_o, of monochromatic radiation in the lowest order of perturbation, is [1]

$$S(\omega,\omega_\ell) = \frac{I\, g^2}{[\gamma^2 + (\omega_o-\omega_\ell)^2]}\, \delta(\omega-\omega_\ell) \quad , \tag{3}$$

which for the non-monochromatic fluctuating field generalizes to [2]

$$S(\omega) = \int S(\omega,\omega_\ell)\; \Gamma(\omega_\ell)\, d\omega_\ell$$

$$= \frac{I\, g^2}{[\gamma^2 + (\omega_o-\omega)^2]}\; \Gamma(\omega_o) \quad . \tag{4}$$

Expression (4) shows that the spectrum of the scattered radiation will consist of peaks at $\omega = \omega_o$ with width γ and the ones determined by the characteristics of the shape of the spectrum of the exciting radiation. In particular for a broad band source $\Gamma(\omega_0) \sim$ constant, the spectrum would consist of a single peak determined by the atomic parameters ω_0 and γ which is in contrast to the monochromatic result (3) where the scattered spectrum simply reflects the spectrum of the exciting source. This is, of course, the celebrated result of Heitler [1] and verified by Gibbs and Venkatesan [3] few years back.

Higher order processes such as two photon absorption depend nonlinearly on intensity of the exciting monochromatic source and that is when the effects of the source statistics first start appearing. For example a Gaussian random process has the property

$$\langle I^n \rangle = n! \langle I \rangle^n \qquad (5)$$

which implies that n^{th} order process in a Gaussian monochromatic field would be n! times more efficient than that in a deterministic field [4]; provided the average intensity is same. Experiments[5] on multiphoton processes have verified such kinematical enhancement factors. If the field is not monochromatic, then the situation is far from simple. In such a case the spectral characteristics of arbitrary order [4,6] start entering. Expressions like (5) also assume that the perturbation theory is valid. Such an assumption starts breaking down when a resonance is involved and when the saturation effects are important. To describe such situations, we have to have a complete stochastic description of the exciting source since the physical quantities will depend on the time correlation functions of field of arbitrarily high order. This itself leads to difficulties as there are very few models in the theory of random processes where the correlations of arbitrary orders are known. These include:-

(i) The complex field $\vec{E}(t)$ is a Gaussian random processes.

(ii) The complex field $\vec{E}(t)$ has a time independent deterministic amplitude but whose phase $\phi(t)$ is such that its derivative is a Gaussian delta correlated random process [4].

(iii) The field $\vec{E}(t)$ is a real dichotomic Markov process [cf.7].

(iv) The field $\vec{E}(t)$ has a time independent frequency ω_ℓ with ω_ℓ being a Gaussian random variable.

The model (iv) is relatively simple because the calculations for any nonlinear process can be done assuming ω_ℓ is fixed and then the end result can be averaged with respect to the probability distribution of ω_ℓ. The situation is similar to the Doppler broadening which one encounters in optical resonance in atomic vapours since in case of Doppler broadening the effective frequency is $\omega_\ell - \vec{k}.\vec{v}$ where $\vec{v}$ is a Gaussian random variable. The subsequent discussion is divided in two parts - one dealing with the perturbative results and the other dealing with the exact results for the models mentioned above.

II. Cross Sections for NonLinear Processes Expressed in Terms of the Convolutions involving the Spectral Tensors of the Field:

Let us first assume that the exciting sources are such that their frequencies are far detuned from any of the atomic resonances, so that one can use the nth order perturbation theory to describe n^{th} order nonlinear process. In the steady state, the nth order nonlinear response of the system can be written as

$$P_{\mu}^{(n)}(\omega) = \int\cdots\int d^{n}\{\omega\}\,\delta(\omega-\sum\omega_i)\,\chi^{(n)}_{\mu\{\alpha_n\}}(\{\omega\}) \prod_{i=1}^{n} E_{\alpha_i}(\omega_i) \quad , \tag{6}$$

where $\chi^{(n)}_{\mu\{\alpha_n\}}(\{\omega\})$ stands for the nth order nonlinear susceptibility $\chi^{(n)}(\omega_1, \omega_2, \omega_3 \ldots \omega_n)$. Here $P(\omega)$ and $E(\omega)$ are respectively, the Fourier transform of $P(t)$ and $E(t)$. If E_{α} is stochastic in nature, then $P_{\mu}^{(n)}(\omega)$ also becomes stochastic. The cross section for the nth order process will be given in terms of a functional involving the polarization fluctuations, the spectrum of such fluctuations will be given by

$$\langle P_{\mu}^{(n)}(\omega)\, P_{\nu}^{(n)}(\omega')\rangle = \delta(\omega+\omega')\, S_{\mu\nu}^{(P)}(\omega) \; ,$$

$$S^{P}_{\mu\nu}(\omega) = \int\cdots\int d^{n}(\omega_i)\, d^{n}(\omega_i')\,\delta(\omega-\sum_i\omega_i)\,\delta(\omega+\sum\omega_i')$$
$$\chi^{(n)}_{\mu\{\alpha_n\}}(\{\omega_i\})\,\chi^{(n)}_{\nu\{\beta_n\}}(\{\omega_i'\})\langle\prod_{i=1}^{n} E_{\alpha_i}(\omega_i)\prod_{i=1}^{n} E_{\beta_i}(\omega_i')\rangle \tag{7}$$

The first line of (7) follows from the stationarity and the Wiener-Khintchine theorem. Relation (7) shows that the cross section for an nth order process will depend on the $2n^{th}$ order correlation function of the field[4,6]. As generally happens in the study of the nonlinear response, one uses sources with different frequencies and so a reasonable assumption to use will be the uncorrelated nature of sources. Then the $2n^{th}$ order correlation can be expressed in terms of the lower ordered correlations of the fields produced by different sources. Such higher order correlations are known only for certain specific models of the sources.

Expression (7) has been recently used to predict the existence of fluctuation induced resonances[8] in the field of a random pump in the context of four wave mixing. For the purpose of calculating the nonlinear gain one has to proceed differently for the gain is directly related to the induced nonlinear polarization. For example the rate of absorption of energy σ from an external field $\vec{E}^{(s)} = 2\mathrm{Re}\,\vec{E}^{(s)}e^{-i\omega_s t}$ per unit incident flux is

$$\sigma = \frac{4\pi\omega_s}{c\,|E^{(s)}|^2}\,\mathrm{Im}\,(\vec{P}^{(s)}\cdot\vec{E}^{(s)*}) \quad , \tag{8}$$

where $\vec{P}^{(s)}$ is the nonlinear polarization induced at ω_s. As an application of (8), consider the case of stimulated Raman gain in the field of a random pump wave $\vec{E}^{(\ell)}(t)$. Writing $\vec{E}(t) = \vec{E}^{(\ell)}(t) + \vec{E}^{(s)}(t)$, using

$$\langle E_\mu^{(\ell)}(\omega_1)\ E_\nu^{(\ell)}(\omega_2)\rangle = \delta(\omega_1+\omega_2)\ \Gamma_{\mu\nu}(\omega_1)\ , \tag{9}$$

and (6), we find that (8) reduces to

$$\sigma = \frac{12\pi\omega_s}{c|E^{(s)}|^2}\ \mathrm{Im}\int \chi^{(3)}_{\mu\alpha\beta\gamma}(\omega_1, -\omega_1, +\omega_s)\ \Gamma_{\alpha\beta}(\omega_1)\ \varepsilon_\gamma^{(s)}\ \varepsilon_\mu^{(s)*}\ d\omega_1 \tag{10}$$

Using the known form of $\chi^{(3)}$ [9] and of the spectral shape of the pump, the nonlinear gain can be obtained. The integrals can be done analytically for the Lorentzian model of the pump i.e. for

$$\Gamma_{\alpha\beta}(\omega_1) = \frac{\gamma_\ell/\pi}{(\omega_1-\omega_\ell)^2+\gamma_\ell^2}\ I_{\alpha\beta}^{(\ell)}\ . \tag{11}$$

For Raman scattering in a three level model [9] with g and f representing the initial and final states and n the intermediate state, we find that the integral in (10) reduces to ($\vec{P}_{ij}$ represents the dipole matrix element)

$$(\vec{P}_{fn}\cdot\vec{\varepsilon}^{(s)*})\ |\vec{P}^*_{ng}\cdot\vec{\varepsilon}^{(\ell)}|^2\ (\vec{P}^*_{nf}\cdot\vec{\varepsilon}^{(s)})\ (\omega_s-\omega_{nf}+i\,\Gamma_{nf})^{-1}$$

$$\Big\{(\omega_\ell-\omega_s-\omega_{fg}-i\,\Gamma_{fg}-i\gamma_\ell)^{-1}\ (\omega_\ell-\omega_{ng}-i\,\Gamma_{ng}-i\gamma_\ell)^{-1}$$

$$-\ 2\,\frac{(\Gamma_{ng}+\gamma_\ell)}{\Gamma_{nn}}\ \left[(\omega_\ell-\omega_{ng})^2+(\Gamma_{ng}+\gamma_\ell)^2\right]^{-1}\Big\}$$

which clearly shows the effect of γ_ℓ on various relaxation coefficients. The linewidth of the Raman transition is now $(\Gamma_{fg}+\gamma_\ell)$. An exactly similar result holds for spontaneous Raman scattering[10]. It is obvious that similar considerations will apply to higher order gain coefficients for example to Hyper Raman gain which can be shown to depend on $\chi^{(5)}$ and the fourth order correlation function of the pump wave.

III. Methods of Solution for Specific Models :

The general problem, with which one has to deal, can be written in the form

$$\frac{\partial\rho}{\partial t} = L_0\rho + E(t)\,L_+\rho + E^*(t)\,L_-\rho\ , \tag{12}$$

where E(t) is the random field. The physical quantities will be given by the ensemble average of ρ and of the two time correlation function of the dipole moment operators. In the interaction picture, one can write (12) as

$$\frac{\partial \rho}{\partial t} = iE(t)\, L_+(t)\, \rho + iE^*(t)\, L_-(t)\, \rho \ , \tag{13}$$

which in some cases may be reduced to

$$\frac{\partial \rho}{\partial t} = i\,\mathcal{E}(t)\, \mathcal{L}(t)\rho, \tag{14}$$

which has a formal solution in terms of the time ordering operator

$$\rho(t) = T\, e^{i\int^t \mathcal{E}(\tau)\mathcal{L}(\tau)d\tau}\, \rho(o) \tag{15}$$

A. Gaussian model for E(t):

Even if $\mathcal{E}(t)$ is Gaussian, we can not evaluate the ensemble averages of (15) due to the appearance of the time ordering operator in (15). Only if $\mathcal{L}(\tau)$ is such that

$$[\mathcal{L}(\tau_1),\ \mathcal{L}(\tau_2)] = 0\ , \tag{16}$$

then (15) simplifies to

$$\langle \rho(t) \rangle = \exp[-\frac{1}{2}\int_o^t dt_1 \int_o^t dt_2 \langle \mathcal{E}(t_1)\ \mathcal{E}(t_2)\rangle \mathcal{L}(t_1)\mathcal{L}(t_2)]\, \rho(o) \tag{17}$$

As an example of the type of situation (16) consider optical Bloch equations in a fluctuating electric field for the special case when the transverse and longitudinal relaxation times are equal

$$\frac{\partial}{\partial t}\begin{bmatrix} \langle s^+ \rangle \\ \langle s^- \rangle \\ \langle s^z \rangle \end{bmatrix} = \begin{bmatrix} i\Delta - \frac{1}{T_2} & 0 & +2i\alpha^*(t) \\ 0 & -i\Delta - \frac{1}{T_2} & -2i\,\alpha(t) \\ i\,\alpha(t) & -i\alpha^*(t) & -\frac{1}{T_1} \end{bmatrix} \begin{bmatrix} \langle s^+ \rangle \\ \langle s^- \rangle \\ \langle s^z \rangle \end{bmatrix} + \begin{bmatrix} 0 \\ 0 \\ \eta/T_1 \end{bmatrix} \tag{18}$$

The variables $\langle s^\pm \rangle$, $\langle s^z \rangle$ correspond respectively to the atomic polarization and inversion. In the special case when $\Delta=0$, $T_1=T_2$, $\alpha(t)$=real, the time evolution operator $U(t,\tau)$ reduces to

$$U(t,\tau) = e^{-\frac{(t-\tau)}{T_2} + i\int_\tau^t \alpha(t')dt'\, A}\ , \quad A = \begin{bmatrix} 0 & 0 & 2 \\ 0 & 0 & -2 \\ 1 & -1 & 0 \end{bmatrix} \tag{19}$$

and therefore the ensemble average of U can be obtained in closed form

$$\langle U(t,\tau) \rangle = e^{-\frac{t-\tau}{T_2} - \frac{1}{2}A^2 \int_\tau^t dt' \int_\tau^t dt'' \langle \alpha(t')\ \alpha(t'')\rangle} \ . \tag{20}$$

However some of these assumptions are quite restrictive and some essential physics is lost by assuming $\alpha(t)$ to be a real field. For example, the character of Stark splittings changes. To see this in its simplest form, consider

$$I(\omega) = \delta(\omega - \omega_\ell - \alpha) + \delta(\omega - \omega_\ell + \alpha) \tag{21}$$

If α is a real Gaussian random variable, then

$$\langle I(\omega)\rangle = \frac{2}{\sqrt{\pi\langle\alpha^2\rangle^{1/2}}} e^{-(\omega-\omega_\ell)^2/\sqrt{\langle\alpha^2\rangle}} \tag{22}$$

showing that the ac Stark effect is completely smeared out. However in the same problem if α had been treated as a complex variable, then

$$I(\omega) = \delta(\omega-\omega_\ell-|\alpha|) + \delta(\omega-\omega_\ell+|\alpha|) \tag{23}$$

The averaging of I over the complex Gaussian distribution leads to

$$I(\omega) = \frac{2|\omega-\omega_\ell|}{\langle|\alpha|^2\rangle} e^{-(\omega-\omega_\ell)^2/\langle|\alpha|^2\rangle} \tag{24}$$

thus preserving ac stark effect.

In the general case, since (16) is not valid, the progress has been quite slow even when E(t) is a complex Gaussian random process. However if E(t) is also Markovian, then it is possible to use the methods from the theory of Markov processes to obtain continued fraction expansions for the ensemble average of the physical quantities : we will now briefly present this formulation, the details can be found in the work of Lambropoulos and coworkers[11]. Let us denote the set $\langle S^+\rangle, \langle S^-\rangle, \langle S^z\rangle$ by ψ, and let $P(\psi, \varepsilon)$ be the joint distribution for ψ and the field ε. The field $\varepsilon(t)$ obeys the Fokker-Planck equation

$$\frac{\partial P(\varepsilon)}{\partial t} = +2b\frac{\partial}{\partial I}(I-I_o)P + 2bI_o\frac{\partial^2}{\partial I^2}IP + \frac{bI_o}{2I}\frac{\partial^2 P}{\partial\phi^2} = KP, \quad \varepsilon = \sqrt{I}\,e^{-i\phi} \tag{25}$$

so that

$$\langle\varepsilon(t)\,\varepsilon^*(t')\rangle = I_o\,e^{-b|t-t'|} \tag{26}$$

Then a simple exercise using (18) and (25) shows that ($\alpha(t) = \alpha_o\,\varepsilon(t) = \alpha_o\sqrt{I}\,e^{-i\phi}$)

$$\begin{aligned}\frac{\partial P(\varepsilon,\psi)}{\partial t} = &-\frac{\partial}{\partial\psi_1}\left[(i\Delta-\tfrac{1}{T_2})\psi_1 P + 2i\alpha_o^*\varepsilon^*\psi_3 P\right] \\ &+\frac{\partial}{\partial\psi_2}\left[(+i\Delta+\tfrac{1}{T_2})\psi_2 P + 2i\alpha_o\varepsilon\psi_3 P\right] \\ &-\frac{\partial}{\partial\psi_3}\left[i\alpha_o\varepsilon\psi_1 P - i\alpha_o^*\varepsilon^*\psi_2 P - \frac{\psi_3}{T_1}P\right] + KP \\ &+\frac{\eta}{T_1}P\end{aligned} \tag{27}$$

This multi-dimensional Fokker-Planck equation is not known to have any analytical solution. The physical quantities which we need to know are

$$\begin{aligned}\langle\psi\rangle &= \int P(\varepsilon,\psi)\,\psi\,d^2\varepsilon\,d\psi \\ &= \int \chi\,d^2\varepsilon\end{aligned} \tag{28}$$

The elements χ obviously satisfy (M being the square matrix of (18))

$$\frac{\partial \chi}{\partial t} = M\chi + \begin{bmatrix} 0 \\ 0 \\ \frac{\eta}{T_1} \end{bmatrix} + K\chi , \tag{29}$$

where M is linear in ε . The next step in the calculation consists of expanding χ in terms of the eigenfunctions of K which are explicitly known

$$K\Phi_{nm} = \lambda_{nm}\Phi_{nm}, \quad \Phi_{nm} = \frac{e^{in\phi}}{2\pi I_o}\left(\frac{m!}{(m+|n|)!}\right)^{\frac{1}{2}}$$

$$e^{-x}\, x^{|n|/2}\, L_m^{|n|}(x), \quad x = I/I_o, \quad \lambda_{nm} = b(2m+|n|) . \tag{30}$$

Once χ is expanded in terms of Φ_{nm}, the coupled equations for the expansion coefficients can be obtained in the usual manner. It has been possible to convert some of these recursion relations for the expansion coefficients into continued fractions, which can be then numerically evaluated. Lambropoulos and coworkers[11] have investigated a wide class of problems in optical physics using this technique.

If the field $\varepsilon(t)$ is Gaussian but nonmarkovian, then no general techniques appear to be known for the solution to the optical resonance equations in such fields. However some special cases can be handled. For example the rate equation say for atomic population obeys the equation of the form

$$\dot{n} = -a - c\,|\varepsilon(t)|^2 n - bn \tag{31}$$

Thus the ensemble average of n(t) requires the knowledge of the functional

$$F = \left\langle e^{-c\int_0^t |\varepsilon(\tau)|^2 d\tau} \right\rangle \tag{32}$$

Using the Karhunen-Lo'eve expansion for the Gaussian process (which can be markovian or non-markovian), one can show that

$$F = \prod_k (1 + c\,\lambda_k)^{-1} , \tag{33}$$

where λ_k's are the eigenvalues of the integral equation

$$\int_0^t \langle \varepsilon^*(t)\, \varepsilon(t+\tau)\rangle \Phi_k(\tau)d\tau = \lambda_k\, \Phi_k(t) . \tag{34}$$

For Lorentzian spectrum (markovian case) the functional F can be evaluated in closed form [12]

$$F = e^{bt}\left[\cosh(\beta bt) + \frac{1}{2\beta}(\beta^2+1)\sinh(\beta bt)\right], \quad \beta^2 = \left(1 + \frac{2I_oc}{b}\right) \tag{35}$$

whereas for other types of spectrum such as rectangular, F can only be evaluated numerically[12] . This method has been applied to some problems in Raman scattering and multiphoton ionization[13] .

B. Random Telegraphical Signal Model for E(t) :

Another theoretical model of E for which the optical resonance equations are exactly soluble corresponds to

$$E(t) = \varepsilon e^{-i\beta\Phi(t) - i\omega_\ell t} , \tag{36}$$

where $\Phi(t)$ is a dichotomic markov process

$$\langle\Phi(t)\rangle = 0, \langle\Phi(t)\,\Phi(t')\rangle = e^{-2\Gamma|t-t'|} , \Phi^2 = 1 \tag{37}$$

leading to

$$E(t) = \varepsilon\cos\beta - i\,\varepsilon\,\sin\beta\,\Phi(t) \tag{38}$$

The auto correlation function of E(t) has an interesting form

$$\langle E(t)\ E^*(t')\rangle = \varepsilon^2 [\cos^2\beta + \sin^2\beta\ e^{-\Gamma|t-t'|}] e^{-i\omega_\ell(t-t')} \tag{39}$$

i.e. the spectrum contains a coherent component and a Lorentzian, the relative strength of the two contributions depends on the parameter β . In view of (38), the original density matrix equation (12) can be written in the component form as a matrix equation

$$\frac{\partial\Psi}{\partial t} = (C_0 + C_1\ \Phi(t))\Psi + g , \tag{40}$$

where both C_0 and C_1 depend on the strength of the applied field. As has been shown elsewhere[14] this model is exactly soluble i.e. ensemble averages of Ψ and of the correlations can be calculated in closed form for example $\langle\Psi\rangle$ obeys the equation

$$\frac{\partial\langle\Psi\rangle}{\partial t} = C_0\langle\Psi\rangle + \int_0^t d\tau\ e^{-2\Gamma(t-\tau)}\ C_1\ e^{C_0(t-\tau)}\ C_1\langle\Psi(\tau)\rangle + g \tag{41}$$

The detailed consequences of this for $\beta = \pi/2$ are given in [7] , however the more interesting case $\beta \neq \pi/2$ is under investigation. It may be noted that the variables Ψ are non-markovian for this model.

C. Phase Diffusion Model :

We next consider the model of the field where

$$E(t) = \varepsilon e^{-i\omega_\ell t - i\Phi(t)} , \dot{\Phi} = \mu(t) ,$$

$$\langle\mu(t)\rangle = 0, \langle\mu(t)\ \mu(t')\rangle = 2\gamma_c\ \delta(t-t') \tag{42}$$

and μ is Gaussian. For this model the general optical resonance equations are exactly soluble [15-17]. Agarwal and coworkers[15,16] have used this model in the context of a large class of optical resonance phenomena such as in resonance fluorescence from two level and three level systems, optical Hanle effect, modulation studies, coherent antistokes Raman scattering, four wave mixing, parametric amplification etc. Rather than give these various applications, we only present the elements of the method used to obtain exact solutions.

A major simplification in optical resonance problems occurs due to the use of the rotating wave approximation i.e. the interaction part of the Hamiltonian is approximated as

$$\mathcal{H} = \sum_{i>j} g_{ij}\, e^{-i\omega_{ij}t - i\,\Phi_{ij}(t)} |i\rangle\langle j| + \text{c.c.} \tag{43}$$

and the sum in (43) is over all the allowed transitions. The notation used in (43) implies that the level $|i\rangle$ is above the level $|j\rangle$ by an amount Ω_{ij}. The field with frequency ω_{ij} and phase Φ_{ij} couples only the transition $|i\rangle \leftrightarrow |j\rangle$. Thus the density matrix equation will be

$$\frac{\partial \rho}{\partial t} = L_0\rho - i\left[\sum_{i>j} g_{ij}\, e^{-i\omega_{ij}t - i\Phi_{ij}(t)} |i\rangle\langle j| + \text{h.c.}\ ,\ \rho\right] \tag{44}$$

where L_0 stands for other coherent interactions as well as other sources of relaxation such as due to collisions, radiative emission etc. The basic idea is to transform (44) into an equation of the form

$$\frac{\partial \tilde{\rho}}{\partial t} = L_0\tilde{\rho} - i\left[\sum_{i>g} g_{ij} |i\rangle\langle j| + \text{h.c},\ \tilde{\rho}\right] + \sum \mu_{ij}(t)\, L_{ij}\tilde{\rho} \tag{45}$$

so that the Gaussian delta correlated random processes appear linearly in the dynamical equations. Such dynamical equations are exactly soluble and we quote the result [15]. Write the basic equation as

$$\dot{x}_i = \sum_j (M_{ij}\, x_j + \mu_{ij}(t)\, x_j)\ , \tag{46}$$

$$\langle \mu_{ij}\rangle = 0,\ \langle \mu_{ij}(t)\ \mu_{kl}(t')\rangle = 2Q_{ijkl}\ \delta(t-t') \tag{47}$$

then

$$\langle \dot{x}_i\rangle = \sum_j M_{ij}\langle x_j\rangle + \sum_{jk} Q_{ikkj}\langle x_j\rangle = \langle A_i\rangle \tag{48}$$

The second moments also satisfy the closed set of equations

$$\frac{\partial}{\partial t}\langle x_k x_l\rangle = \langle A_k x_l\rangle + \langle A_l x_k\rangle + \sum_{ij} Q_{kilj}\langle x_i x_j\rangle \tag{49}$$

The variables $x_i^{'s}$ are also markovian, so that the time correlations can be obtained from (48)

$$\frac{d}{dt}\langle x_i(t)\ x_k(t')\rangle = \sum_j (M_{ij} + \sum_k Q_{ikkj})\ \langle x_j(t)\ x_k(t')\rangle\quad . \tag{50}$$

Application of (48) to (45) would then lead to

$$\frac{\partial\langle\tilde{\rho}\rangle}{\partial t} = L_0\langle\tilde{\rho}\rangle - i\left[\sum_{i>j} g_{ij}|i\rangle\langle j| + \text{h.c.},\ \langle\tilde{\rho}\rangle\right]$$

$$+ \sum Q_{ijkl}\, L_{ij}\, L_{kl}\langle\tilde{\rho}\rangle. \tag{51}$$

In order to see how the transformation from (44) to (45) is done and what is the physics of this new density matrix $\tilde{\rho}$, consider the case of resonant Raman scattering in fields of arbitrary intensity. As before let a laser of frequency ω_ℓ [ω_s]phase Φ_ℓ [Φ_s]act between the ground level $|g\rangle$[final level $|f\rangle$] and the resonant intermediate level $|n\rangle$. In such a case ρ and $\tilde{\rho}$ are related by the canonical transformation such that

$$\tilde{\rho}_{\alpha\alpha} = \rho_{\alpha\alpha}; \quad \tilde{\rho}_{gn} = e^{-i\Phi_\ell - i\omega_\ell t} \rho_{gn}$$

$$\tilde{\rho}_{fn} = e^{-i\Phi_s - i\omega_s t} \rho_{fn}, \quad \tilde{\rho}_{gf} = e^{-i(\Phi_\ell - \Phi_s) - it(\omega_\ell - \omega_s)} \tag{52}$$

Use of (51), then shows that the equations of motion for $\langle\tilde{\rho}\rangle$ can be obtained from those in the absence of any phase fluctuations, if we change the relaxation coefficients in the off-diagonal elements as follows

$$\Gamma_{ng} \rightarrow \Gamma_{ng} + \gamma_\ell, \; \Gamma_{nf} \rightarrow \Gamma_{nf} + \gamma_s, \quad \Gamma_{fg} \rightarrow \Gamma_{fg} + (\gamma_\ell + \gamma_s - 2\gamma_{\ell s}) \tag{53}$$

where $\gamma_{\ell s}$ denotes the cross correlation between the lasers i.e.

$$2\gamma_{\ell s}\,\delta(t-t') = \langle \Phi_\ell(t)\,\Phi_s(t') \rangle \tag{54}$$

This simple substitution rule [cf.15,18] enables one to obtain the characteristics of the Raman system for example Raman gain which can be shown to be related to Im $\tilde{\rho}_{fn}$. An important point worth emphasizing is that in case the same laser is used on both the transitions, then $\gamma_\ell + \gamma_s - 2\gamma_{\ell s} = 0$ and hence Γ_{fg} remains unchanged. This has important consequences on the trapping of the population in the state $|f\rangle$[19]. This cross correlation has been successfully used by Thomas et al [20] in the observation of Ramsay fringes using a resonant Raman transition which will probably have important implications on frequency standards. For the description of the nonlinear optical processes, such as four wave mixing [6,8] which are characterized by the square of the induced polarization, we have to use equations like (49). Note further that if one wants to calculate the dipole moment e.g. ρ_{fn}, then we can not use $\tilde{\rho}_{fn}$, but we have to make another transformation $\tilde{\rho} \rightarrow \tilde{\tilde{\rho}}$ such that

$$\tilde{\tilde{\rho}} = \tilde{\rho}\, e^{i\Phi_s} \tag{55}$$

Equations for $\tilde{\tilde{\rho}}$ would be similar to (45) so that (48) can be applied directly. Now the equations for the diagonal elements of $\langle\tilde{\tilde{\rho}}\rangle$ will depend on γ_ℓ, γ_s etc. The dipole moment ρ_{fn} is needed to evaluate the spectrum of the spontaneously emitted radiation; which is related to the two time correlation function $\langle\langle (|n\rangle\langle f|)_t \; (|f\rangle\langle n|)_{t'}\rangle$. Such two time correlations can be obtained by using the equations for $\langle\tilde{\tilde{\rho}}\rangle$ and the markov property of the variables $\tilde{\tilde{\rho}}_{\alpha\beta}$. Note also that the transformation (55) is not adequate if we want to calculate the dipole moment ρ_{gn} - in this case we should use

$$\bar{\rho} = \tilde{\rho}\, e^{i\Phi_\ell} \tag{56}$$

So far we have not put in any propagation effects. It turns out that even the propagation effects can be incorporated in the optical resonance equations in the fluctuating fields, which then could be solved for the phase diffusion model. To see this, let us consider the equations [cf.21] for the stimulated Raman scattering under the conditions that the saturation effects are not important :

$$\frac{\partial \mathcal{E}_s}{\partial z} + \frac{1}{v_s}\frac{\partial \mathcal{E}_s}{\partial t} + \Gamma_s \mathcal{E}_s = \sigma_1 \, Q^* |\mathcal{E}_\ell| \, e^{-i\Phi_\ell\left(t - \frac{z}{v_\ell}\right)}$$

$$\frac{\partial Q^*}{\partial t} + \frac{Q^*}{T_2} = \sigma_2 |\mathcal{E}_\ell| \mathcal{E}_s \, e^{i\Phi_\ell\left(t - \frac{z}{v_\ell}\right)} \tag{57}$$

where $\mathcal{E}_s$ is the slowly varying envelope of the stokes field and Q is the Raman mode. A simple change of variables

$$z' = z, \quad \tau = t - z/v_s \quad . \quad P = Q^* e^{-i\Phi_\ell} \tag{58}$$

brings the above equations in the form

$$\frac{\partial \mathcal{E}_s}{\partial z'} + \Gamma_s \mathcal{E}_s = \sigma_1 |\mathcal{E}_\ell| P \ ,$$

$$\frac{\partial P}{\partial \tau} + \frac{P}{T_2} + i\mu(\tau - \beta z') P = \sigma_2 |\mathcal{E}_\ell| \mathcal{E}_s \ , \quad \beta = \frac{1}{v_s} - \frac{1}{v_\ell} \ . \tag{59}$$

The Gaussian delta correlated noise appears in (59) in the multiplicative form and we can solve the equations for first and second moments by using (48), (49) and the Laplace transform of (59) with respect to the z' coordinate. The results so obtained are found to agree with those derived by using other methods [21].

Elliott et al[22]have considered a modification of the laser by putting a modulator so that the phase fluctuations can be controlled and they are planning extensive studies of the optical resonance phenomena in fluctuating fields. In their set up, one has

$$\langle \mu(t) \ \mu(t') \rangle = \frac{\mu^2}{\pi} \ \frac{\sin B/2\,(t-t')}{(t-t')} \tag{60}$$

which for large B, goes over to a delta correlated process. For small B, $\langle \mu(t) \ \mu(t') \rangle \sim \mu^2 B/2\pi$ which corresponds to an external frequency ω_ℓ being a Gaussian random variable and thus the situation is similar to Doppler broadening. For intermediate values of B no explicit solutions are known.

Finally we would like to mention that the methods discussed above are equally applicable[23]to study the effect of the velocity changing collisions on optical line shapes. In the classical model of weak collisions[24]the effective field acting on the atom can be written as

$$E(t) = \mathcal{E} e^{-i\omega_\ell t + i\vec{k}_\ell \cdot \vec{R}(t)} \ , \tag{61}$$

where $\vec{R}(t)$ is the position of the atom, which is given by the familiar Brownian motion equations

$$\dot{\vec{R}} = \vec{v} \quad , \quad \dot{\vec{v}} + \gamma_v \vec{v} = \vec{F}(t) \quad ,$$

$$\langle F_\alpha(t) F_\beta(t') \rangle = 2 \Gamma_v \delta(t-t') \delta_{\alpha\beta} \quad . \tag{62}$$

Notice the similarity of this situation to the phase diffusion model, the correspondence is exact in the limit of large γ_v . Hence the results obtained in the context of the phase diffusion model could be directly used to get the corresponding results for the case of velocity changing collisions. For example for the case of Raman gain, we will have[23]in place of (53)

$$\Gamma_{ng} \rightarrow \Gamma_{ng} + \hat{k}_\ell^2 \tilde{\Gamma}_v \quad , \quad \Gamma_{nf} \rightarrow \Gamma_{ng} + \hat{k}_s^2 \tilde{\Gamma}_v \quad ,$$

$$\Gamma_{fg} \rightarrow \Gamma_{fg} + \tilde{\Gamma}_v (\hat{k}_\ell - \hat{k}_s)^2 \quad , \quad \tilde{\Gamma}_v = \Gamma_v / \gamma_v^2 \tag{63}$$

where $\hat{k}_\ell$ and $\hat{k}_s$ are, respectively, the unit vectors in the direction of propagation of the pump and Stokes radiation. The case of finite γ_v can also be handled by using methods similar to those discussed in connection with Gaussian model for the appropriate $\tilde{\rho}$ equation involves the general Gaussian stochastic variable v(t).

References :

1. W. Heitler, Quantum Theory of Radiation (Oxford, N.Y. 1954) Sec.20.
2. cf. M.G. Raymer and J. Cooper, Phys. Rev. A20, 2238 (1979).
3. H.M. Gibbs and T.N.C.Venkatesan, Opt. Commun. 17, 87 (1976).
4. G.S. Agarwal, Phys. Rev. A1, 1445 (1970); B.R.Mollow, Phys. Rev., 168 1418 (1968).
5. G. Mainfray in "Multiphoton Processes" eds. J.H. Eberly and P.Lambropoulos (John Wiley, New York 1978) p.253.
6. G.S. Agarwal and S. Singh, Phys. Rev. A25, 3195 (1982).
7. G.S. Agarwal, Z. Physik B33, 111 (1979).
8. G. S. Agarwal and C.V. Kunasz, Phys. Rev. A27, 996 (1983).
9. Y.R. Shen, Phys. Rev. B9, 622 (1974); N. Bloembergen, H.Lotem and R.T. Lynch Jr., Ind. J. Pure Appl. Phys. 16, 151 (1978).
10. G.S. Agarwal and S.S. Jha, J. Phys. B12, 2655 (1979).
11. P. Zoller, Phys. Rev. A19, 1151 (1979); ibid A20, 1019 (1979); S.N.Dixit, P.Zoller and P. Lambropoulos, Phys. Rev. A21, 1289 (1980); A.T. Georges, Phys. Rev. A21, 2034 (1980); A.T. Georges and P.Lambropoulos, Phys. Rev. A20, 991 (1979).
12. S.K. Srinivasan, S. Sukavanam and E.C.G.Sudarshan, J. Phys. A6, 1910 (1973) ; M.L. Mehta and C.L. Mehta, J. Opt. Soc. Am. 63, 826 (1973).
13. G.S. Agarwal, Opt. Commun. 35, 267 (1980); M. Lewenstein, P. Zoller and J.Mostowski, J. Phys. B16, 563 (1983); A.T. Georges, Opt. Commun. 41, 61 (1982).
14. G.S. Agarwal, this volume.
15, G.S. Agarwal, Phys. Rev. Lett. 37, 1383 (1976); Phys. Rev. A18, 1490 (1978).
16. P. Anantha Lakshmi and G.S. Agarwal, Phys. Rev. A23, 2553 (1981); R. Saxena and G.S. Agarwal, Phys. Rev. A25, 2123 (1982); G.S. Agarwal and P.A. Narayana, Opt. Commun. 30, 364 (1979).
17. P. Zoller, in Laser Physics eds. D.F. Walls and J.D. Harvey (Academic, Sydney 1980); K. Wodkiewicz, Phys. Rev. A19, 1686 (1979).
18. J.H. Eberly, Phys. Rev. Letters 37, 1386 (1976).
19. B.J. Dalton and P.L. Knight, J. Phys. B15, 3997 (1982).
20. J.E. Thomas, P.R. Hemmer, S. Ezekiel, C.C. Leiby, R.H. Picard and C.R. Willis, Phys. Rev. Lett. 48, 867 (1982).

21. M.G. Raymer, J. Mostowski and J.L. Carlsten, Phys. Rev. A19, 2304 (1979).
22. D.S. Elliott, R. Roy and S.J. Smith, Phys. Rev. A26, 12 (1982).
23. G.S. Agarwal, to be published.
24. L. Galatry, Phys. Rev. 122, 1218 (1961); S.G. Rautian and I.I.Sobelman, Sov. Phys. Uspekhi 9, 701 (1967).

STOCHASTIC MODELLING OF RELAXATION EFFECTS IN LINE SHAPES

S. Dattagupta
School of Physics, University of Hyderabad
Hyderabad 500 134, India

1. Introduction

Relaxation phenomena are best studied by spectroscopic techniques. The central formula for the line shape, i.e., the intensity as a function of frequency can be written as the real part of the Laplace transform of a certain correlation function :

$$I(\omega) = \mathrm{Re}\int_0^\infty dt \exp(-st) \langle d^+(o) \exp(-i\vec{k}.\vec{r}(o))\, d(t) \exp(i\vec{k}.\vec{r}(t))\rangle \quad (1.1)$$

Here the Laplace transform variable s is complex whose imaginary part is the frequency ω. The real part of s is a certain width (assumed known in the present context) which may arise from instrumental resolution as well as natural causes such as the finite life time of decay. The operator d represents the transition operator, $\vec{k}$ denotes the wave vector of the emitted or absorbed radiation, and $\vec{r}$ indicates the centre of mass coordinate of the radiating system e.g. an atom, a molecule or a nucleus, as the case may be. The angular bracket in (1.1) has the usual meaning.

$$\langle \ldots \rangle = \mathrm{Tr}\,(\rho \ldots), \quad (1.2)$$

where $\rho = \exp(-\beta\mathcal{H})/\mathrm{Tr}\,[\exp(-\beta\mathcal{H})]$, $\beta = (k_B T)^{-1}$, (1.3)

$\mathcal{H}$ being the Hamiltonian of the entire system. The time-dependence of various operators in (1.1) are given by

$$d(t) = \exp(i\mathcal{H}t)\, d(o) \exp(-i\mathcal{H}t)$$

$$\vec{r}(t) = \exp(i\mathcal{H}t)\, \vec{r}(o) \exp(-i\mathcal{H}t),$$

where we have set $\hbar = 1$.

A wide variety of line shape problems can be tackled, starting from (1.1). Examples are :

(i) Atomic spectroscopy (infrared or microwave) :

In this case, d is the dipole moment operator for the atom. Most infrared or microwave studies (which are of concern to us here) are carried out in liquids or gases where $\vec{r}$ is a continuous variable that may be treated classically [1]. Thus

$$\vec{r}(t) = \vec{r}(o) + \int_0^t \vec{v}(t')\, dt', \quad (1.4)$$

where $\vec{v}$ is the velocity. The line shape expression then takes the form

$$I(\omega) = \mathrm{Re}\int_0^\infty dt \exp(-st) \langle d^+(o)\, d(t) \exp(-ik\int_0^t v(t')dt')\rangle, \quad (1.5)$$

as only the component of $\vec{v}$ along the direction of $\vec{k}$ enters into consideration.

(ii) Molecular spectroscopy (infrared or Raman) :

Here one investigates the vibrational, translational and rotational motion of molecules. The transition operator d is therefore the molecular dipole moment in the infrared case while it is the polarizability in the case of Raman scattering[2]. In most instances, the coupling between vibrational, translational and rotational motions is so weak that the respective contributions can be separated. Furthermore, we shall restrict our attention to vibrational and rotational relaxation only. One may now treat two distinct cases.

(a) Vibrational relaxation:

Here one analyzes a vibrational line and through it the effect of dynamics on molecular vibrations occasioned by, say, phonons as in solids, or collisional effects as in gases or liquids [3] . The transition operator d is now a normal mode coordinate Q of molecular vibration, and hence the vibrational line shape reads

$$I(\omega) = \mathrm{Re} \int_0^\infty dt \exp(-st) \langle Q^+(o) Q(t) \rangle \quad . \tag{1.6}$$

(b) Rotational relaxation:

This case is more relevant to liquids and gases. Although one looks again at a transition between molecular vibrational levels, one is now interested in the effect of rotational motion on the vibration-rotation bands. In most instances the molecule concerned would have atleast one axis of symmetry. The rotational contribution to both the infrared and depolarized Raman line shapes is then given by [4]

$$I(\omega) = \mathrm{Re} \int_0^\infty dt \exp(-st) \langle \langle \ell\, m{=}o | \exp [-i\vec{L} . \int_0^t \vec{\Omega}(t')dt'] | \ell\, m{=}o \rangle \rangle , \tag{1.7}$$

where vibrations parallel to the symmetry axis are considered. Here $\vec{L}$ is the angular momentum operator and $\vec{\Omega}$ the angular velocity of the molecule. It turns out that ℓ is 1 for infrared spectroscopy and is 2 for Raman scattering.

(iii) Nuclear Spectroscopy :

(a) Mössbauer effect:

Here one observes resonant emission or absorption of a gamma ray by a nucleus. The transition operator d can be traced to the interaction between the nucleus and the surrounding radiation field of the gamma ray. For magnetic dipole transitions, the matrix elements of d are proportional to certain Clebsch-Gordan coefficients. In a viscous liquid, the Mössbauer line shape is given by an expression similar to (1.5) where $\vec{k}$ is the wave vector of the gamma ray and $\vec{v}$ is the velocity of the nucleus. On the other hand, in a solid (where most applications of Mössbauer spectroscopy are made), (1.1) is a more appropriate expression where $\vec{r}$ denotes the discrete position of the nucleus in a lattice [5] .

(b) Nuclear magnetic resonance (NMR):

Here the $\vec{k}$-dependence of the kind indicated in (1.5) can be ignored in most cases. The transition operator d is now given by the component of the nuclear magnetic moment along the direction of the radio frequency magnetic field[6] .

(c) Muon spin rotation (μSR):

The μSR line shape is obtained from an expression which is very similar to that for the NMR signal except that it is recorded in the time domain while NMR is observed as a function of the frequency of the time-varying field [6] . The transition operator d is now a Pauli spin angular momentum along the initial direction of polarization of the muon.

In all the different cases mentioned above, there is an underlying common theme. One is almost always looking at the time-dependence of a transition operator $[d^+ \exp(-i\vec{k}.\vec{r})]$ governed by the time-evolution operator

$$\mathcal{U}(t) = \exp(i\mathcal{H}^{\times} t) \quad , \tag{1.8}$$

where $\mathcal{H}^{\times}$ is the Liouville operator associated with $\mathcal{H}$, the Hamiltonian of a many body system. The point of view adopted in the stochastic theory of line shape is that the interactions present in the surroundings of the radiating system produce explicitly time dependent forces at the radiating system[7] . The surrounding many body system is therefore viewed as a heat bath whose role is to drive random fluctuations into the radiating system. Such ideas are of course quite familiar in the context of the Brownian motion, a subject of extensive discussion in the present volume. How this kind of stochastic modelling makes physical sense in line shape problems would hopefully become clear in the subsequent sections where we deal with individual cases separately. We shall motivate the development of the general theory, first in terms of a simple example (§ 2.1), and then keep adding new features, within, however, a common theoretical framework. Finally, we shall deal with a certain model of the socalled transition probability matrix which would allow us to provide analytic expressions for the line shape. We shall see how the particular form chosen for the transition matrix accounts for distinct kinds of physical phenomena in the context of different experimental situations outlined above.

2. Dephasing in vibrational relaxation :

In simple physical terms, vibrational dephasing can be understood easily for dilute gaseous molecules for which the most dominant collision events are of the binary type. The vibrational wave function of a molecule is determined by the occupation number of the various normal modes and an overall phase factor. Now, elastic collisions with other molecules cause phase shifts resulting in fluctuations of the vibrational frequency. This effect is known as vibrational dephasing.

In a dense liquid or a solid many body effects make the description of vibrational dephasing much more complicated than that in the gas phase. However, in many a situation involving solute molecules in a solvent medium, the gas type binary collision picture provides a reasonable model [3]. The line shape can then be written from (1.6) as

$$I(\omega) = \mathrm{Re} \int_0^\infty dt \exp(-st)\, \mathrm{Tr}_v \left\{ \rho_v Q^+(o) \left[\langle \exp_T (i \int_0^t \mathcal{H}^\times (t')dt') \rangle Q(o) \right] \right\}, \quad (2.1)$$

where, in the stochastic picture, the vibrational part of the Hamiltonian $\mathcal{H}$ is assumed to be randomly time-dependent. Here $\mathrm{Tr}_v \{ \ldots \}$ denotes a trace over the vibrational states characterized by the density matrix ρv, $\exp_T(\ldots)$ implies a suitable time-ordering of the operators and $\langle \ldots \rangle$ indicates an average over the stochastic properties of $\mathcal{H}(t)$.

We shall, for the sake of simplicity, be concerned here with only the ground and first excited vibrational levels. The two lower most vibrational levels can then be represented by the eigenstates of the z component S_z of a pseudo-spin operator ($\vec{S} = \frac{1}{2}$). The vibrational part of the Hamiltonian may be modelled as [8]

$$\mathcal{H}(t) = [\omega_o + \Delta\omega(t)] \, S_z \, , \quad (2.2)$$

where ω_o specifies the static part of the vibrational frequency and $\Delta\omega(t)$ the fluctuations with zero mean. Using the fact that in the pseudo spin language the transition operators may be represented by raising (or lowering) angular momentum operators, (2.1) may be simplified to

$$I(\omega) = Z \, \mathrm{Re} \int_0^\infty dt \exp[-(s-i\omega_o)t] \langle \mathcal{U}(t) \rangle \, , \quad (2.3)$$

where $$\mathcal{U}(t) = \exp(i \int_0^t \Delta\omega(t')\, dt') \, , \quad (2.4)$$

Z being a temperature-dependent prefactor.

2.1 Stochastic Theory Calculation of $\langle \mathcal{U}(t) \rangle$:

We assume that the system is subject to certain fluctuations (or 'collisions', in a generalized sense) at random instants of time which make the frequency jump instantaneously from one to the other of the n possible forms. For a continuous stochastic process, we can take the limit $n \longrightarrow \infty$ at a suitable stage, as would be made apparent later on. The instants of time at which the collisions occur are assumed to be randomly distributed with a Poisson distribution, and in between two successive collisions, the system is assumed to be unperturbed [9]. In the linear vector space spanned by the n values of $\Delta\omega(t)$, the time development operator of (2.4) can be viewed as a matrix which is constructed as

$$\mathcal{U}(t) = e^{i\Omega t_1} \mathcal{T} e^{i\Omega(t_2-t_1)} \mathcal{T} \cdots\cdots \mathcal{T} e^{i\Omega(t-t_s)} \quad (2.5)$$

where $\mathcal{T}$ is a transition probability matrix whose elements specify the probabilities of jump of $\Delta\omega(t)$ from one value to another. The frequency matrix Ω is diagonal with elements given by the various values of $\Delta\omega(t)$, viz., $\Delta\omega_1, \Delta\omega_2 \ldots\ldots \Delta\omega_n$. Denoting by $|a)$, $|b)$, etc., the stochastic states, such that a=1,2....n correspond to the frequency values $\Delta\omega_1, \Delta\omega_2, \ldots\ldots, \Delta\omega_n$, respectively, we may write Ω as

$$\Omega = \sum_j \Delta\omega_j \; F_j \quad , \tag{2.6}$$

where

$$(a|\,F_j\,|b) = \delta_{aj} \; \delta_{ab} \quad , \tag{2.7}$$

Thus

$$(a|\Omega|\,b) = \Delta\omega_a \; \delta_{ab} \quad . \tag{2.8}$$

If λ specifies the mean rate of collisions, then the Laplace transform of $\langle \mathcal{U}(t) \rangle$ can be shown to be given by [9]

$$\langle \mathcal{U}(s) \rangle = \sum_{ab} p_a \; (a\,|\,[\, s - i\omega_o - i\Omega - \lambda(\mathcal{J} - 1)\,]^{-1}\,|\,b) \quad , \tag{2.9}$$

where p_a is the a-priori probability of the occurrence of the stochastic state $|a)$. We shall treat later in § 8.1 a specific form of the matrix $\mathcal{J}$, and based on that, obtain an explicit solution to the vibrational dephasing problem.

Before concluding this Section, it may be mentioned that the ideas developed here are rather similar to the Continuous Time Random Walk models[10]. Also, (2.9) is equivalent to the stochastic Liouville equation discussed elsewhere in this volume [11].

2.2 <u>Dephasing cum depopulation</u> :

As mentioned earlier, dephasing is essentially due to elastic collisions. Some of the collisions, however, are expected to be inelastic which may lead to direct transitions between the vibrational levels so as to induce population or energy relaxation. In the process called 'depopulation', the vibrational energy is transferred to translational, rotational or different vibrational degrees of freedom. Depopulation, like dephasing, can also be discussed within the same stochastic model framework outlined above. The Hamiltonian in (2.2) has to be now extended to

$$\mathcal{H}(t) = [\omega_o + \Delta\omega(t)\,]\, S_z + \sum_{i=1}^{n} (\vec{S}.\vec{h}_i) \;\; \delta(t-t_i) \tag{2.10}$$

where t_i (i = 1,2,.....n) are the instants at which the collisions occur and h_i is an 'effective' field the molecule experiences during a collision [8]. The additional term which contains the raising and lowering operators S_+ and S_- can cause transitions between the vibrational energy states giving rise to depopulation. The field $\vec{h}_i$, in the present formulation, is of course a random variable which should be averaged over in the final expression for the line shape.

In order to incorporate the simultaneous presence of vibrational dephasing and and depopulation within the same stochastic model, the transition matrix $\mathcal{J}$ now should have an additional feature. It should be replaced by $\mathcal{J}_1 \; \mathcal{J}_2$ where $\mathcal{J}_1$ has the same meaning as $\mathcal{J}$ while $\mathcal{J}_2$ is given by [8]

$$\mathcal{J}_2 = \exp(i\vec{h}_i \cdot \vec{S}^{\times})_{av} , \qquad (2.11)$$

where $(...)_{av}$ denotes an average over the type of the field $\vec{h}_i$ the molecule experiences at the time t_i. It is perhaps appropriate here to mention the limitations of the model described by (2.11). Each collision, in at least so far as causing direct transitions between the two levels, is assumed to have an averaged effect only. This means that the fields $\vec{h}_i$ are taken to be uncorrelated from one collision to the next. Further, since the interaction $\vec{h}_i.\vec{S}$ is hermitian, the transitions induced by it between the eigenstates of S_z are equally probable in either direction, and hence finite temperature effects are ignored in the model.

3. Collision broadening in atomic spectroscopy :

We turn our attention next to the problem of collision broadening of atomic spectral lines in gases. Although this phenomenon is studied in an altogether different physical context, the mathematical treatment of it is quite similar to that of vibrational relaxation considered in § 2 [12]. Generally speaking, collision broadening is regarded as an effect which arises from the simultaneous perturbations of the upper and the lower levels of a transition due to the interaction between the radiating atom called the emitter and the rest of the atoms in a buffer gas called the perturbers. The emission spectrum may consist of several lines originating from a completely or partially lifted degeneracy of the energy levels involved in the transition. The interaction during a collision modulates the wavefunction of the emitter in its excited and ground states, which makes the frequencies of transitions randomly time-dependent. This frequency modulation effect is therefore quite akin to vibrational dephasing discussed earlier. In addition to causing frequency modulation, a collision can also induce a direct (inelastic) transition between the two levels. To illustrate, if the perturber is electrically charged, it may give rise to a 'pulse' of an electric field as it collides with the emitter. This electric field then couples directly with the dipole moment of the emitter (the Stark effect). The resultant effect is (curiously) known as phase modulation in the literature on collision broadening although it is very similar to vibrational depopulation as opposed to dephasing (§ 2.2). It is evident that the simultaneous occurrence of frequency and phase modulations can be accounted for, in the pseudo-spin language, in terms of the stochastic model Hamiltonian given in (2.8), and the analysis of collision broadening of atomic lines would be very similar to that of vibrational relaxation in molecular spectroscopy.

In discussing frequency and phase modulations, we have, so far, not considered another source of broadening, namely the Doppler broadening owing to the thermal velocity of the emitter, and the effect on it of the change in the velocity of the emitter due to a collision. It turns out, velocity changes can cause an appreciable reduction in the Doppler broadening contribution. The simultaneous effects of frequency, phase and velocity modulations can again be treated by way of a further extension of the formalism outlined in § 2[19] . The Hamiltonian is still given by (2.10). However, the time evolution operator now is to be written as

$$\mathcal{U}(t) = \exp_T \left[i \int_0^t dt' \, (\mathcal{H}^{\times}(t') - k\, v\, (t')) \right] , \tag{3.1}$$

and the line shape is to be calculated from (1.5).

We may point out that in the problem at hand we have a combination of both discrete ($\Delta\omega(t)$) and continuous ($v(t)$) stochastic processes. Of course, in some cases, $\Delta\omega(t)$ can also be a continuous process. Following the matrix method indicated in § 2, we now have

$$\mathcal{U}(s) = \sum_{ab} p_a \int p(v_o)\, dv_o\, dv_1 \, (a v_o | [s - i\omega_o - i\Omega + ikV - \lambda(\mathcal{T} - 1)]^{-1} | b v_1) , \tag{3.2}$$

where the stochastic space has been suitably extended to accomodate the velocity variables, and V is a 'velocity matrix' (similar to Ω) which is diagonal and whose elements are the allowed values of the velocity. Similarly, the transition matrix $\mathcal{T}$ has to be expanded in order to account for frequency, phase as well as velocity modulations occurring in the same collision. Thus we may write

$$\mathcal{T} = \mathcal{T}_1 \mathcal{T}_2 \mathcal{T}_3 , \tag{3.3}$$

where $\mathcal{T}_1$ and $\mathcal{T}_2$ have the same interpretations as before whereas $\mathcal{T}_3$ is a matrix whose elements give the probabilities of transitions of the velocity from one value to another. We shall describe later a specific model for the transition matrix $\mathcal{T}$(§8.2).

4. Molecular rotations in liquids and gases :

We consider here molecular rotations in dense phases, introduced in § 1(ii)(b), as can be investigated by Raman and infrared spectroscopy. The most extensively used model for the analysis of rotational line shapes is the extended diffusion models (EDM) [13] . In these, the 'tagged' molecule is assumed to undergo free rotations unless interupted by random instantaneous collisions which may alter its state of rotation. We shall show here how the EDM fit into the general stochastic theory framework we have built up [4] .

For the purpose of keeping the mathematical discussion simple we shall consider a linear molecule. The rotational line shape is given by (1.7) where the time-development operator can be expressed as

$$\mathcal{U}(t) = \exp \left[-i \vec{L} \cdot \int_0^t \vec{\Omega}\, (t')\, dt' \right] . \tag{4.1}$$

In the EDM, the angular velocity $\vec{\Omega}$ is characterized by three stochastic variables : the components ω_x and ω_y in the plane normal to the plane of rotation, and the azimuthal angle ϕ which specifies the plane of rotation. In between collisions, $\{\omega_x, \omega_y$ and $\phi\}$ are assumed to have a set of fixed values (governed, of course, by certain probability distributions) which get randomized due to collisions. Here, once again, one is dealing with an admixture of continuous (ω_x and ω_y) and discrete (ϕ) stochastic variables. If t_i's denote the instants of collisions as before, we may write

$$\vec{\Omega}(t) = \hat{i}\ \omega_x(t) + \hat{j}\omega_y(t) + \hat{k} \sum_{i=1}^{n} \Phi_i\ \delta(t-t_i)$$

$$= \vec{\omega}(t) + \hat{k} \sum_{i=1}^{n} \Phi_i\ \delta(t-t_i)\ , \qquad (4.2)$$

where $\vec{\omega}(t)$ denotes the angular velocity vector in the XY-plane. Proceeding as before, we obtain

$$< \mathcal{U}(s) > = \frac{1}{(2\pi)^2} \iint p(\omega_0) d\vec{\omega}_0\, d\vec{\omega}_1\, (\vec{\omega}_0 | [s + i \vec{L} . \boldsymbol{\omega} - \lambda (\mathcal{J}_1 \mathcal{J}_2 - 1)]^{-1} | \vec{\omega}_1) \qquad (4.3)$$

where $\boldsymbol{\vec{\omega}}$ is a diagonal matrix (analogue of Ω and V of § 2 and § 3 respectively) whose elements are the allowed values of the angular velocity vector $\vec{\omega}$ in the xy plane, $\mathcal{J}_1$ is the transition probability matrix whose elements yield the probabilities of 'jumps' from one $\vec{\omega}$ to another, while $\mathcal{J}_2$ is given by (cf., (2.11))

$$\mathcal{J}_2 = (\exp (i \Phi L_z))_{av}\ . \qquad (4.4)$$

5. <u>μSR : dipolar relaxation in a solid</u> :

In this section we discuss an example from hyperfine spectroscopy involving the muon spin rotation in a paramagnetic solid. For a μ^+ occupying an interstitial site in a lattice, the effect of its (magnetic) interaction with the environment is to alter its state of spin polarization which can be measured via the direction of emission of the positron. The most dominant mechanism for muon depolarization in a paramagnetic solid is due to the dipolar interaction between the muon spin and the spins of the surrounding nuclei. Now, as the muon undergoes thermally activated jumps from one interstitial site to another, the local dipolar field changes at random. Hence, the dipolar Hamiltonian will again have stochastic elements, and may be modelled as [6]

$$\mathcal{H}(t) = \vec{H}(t) . \vec{S}\ , \qquad (5.1)$$

where $\vec{S}$ ($S=\frac{1}{2}$) is the muon spin and $\vec{H}(t)$ the <u>effective</u> dipolar field. In contrast to the cases discussed earlier (see, for example, (2.2) the Hamiltonian in (5.1) <u>does not</u> commute with itself at different times because the field $\vec{H}(t)$ changes its direction (as well as magnitude) from time to time. Hence the quantum and stochastic variables get entangled in the present case. Following the treatment outlined before, the Laplace transform of the averaged time-development operator is now given by

$$<\mathcal{U}(s)> = \sum_{a,b=1}^{n} p_a\ (a | [s - i \sum_{j=1}^{n} V_j^{\times} F_j - \lambda(\mathcal{J} - 1)]^{-1} | b)\ , \qquad (5.2)$$

where the symbols have the same meaning as before, while

$$V_j^{\times} = \vec{H}_j . \vec{S}^{\times}\ , \qquad (5.3)$$

$\vec{H}_j$ being the jth form of the dipolar field and $\vec{S}^{\times}$ the Liouville operator associated with $\vec{S}$.

6. Relaxation cum diffusion in Mössbauer spectra :

Our final physical example, in which stochastic modelling is useful, concerns the study of dynamics in solids or viscous liquids using the Mössbauer effect. In particular, we are interested in a situation in which the Mössbauer nucleus, due to its diffusive motion, jumps into a new environment where it finds either a different electric field gradient, or a different dipolar field as in the μSR case [14]. A new environment means a new Hamiltonian for the nucleus (cf. § 5) which causes relaxation effects. On the other hand, the diffusive motion itself causes a line broadening, a feature that is absent in the μSR case.

As mentioned before the Mössbauer line shape is given by (1.1) where d is a nuclear operator which is responsible for the emission or absorption of a gamma ray of frequency ω and natural line width Γ (i.e. $s = -\omega + \Gamma/2$), and $\vec{r}$ is the coordinate of the centre of mass of the nucleus. In order to treat diffusion it is convenient to recast (1.1) as [14]

$$I(\omega) = \mathrm{Re} \int_0^\infty dt \exp(-st) \int d\vec{R} \exp(i\vec{k}.\vec{R}) \mathrm{Tr}[\rho d^+(o)[\langle \mathcal{U}(\vec{R},t)\rangle d(o)]] \qquad (6.1)$$

where the operator $\langle \mathcal{U}(\vec{R},t)\rangle$ describes, in an average sense, the position and time development of the quantum state of the Mössbauer atom, the average being taken over the random process under consideration, viz., diffusion in the present case. The 'trace' in (6.1) indicates an average over the angular momentum states of the nucleus.

In evaluating $\langle \mathcal{U}(\vec{R},t)\rangle$ one adopts a suitable generalization of the stochastic model presented in § 2.1 [14]. The physical idea is that the effect of the heat bath in which the nucleus is embedded, is to subject the nucleus to sudden 'collisions' at random instants of time. Each such collision induces the nucleus to jump to a new site causing an instantaneous change in phase of the emitted gamma ray. At the same time, the jump takes the nucleus to a new environment where it 'sees' a different Hamiltonian. In between collisions, the nucleus is assumed unperturbed. As before we would like to picture $\mathcal{U}(\vec{R},t)$ as a matrix in the stochastic space (its dimension is $n \times n$ if n is the number of different forms of the Hamiltonian at different sites) which may be constructed as

$$\mathcal{U}(\vec{R},t) = \exp(i\mathcal{H}_o^\times t_1)\, g(\vec{o},\vec{R}_1)\, \mathcal{J}_1 \exp(i\mathcal{H}_o^\times (t_2-t_1))$$

$$\times g(\vec{R}_1,\vec{R}_2)\, \mathcal{J}_1 \exp(i\mathcal{H}_o^\times (t_3-t_2)) \times \ldots g(\vec{R}_{s-1},\vec{R})\, \mathcal{J}_1 \exp(i\mathcal{H}_o^\times (t-t_s)) \qquad (6.2)$$

where

$$\mathcal{H}_o^\times = \sum_{j=1} V_j^\times F_j . \qquad (6.3)$$

$V_j^\times$ is a quantum mechanical Liouville operator associated with the jth form of the Hamiltonian V_j (cf. (5.2)), the matrices F_j and $\mathcal{J}_1$ have been introduced before (the elements of $\mathcal{J}_1$, in the present case, yield the probabilities of jumps of the Hamiltonian from one given form to another), and $g(\vec{R}_1,\vec{R}_2)$ is the probability that a jump takes the nucleus from $\vec{R}_1$ to $\vec{R}_2$. The Laplace-Fourier transform of the time-development operator can be shown to have the structure [14]

$$\langle \mathcal{U}(\vec{k},s) \rangle = \sum_{ab} p_a \, (a \,|\, [\, s - i\, \mathcal{H}_o^{\times} - \lambda(g(\vec{k})\, \mathcal{J}_1 - 1)\,]^{-1} \,|\, b) \,, \qquad (6.4)$$

where

$$g(\vec{k}) \equiv \int d\vec{R} \; \exp(i\vec{k}.\vec{R}) \; g(\vec{R}) . \qquad (6.5)$$

7.1 Mathematical meaning of the transition probability matrix $\mathcal{J}$

We have so far considered a wide variety of spectroscopy problems (in which relaxation effects are important) and cast them in a common theoretical framework. We would like now to indicate the mathematical basis of the treatment given here in the context of the formal theory of stochastic processes covered in great detail in the present volume. To do this, it is important to examine the structure of $(a|P(t)|b)$ which defines the conditional probability that the stochastic state is $|b)$ at time t, given that the state was $|a)$ at t=o. In terms of the model outlined in § 2.1, it is easy to see that the P-matrix is given by [15]

$$P(t) = \sum_{s=o}^{\infty} \int_o^t dt_s \ldots \int_o^{t_2} dt_1 \, \exp(-\lambda t_1) \, (\lambda \mathcal{J}) \, \exp(-\lambda(t_2-t_1)) \ldots (\lambda \mathcal{J}) \exp(-\lambda(t-t_s)) \qquad (7.1)$$

where λ is the mean rate of collisions, as defined before. Equation (7.1) implies that

$$P(t) = \exp(W t) \,, \qquad (7.2)$$

where

$$W \equiv \lambda(\mathcal{J} - 1) \,. \qquad (7.3)$$

Equation (7.2) is the solution of the Chapman-Kolmogorov-Smoluchowski equation that defines a stationary Markov process [16] :

$$(a \,|\, P(t) \,|\, b) = \sum_c (a \,|\, P(\tau) \,|\, c)\, (c \,|\, P(t-\tau) \,|\, b) . \qquad (7.4)$$

The differential form of (7.4), of course, yields the familiar master equation for the probability :

$$\frac{d}{dt} (a \,|\, P(t) \,|\, b) = \sum_c (a|P(t)|c)\, (c|W|b) \,. \qquad (7.5)$$

The continuous version of (7.5) (as would be needed in describing velocity modulations in collision broadening (§ 3)) reads

$$\frac{d}{dt} (a \,|\, P(t) \,|\, b) = \int dc \, (a \,|\, P(t) \,|\, c) \, (c \,|\, W \,|\, b) \,. \qquad (7.6)$$

We also have from (7.2),

$$W = \left(\frac{dP}{dt} \right)_{t=o} , \qquad (7.7)$$

and therefore, $(a|W|b)$ specifies the probability per unit time of an instantaneous jump of the stochastic state from $|a)$ to $|b)$. As the total probability is conserved,

(7.7) leads to

$$\sum_b (a|W|b) = 0 \quad . \tag{7.8}$$

In addition to (7.8), the W-matrix must satisfy the condition of detailed balance for a system in thermal equilibrium :

$$p_a (a|W|b) = p_b (b|W|a) \tag{7.9}$$

The transition probability matrix, on the other hand, satisfies (cf. (7.3), (7.8) and (7.9))

$$\sum_b (a|\mathcal{J}|b) = 1$$

and

$$p_a (a|\mathcal{J}|b) = p_b (b|\mathcal{J}|a) \quad . \tag{7.10}$$

Here p_a, as has been defined before, is the a priori probability of the occurrence of the stochastic state $|a)$.

7.2 $\mathcal{J}$ - matrix in the strong collision approximation :

It is clear that in order to derive line shape expressions it is necessary to introduce suitable models for the transition matrix $\mathcal{J}$, restricted of course by the requirements of (7.10). For the purpose of fixing our ideas we shall use here the jargon of collision broadening, although the following discussion applies equally well to other kinds of fluctuations in other areas of spectroscopy which have been considered earlier.

In the context of collision broadening, the stochastic variable is, say, the velocity, which is a continuous process. If the collisions have no effect on velocity, then $\mathcal{J} = 1$. Next comes the weak-collision approximation in which each collision is assumed to alter the pre-collision value of the velocity only infinitesimally [17]. This leads to the Fokker-Planck equation for $(v_o|P(t)|v)$ with the initial condition $(v_o|P(o)|v) = \delta(v-v_o)$. As the Fokker-Planck equation has been the topic of a great amount of discussion in the present volume, we shall consider here the other extreme form for $\mathcal{J}$ which is based on what is variously called the Kubo - Anderson process [18] the random phase approximation [19] or the strong collision approximation [17]. In this it is assumed that the distribution 'equilibrates' so rapidly that it loses all memory of the pre-collision value of the variable. In other words,

$$(v_o|\mathcal{J}|v) = p(v) \quad , \tag{7.11}$$

the right hand side being the only v_o-independent form that satisfies the continuous version of (7.10). Substituting (7.11) in (7.3) and carrying out the exponentiation required by (7.2), the solution obtained is

$$(v_o|P(t)|v) = \delta(v-v_o) \exp(-\lambda t) + p(v) [1-\exp(-\lambda t)] \quad . \tag{7.12}$$

The discrete version of (7.12), as would be needed, say, in problems concerning vibrational dephasing or Mössbauer study in solids, would have the obvious structure

$$(a|P(t)|b) = \delta_{ab}\exp(-\lambda t) + p_b\,[1-\exp(-\lambda t)] \quad . \tag{7.13}$$

It may be noted also that in the special case of a discrete two-state stochastic process with equal a-priori probabilities (i.e. $p_a = \frac{1}{2}$), the strong collision approximation implies that

$$(a|\mathcal{J}|b) = \frac{1}{2} \quad . \tag{7.14}$$

Hence, from (7.3), W which is now a 2x2 matrix, has the structure

$$W = \lambda \begin{bmatrix} -1/2 & 1/2 \\ 1/2 & -1/2 \end{bmatrix} . \tag{7.15}$$

The process described by (7.15) is called a telegraphic or a dichotomic Markov process [18].

8. Line shape expressions in the strong collision model :

In recent years there have been numerous applications of the strong collision model to the analysis of line shapes in the context of various spectroscopy experiments. Here we shall not go into the details of these applications but merely point out the physical context in which the strong collision model is applied to individual cases mentioned before. The discussion will allow us to provide a somewhat unified stochastic theory picture of a variety of spectroscopic studies.

8.1 Vibrational dephasing :

As we have discussed before, vibrational relaxation is viewed to occur as a result of fluctuations in the environment of the active molecule. As the 'environment' is basically a system with a large number of degrees of freedom, the fluctuations may be taken to be short-lived and hence modelled as instantaneous 'collisions'. Each collision perturbs the active molecule and causes the frequency separation $\Delta\omega$(of the two lines involved in the transition) to 'jump' instantaneously from one value to another. We may model this jump process in the strong collision approximation, and obtain an analytic expression for the line shape, as indicated below.

From (2.1) and (2.3), it is evident that the vibrational line shape is given by Re $\langle \mathcal{U}(s) \rangle$ where u(s) can be expressed as (cf.(2.9))

$$\mathcal{U}(s) = [s-i\omega_o-i\Omega - \lambda(\mathcal{J}-1)]^{-1} = \mathcal{U}_o(s+\lambda)\,[1+\lambda\mathcal{J}\,\mathcal{U}(s)] \quad , \tag{8.1}$$

where

$$\mathcal{U}_o(s+\lambda) = (s-i\omega_o - i\Omega + \lambda)^{-1} \quad . \tag{8.2}$$

Now, the 'stochastic average' of u(s), defined as

$$\langle\mathcal{U}(s)\rangle = \sum_{ab} p_a \;(a|\mathcal{U}(s)|b) \quad , \tag{8.3}$$

may be written from (8.1) as [19]

$$\langle \mathcal{U}(s) \rangle = \langle \mathcal{U}_o(s+\lambda) \rangle + \lambda \sum_{abcd} p_a \, (a|\mathcal{U}_o(s+\lambda)|c) \, (c|\mathcal{J}|d) \, (d|\mathcal{U}(s)|b) \, , \tag{8.4}$$

where we have used the closure property of stochastic states. Using next the strong collision form of the $\mathcal{J}$ -matrix (cf. (7.11)) in (8.4), we obtain

$$\langle \mathcal{U}(s) \rangle = \langle \mathcal{U}_o(s+\lambda) \rangle + \lambda \sum_{ac} p_a \, (a|\, \mathcal{U}_o \, (s+\lambda)\,|\, c) \sum_{db} p_d \, (d|\mathcal{U}(s)\,|\, b)$$

$$= \langle \mathcal{U}_o(s+\lambda) \rangle + \lambda \langle \mathcal{U}_o(s+\lambda) \rangle \langle \mathcal{U}(s) \rangle$$

or,

$$\langle \mathcal{U}(s) \rangle = [\langle \mathcal{U}_o(s+\lambda) \rangle^{-1} - \lambda]^{-1} \quad . \tag{8.5}$$

Equation (8.5) expresses the Laplace-transform of the time-evolution operator (or the Green function) in terms of the 'free propagator' only. The structure of (8.5) is suggestive as to why the present approximation is also referred to as the random phase approximation.

From (8.2) and the definition of the stochastic average as in (8.3), as well as the expressions for Ω and F_j (cf. (2.6) and (2.7)), we have

$$\langle \mathcal{U}_o(s+\lambda) \rangle = \sum_{j=1}^{n} p_j \, (s-i\omega_o-i\,\Delta\omega_j + \lambda)^{-1} \tag{8.6}$$

Now, in the context of vibrational dephasing in liquids, there is a large number of possible frequency values. Hence the central limit theorem applies and we may replace the summation in (8.6) by an integral

$$\langle \mathcal{U}_o(s+\lambda) \rangle = \int_{-\infty}^{\infty} d(\Delta\omega) \; p(\Delta\omega) \; (s-i\omega_o-i\Delta\omega + \lambda)^{-1} \quad , \tag{8.7}$$

where

$$p(\Delta\omega) = (4\pi\sigma^2)^{\frac{1}{2}} \; \exp\,[-\,(\Delta\omega)^2/4\sigma^2\,] \; . \tag{8.8}$$

Carrying out the integral in (8.7) with the aid of (8.8),

$$\langle \mathcal{U}_o \; (s+\lambda) \rangle = \frac{\sqrt{\pi}}{2\,\sigma} \exp(x^2) \; \mathrm{erfc}(x) \; , \tag{8.9}$$

where erfc(x) denotes the complimentary error function of its argument x defined as

$$x \equiv (s-i\omega_o + \lambda)/_{2\sigma} \; . \tag{8.10}$$

Substitution of (8.9) into (8.5) completes the solution to the vibrational line shape in the strong collision model. We refer the reader to [8] for application of these results to experimental data.

8.1.a <u>Dephasing cum depopulation:</u>

As discussed before, the simultaneous occurrence of dephasing and depopulation can be taken into account by including $\mathcal{J}_2$ (see (2.11)) in (2.9). It turns out that the line shape expression is now a simple generalization of (8.5) [8] :

$$\langle \mathcal{U}(s) \rangle = [\langle \mathcal{U}_o(s+\lambda) \rangle^{-1} - \lambda(1-y)]^{-1} \quad , \tag{8.11}$$

where the parameter y depends on the averages of certain functions of the magnitude and direction of the effective field $\vec{h}_i$:

$$y = [\sin^2(h_i/2)\ (2-\sin^2\theta_i)]_{av} + i(\sinh_i \cos\theta_i)_{av} \quad . \tag{8.12}$$

The azimuthal angle ϕ_i specifying the orientation of $\vec{h}_i$ does not appear in (8.12) as $\vec{h}_i$ is assumed to be cylindrically symmetric. We may point out that the result in (8.11) implies a 'statistical dependence' of dephasing and depopulation on each other. This means that the total line widths cannot be taken to be the sum of individual linewidths (due to dephasing and depopulation separately).

8.2 Collision broadening:

We turn our attention now to the topic of § 3. For the sake of simplicity, we assume the frequencies of transition to have only two distinct values (i.e. $\Delta\omega_j = \pm\Delta\omega$) which jump from one to the other following a telegraphic process (cf.(7.14)). On the other hand, as far as the velocity changes are concerned, the strong collision model applies. Hence from (3.3) the transition probability matrix in the combined space of frequency and velocity variables can be written as [12]

$$\begin{aligned}(a\ v_o | \mathcal{J} | b\ v) &= \mathcal{J}_2\ (a | \mathcal{J}_1 | b)\ (v_o | \mathcal{J}_3 | v) \\ &= \frac{1}{2}\ (1-y)\ p(v) \quad ,\end{aligned} \tag{8.13}$$

when y is given by (8.12). Here we have used also (7.11) and (7.14). Now, for sufficiently dilute gases in which the only important collisions are binary the stationary distribution p(v) may be taken to be a Maxwellian :

$$p(o) = (2\pi\langle v^2\rangle)^{-\frac{1}{2}} \quad \exp(-\ v^2/2\langle v^2\rangle) \quad , \tag{8.14}$$

where

$$\langle v^2\rangle = \int_{-\infty}^{\infty} v^2\ p(v)\ dv \quad . \tag{8.15}$$

going back to (3.2) and using the same mathematical development as in § 8.1, we can show that the line shape has the same structure as in (8.11), except now [12] ,

$$\langle \mathcal{U}_o(s+\lambda)\rangle = \frac{1}{2}\int dv\ p(v) \sum_{j=1}^{2} (s+\lambda-i\omega_o+i\,\Delta\omega_j+ikv)^{-1} \quad . \tag{8.16}$$

If we neglect the frequency modulation (i.e. $\Delta\omega_j$=o), (8.16) yields the line shape expression derived by Rautian and Sobel'man in the strong collision model, using however the Boltzmann kinetic equation for the velocity distribution function [17] .

8.3 Molecular rotations:

The rotational line shape is given by (see (1.7) and (4.3))

$$I\ (\omega) = \mathrm{Re}\ \langle \ell\ m{=}o\ |\ \langle\mathcal{U}(s)\rangle\ |\ \ell\ m{=}o\rangle \quad , \tag{8.17}$$

where $\langle\mathcal{U}(s)\rangle$ is obtained from

$$\mathcal{U}(s) = [s+i\ \vec{L}.\vec{\omega} - \lambda\ (\mathcal{J}_1\mathcal{J}_2 - 1)]^{-1} \tag{8.18}$$

We may write (cf. (8.1)

$$\mathcal{U}(s) = \mathcal{U}_o(s+\lambda)\,[1+ \lambda \mathcal{J}_1 \mathcal{J}_2 \mathcal{U}_o(s+\lambda)+ \lambda \mathcal{J}_1 \mathcal{J}_2 \mathcal{U}_o(s+\lambda)\, \lambda \mathcal{J}_1 \mathcal{J}_2 \mathcal{U}_o(s+\lambda) + \ldots\ldots], \tag{8.19}$$

where $\mathcal{U}_o(s+\lambda) = (s+i\, \vec{L}.\vec{\omega} + \lambda)^{-1}$ (8.20)

Noting that (cf. (4.4)

$$\langle \ell m \mid \mathcal{J}_2 \mid \ell m' \rangle = \frac{1}{2\pi} \int_o^{2\pi} d\phi \langle \ell m \mid \exp (i\phi L_z) \mid \ell m' \rangle = \delta_{mm'}\, \delta_{mo} \tag{8.21}$$

we have from (8.19),

$$\langle \ell m{=}o \mid \mathcal{U}(s) \mid \ell m{=}o \rangle = [\, G_o(s+\lambda)^{-1} - \lambda \mathcal{J}_1 \,]^{-1} \quad , \tag{8.22}$$

where

$$G_o(s+\lambda) = \langle \ell\, m{=}o \mid \mathcal{U}_o(s+\lambda) \mid \ell\, m{=}o \rangle \tag{8.23}$$

Now, recalling that $\vec{\omega}$ in (8.20) is a vector in the xy-plane and noting that an arbitrary rotation about the z axis can bring $\vec{\omega}$ parallel to the y-axis, we may simplify (8.23) to [4]

$$G_o(s+\lambda) = \langle \ell\, m{=}o \mid \tilde{\mathcal{U}}_o(s+\lambda) \mid \ell\, m{=}o \rangle \quad , \tag{8.24}$$

where

$$\tilde{\mathcal{U}}_o\,(s+\lambda) = (s+ i\boldsymbol{\omega} L_y + \lambda)^{-1} \tag{8.25}$$

Combining (8.17) and (8.22), the rotational line shape from (4.3) is given by

$$I_\ell(\omega) = \mathrm{Re} \iint p(\omega_o)\, d\omega_o\, d\omega\, (\omega_o \mid \langle \ell\, m{=}o \mid \mathcal{U}(s) \mid \ell\, m{=}o \rangle \mid \omega), \tag{8.26}$$

where

$$p(\omega) = (I/k_B T)\, \omega \exp (-I\omega^2/2k_B T) \ . \tag{8.27}$$

I being the moment of inertia of the linear molecule. The problem now reduces to specifying a form for the transition matrix $(\omega_o \mid \mathcal{J}_1 \mid \omega)$. It turns out that two distinct models, collectively referred to as the extended diffusion models (EDM), have been considered in the literature[13]. Of these, in the <u>M-diffusion model</u>, the magnitude ω of the angular velocity is assumed to be unaltered in a collision. This implies that [4]

$$(\omega_o \mid \mathcal{J}^M_1 \mid \omega) = \delta(\omega_o - \omega) \quad . \tag{8.28}$$

The line shape, from (8.22 - 8.26) is now given by

$$I^M_\ell(\omega) = \mathrm{Re}\int p(\omega)\, d\omega\, [\, (g_o(s+\lambda))^{-1} - \lambda \,]^{-1} \quad . \tag{8.29}$$

where

$$g_o(s+\lambda) = \langle \ell\, m{=}o \mid (s+\lambda+ i\,\omega L_y)^{-1} \mid \ell\, m{=}o \rangle \quad , \tag{8.30}$$

ω being the eigenvalue of the matrix $\boldsymbol{\omega}$. On the other hand, the <u>J -diffusion model</u> is equivalent to the strong collision approximation discussed here, i.e.,

$$(\omega_o | \mathcal{J}_1^J | \omega) = p(\omega) \quad , \tag{8.31}$$

where $p(\omega)$ is given by (8.27). Following the familiar mathematical steps (see § 8.1), we now have [4]

$$I_{\ell}^{J}(\omega) = \mathrm{Re}\, [\langle g_o (s+\lambda)\rangle^{-1} - \lambda]^{-1} \quad , \tag{8.32}$$

where

$$\langle g_o(s+\lambda)\rangle = \int p(\omega)d\omega \langle \ell m{=}o | (s+\lambda+i\omega L_y)^{-1} | \ell m{=}o\rangle \; . \tag{8.33}$$

Employing the general result of (8.22), other kinds of models which interpolate between the J and M limits have also been treated in the literature [4].

8.4 μSR : dipolar relaxation:

As the μ^+ jumps from one interstitial site to another, the post-jump dipolar field may be assumed to be completely uncorrelated to the pre-jump field. This means that immediately following a jump, the μ^+ loses all the memory of the environment it came from. Hence the strong collision model applies and the μSR line shape from (5.2) can be shown to have the same structure as in (8.5) where now (cf.,(5.3)),

$$\mathcal{U}^o(s+\lambda) = \int d\vec{H}\, p(\vec{H})\, (s+\lambda-i\,\vec{H}.S^{\times})^{-1} \tag{8.34}$$

$p(\vec{H})$ being the distribution of the local field. Applications of these results have been considered in detail elsewhere[20]. Generalizations of the strong collision-like models for studying μ^+-diffusion in the presence of trapping impurities have also been treated in the literature [21]. Furthermore, the strong collision model for discrete jump processes has been applied to the analysis of Electron Paramagnetic Resonance[22], and quadrupolar relaxations in Nuclear Quadrupole Resonance, Mössbauer Effect and Perturbed Angular Correlation of Gamma Rays[6].

8.5 Relaxation cum diffusion in Mössbauer spectra :

Our final application concerns the diffusion of a Mössbauer atom in a highly viscous liquid. As the atom jumps to a new position, it finds itself in a new environment in which the electric field gradient, say, has changed its direction arbitrarily. If the electric field gradients at different sites are randomly oriented, the transition probabilities to all sotchastic states are equal. Accordingly,

$$(a| \mathcal{J}_1 | b) = \frac{1}{n} \quad , \tag{8.35}$$

where n is the number of possible orientations of the field gradient. Evidently, (8.35) is a special case of the strong collision model in which the a-priori probabilities are all equal, i.e.

$$p_a = 1/n \tag{8.36}$$

In the present case, the interaction Hamiltonian (cf.(6.3)) is given by [14]

$$V_j = \omega_o + Q \left[3 \, (I_e^j)^2 - I_e \, (I_e+1) \right] \quad , \tag{8.37}$$

where ω_o is the frequency of the emitted gamma ray (= 14.4 KeV for ^{57}Fe); Q is the quadrupole coupling constant and I_e^j is the spin of the nucleus in its excited state along the jth direction of the field gradient.

The strong collision form for the transition probability matrix in (8.35) allows us to derive (cf. § 8.1)

$$\langle \mathcal{U}(\vec{k},s) \rangle = \left[\langle \mathcal{U}_o(s+\lambda) \rangle^{-1} - \lambda g(\vec{k}) \right] \quad , \tag{8.38}$$

where

$$\langle \mathcal{U}_o(s+\lambda) \rangle = \frac{1}{n} \sum_{j=1}^{n} (s+ \lambda -i \, V_j)^{-1} \quad . \tag{8.39}$$

In the limit when the field gradients are distributed isotropically in space, we may write

$$\langle \mathcal{U}_o(s+\lambda) \rangle = \frac{1}{4\pi} \int d \, \Omega_j \quad (s+\lambda - iV_j)^{-1} \quad , \tag{8.40}$$

where $d \, \Omega_j$ is the elementary solid angle in the jth direction. The isotropy implied in (8.40) allows one to derive(for ^{57}Fe) a simple closed form expression for the line shape (cf. (6.1)) [14] :

$$I(\omega) = \mathrm{Re} \left[(s-i\omega_o + \lambda(1-g(\vec{k})) + \frac{9 \, Q^2}{s+\lambda - i\omega_o} \right]^{-1} \quad . \tag{8.41}$$

We refer the reader to [14] for an elaborate discussion of the result given in (8.41).

In conclusion, we have discussed here several applications of the strong collision model to a variety of line shape problems. As mentioned earlier, the strong collision model is, in some sense, the opposite of the weak collision or Fokker-Planck model. It is these two extreme limits in which (stationary) Markov processes are found to be amenable to analytical studies. The applications of Fokker Planck-type models are numerous, especially in the context of continuous Markov (i.e. diffusion) processes. Many of these applications have been dealt with elsewhere in this volume. The present two lectures provide an altogether different set of applications.

References :

1. A. Ben-Reuven, Adv. Chem. Phys. 33, 235 (1975).
2. R.G. Gordon, in Adv. Magn. Reson. 3, 1 (1968).
3. D.W. Oxtoby, Adv. Chem. Phys. 40, 1 (1979).
4. S. Dattagupta and A.K. Sood, Pramana 13, 423 (1979) and Zeit. Phys. B44,85(1981).
5. S. Dattagupta, in Mossbauer spectroscopy : applications to research in physics, chemistry and biology, ed. B.V. Thosar and P.K. Iyengar, Elsevier, Amsterdam,1983.
6. For a common theoretical formulation of Nuclear Spectroscopy tools including NMR and μSR, see S. Dattagupta, Hyperfine Interactions 11, 77 (1981).
7. M. Blume, Phys. Rev. 174, 351 (1968).
8. A.K. Sood and S. Dattagupta, Pramana 17, 315 (1981).
9. M.J. Clauser and M. Blume, Phys. Rev. B3, 853 (1971); see also [7] and S.Dattagupta, Phys. Rev. B16, 158 (1977).
10. V. Balakrishnan, this volume.

11. J.H. Freed, this volume.
12. S. Dattagupta, Pramana 9, 203 (1977).
13. R.G. Gordon, J. Chem. Phys. 44, 1830 (1966).
14. S. Dattagupta, Phys. Rev. B12, 47 (1975) and ibid 14, 1329 (1976).
15. S. Dattagupta, Nucl. Phys. Solid State Phys. (India) A20, 19 (1977).
16. R. Vasudevan, this volume.
17. S.G. Rautian and I.I. Sobel'man, Sov. Phys. Usp. 9, 701 (1967) (Usp.Fiz.Nauk. 90, 209 (1966)).
18. R. Kubo, J. Phys. Soc. Japan 9, 935 (1954); P.W. Anderson, ibid. 316 (1954); see also A. Brissaud and U. Frisch, J. Math. Phys. 15, 524 (1974).
19. S. Dattagupta and M. Blume, Phys. Rev. B14, 4540 (1974).
20. R.S. Hayano, Y.J. Uemura, J. Imazato, N. Nishida, T. Yamazaki and R. Kubo, Phys. Rev. B20, 850 (1979); see also [6].
21. K.W. Kehr, G. Honig and D. Richter, Z. Phys. B 32, 49 (1978), and S. Dattagupta and B. Purniah, ibid. 46, 331 (1982).
22. S. Dattagupta and M. Blume, Phys. Rev. B 10, 4551 (1974).

BROWNIAN MOTION AND CONDENSED MATTER PHYSICS

CLASSICAL AND QUANTUM DIFFUSION

N. Kumar
Physics Department
Indian Institute of Science
Bangalore 560012, India

I. Classical Brownian Motion

I have been asked to cover diffusion, more specifically quantum diffusion, and I have two one hour lectures in which to do the impossible. I have, therefore, planned my lectures thus. In Lecture I starting now I will begin by summarizing the physical ideas of classical diffusion and then discuss the underlying Brownian motion - from Einstein-Smoluchowski through Ornstein-Uhlenbeck. In doing so I will be graduating from a highly idealized position - Markoff process, namely the Wiener process, to the physically more admissible phase-space Markoff process, namely the Ornstein-Uhlenbeck process, or in other words, from stochastic kinematics to stochastic dynamics. All this should be quite familiar from the 1943 classic paper of Chandrasekhar [1] in the Reviews of Modern Physics, which remains a standard reference on the physics of the problem to this day. Except, however, that I may emphasize the formal stochastic point of view to a greater extent, in keeping with the theme of this School. The limitation of these idealized stochastic processes when applied to a real many-body problem (e.g. self-diffusion in a liquid, say) will be only very briefly commented upon. Towards the end of the first lecture, I will pose the problem of quantizing the Ornstein - Uhlenbeck process in a certain definite sense. I will also mention a rather interesting related problem of stochastic quantization due to Nelson. Then in Lecture II, I will take up the quantum treatment of the Ornstein-Uhlenbeck process through the Frictional Hamiltonian formalism that incorporates the dissipative effects. I will specifically describe our recent work along these lines that reveals certain inconsistencies of the existing 'exact' treatments and seeks to cure these. While Lecture I is in the nature of an overview, Lecture II is intended as a seminar. For reviews, see refs. [2-6].

I.1 Diffusion

When we think of diffusion, what immediately comes to the mind is a dust of particles dispersed through and moving chaotically and perpetually in a medium, e.g., historically it was plant pollen in water, but it could be a colloidal suspension, or, for demonstration purposes, carmine in acetone. At the low space-time resolution of a naked human eye, however, one simply observes a material continuum with a coarse-grained number density $n(\vec{x}\ , t)$ varying slowly in space and time. Furthermore, it is expected that the natural tendency for the system is to run down in the sense that the spatial gradients of $n(\vec{x}, t)$ are equalized as the system evolves in time implying material transport. (That is assuming the system to be thermodynamically stable, chemically non-reacting and free from active transport). This transport current $\vec{j}(\vec{x},t)$ is described by the empirical Fick's Law

$$\vec{j}(\vec{x},t) = -D \vec{\nabla} n(\vec{x},t) \tag{I-1a}$$

for small gradients, defining the diffusion coefficient D. Since the material is locally conserved, one has the continuity equation

$$\frac{\partial n(\vec{x},t)}{\partial t} + \vec{\nabla}\cdot \vec{j}(\vec{x},t) = 0 \tag{I-1b}$$

Combining the two one gets the diffusion equation

$$\frac{\partial n(\vec{x},t)}{\partial t} = D \nabla^2 n(\vec{x},t) \tag{I-1c}$$

This phenomenological equation, depending on the nature of $n(\vec{x},t)$, forms the basis of the transport theory in numerous contexts of physics, chemistry, biology and astrophysics. I will just mention the historical example of Fourier's treatment of heat conduction in solids. To have a feel for numbers, the self-diffusion coefficient D for most liquids is $\sim 10^{-5}\ cm^2\ sec^{-1}$.

Mathematically, (I-1c) is a parabolic differential equation that can be solved for the well posed initial boundary values (i.e. Dirichlet or Neumann conditions and open boundaries) and describes an irreversible evolution into future such that any singularities are smoothed out in the process. The fundamental solution or the propagator $K^o_\infty(\vec{x},t, \vec{x}_o,t_o)$ for an unbounded d-dimensional system is given by

$$K^o_\infty(\vec{x},t;\vec{x}_o,t_o) = \left(\frac{1}{4\pi D(t-t_o)}\right)^{d/2} \exp\left(-|\vec{x}-\vec{x}_o|^2/4D(t-t_o)\right), \quad t \geqslant t_o . \tag{I-2a}$$

It corresponds to a delta-function singularity of unit weight concentrated at $\vec{x}_o$ at the initial epoch t_o. Thus, the solution for an arbitrary initial distribution $n_o(\vec{x}_o,t_o)$ is given by

$$n(\vec{x},t) = \int n_o(x_o,t_o)\, K^o_\infty(\vec{x},t;\vec{x}_o,t_o)\, dx_o , \quad t \geqslant t_o \tag{I-2b}$$

More generally, for an arbitrary spatial boundary ($\partial\Omega$) condition at t=o, the propagator $K^o_\Omega(\vec{x},t;\vec{x}_o,t_o = o)$ is given by

$$K^o_\Omega(\vec{x},t;\vec{x}_o,t_o = o) = \sum_{n=o}^{\infty} e^{-\lambda_n t} \phi_n(\vec{x})\, \phi_n(\vec{x}_o), \quad t \geqslant o , \tag{I-2c}$$

where ϕ_n is the nth eigenfunction associated with (I-1c) and λ_n the corresponding eigenvalue for the prescribed boundary condition. Now two generalizations :

a. Diffusion with advection :

The containing medium may itself be in a state of motion described by a velocity field $\vec{v}(\vec{x},t)$ and the diffusing material is assumed to co-move with it. Thus $\partial/\partial t$ in (I-1c) must be replaced by the substantial derivative giving

$$\frac{\partial n}{\partial t} = -\vec{\nabla}\cdot(n\vec{v}) + D\nabla^2 n \qquad \text{(I-3)}$$

A particularly interesting situation is the one in which the velocity-field $\vec{v}(\vec{x},t)$ is turbulent, rendering (I-3) a stochastic differential equation. For $\vec{v}(\vec{x},t)$ a stochastic process stationary to second order with whitenoise spectrum and corresponding to incompressible flow, one obtains enhanced renormalized, diffusion coefficient. Obvious relevance is to diffusion in a system undergoing pre-transitional hydrodynamical instabilities may be noted. This is why the cup of coffee is stirred!

b. Diffusion with drift in a force-field :

Here one assumes that an external force such as the one due to gravity produces a drift velocity analogous to Ohm's law (and not an acceleration). Then, for a force derivable from a potential $V(\vec{x})$

$$\vec{j}(\vec{x},t) = -D\vec{\nabla}n(x,t) - \mu(\vec{\nabla}V(\vec{x}))\,n(\vec{x},t)$$

$$\frac{\partial n(\vec{x},t)}{\partial t} = D\vec{\nabla}\cdot\left(\vec{\nabla}n(\vec{x},t) + \frac{\mu}{D}\,n(\vec{x},t)\,\vec{\nabla}V(\vec{x})\right) \qquad \text{(I-4)}$$

(Smoluchowski equation)

This can be reduced to the previous case by the substitution $n(x,t)=\exp(-\mu V/2D)\,\sigma(x,t)$ that eliminates the term involving $\vec{\nabla}_n \cdot \vec{\nabla}V$ and makes the equation for σ self-adjoint. One obtains for the propagator now

$$K^V_\Omega(\vec{x},t;\vec{x}_o,t_o) = \exp\left(\mu V(\vec{x}_o)/D\right)\sum_{n=o}^{\infty}\phi_n(\vec{x}_o)\,\phi_n(\vec{x})\,e^{-\lambda_n t} \qquad \text{(I-5)}$$

where ϕ_n is the nth eigenfunction of (I-4) and λ_n the corresponding eigenvalue. Note that the Einstein-Nernst relation $\mu/D = 1/k_BT$ obtains on setting $\partial n/\partial t=o$ in (I-4) for equilibrium.

Equation (I-1) - (I-5) describe the macroscopic coarse-grained behaviour observed at the low resolution of a naked eye - 'the harvest of a naked eye'. Now, should we observe the system under the high resolution of an ultramicroscope, as Robert Brown would, the material continuum will resolve into the individual particles in a state of perpetual and chaotic motion - called Brownian motion after the 19th century botanist. We shall not tarry here to look up the rather amusing history of the subject except to note that hardly a century ago this motion was attributed to some form of 'vital' force inherent in the particle, or to some equally obscure 'intestine' motion, and that when finally early this century Einstein, barely aware of Robert Brown's findings, gave the correct quantitative theory, it was only to adduce evidence for the reality of and the definite finite (non-zero) size for atoms and molecules! His theory was statistical rather than dynamical and that accounted for its success since it circumvented many difficult problems associated with the stochastic differential equations not heard of before Langevin.

I.2 Wiener Process [7,8] :

The correct kinetic theoretic explanation of the Brownian motion is, of course, now well known. The Brownian particle moves under the action of the colliding molecules of the medium which is in a state of thermal agitation - the heat 'bath'. Indeed, recalling that the typical size of a colloidal Brownian particle is $\sim 1\ \mu m$, and there are $\sim 10^{23}$ molecules cm^{-3} in the medium moving at thermal speeds $\sim 10^{5}$ cm sec^{-1}, the particle should suffer $\sim 10^{21}$ collisions per second. Given its large inertia $\sim 10^{-12}$ gm it will integrate a large number of small impulses before moving appreciably. Simple appeal to the law of large numbers should make the successive displacements independent, gaussian random variables. The basic ansatz is thus that of 'stosszahl ansatz' or the 'molecular chaos' which is ultimately related to the large number of bath degrees of freedom interacting weakly with the Brownian particle. Thus it is a pedestrian problem. It has to do with the random walk !

Some of the observed general features of the Brownian motion are :

(i) the particles move independently of one another

(ii) successive displacements separated by macroscopic time intervals are statistically uncorrelated

(iii) mean-squared displacement grows linearly with time $t \to \infty$

(iv) the motion is perpetual.

It should be clear from (i) that the normalized density $n(\vec{x},t)$ discussed above may be identified with the probability density of finding the particle at $\vec{x}$ at time t, assuming, of course, that the Brownian particles are identical. The question then is what the stochastic motion of the Brownian particle should be in order that the probability density evolves as in (I-1) through (I-5). To answer this question we argue with Einstein that there exists a time scale τ such that the displacements separated by time intervals $\gg \tau$ are statistically uncorrelated and yet τ is sufficiently large to ensure large number of collisions before the particle moves appreciably, i.e. $\tau_{macro} \gg \tau \gg \tau_{micro}$. This should certainly be so for large particle inertia, weak coupling to the bath and high temperatures. And now we boldly consider the idealized limit $\tau = o$, This leads us to the following stochastic motion [7] :

$$d\,\vec{X}(t) = d\,\vec{W}(t) \quad , \tag{I-6}$$

with $d\,\vec{W}(t)$'s independent, gaussian, white noise, random variables with mean $\langle d\,W(t)\rangle = o$ and covariance $\langle d\,W_i(t)\ d\,W_j(t)\rangle = 2dD\delta_{ij}dt$, and $d\,\vec{W}(t)$ independent of $\vec{X}(s)$ for $s \leqslant t$ (the Markoff assumption). Such a position Markoff process is called the Wiener process. The Wiener processes constitutes a classical paradigm in the sense of Kuhn and forms the kernel of our thinking about all stochastic processes related to diffusion and time dependent statistical mechanics. Its relation to diffusion is rendered more compelling by the following remarkable theorem. Let $p_t(\vec{x})$ be the probability density for the state space $\vec{x}$. Then, given

(a) the Markoffian property (Chapman - Kolmogoroff condition)

$$p_t \otimes p_s = p_{t+s} \quad , \quad o \leqslant t, \ s \leqslant \infty \quad , \tag{I-7}$$

(b) the persistence - at - a - point, i.e. for $\eta > o$

$$\lim_{t \to o} p_t(\{\vec{x} \mid |\vec{x}| \geqslant \eta\}) = O(t) \tag{I-8}$$

and

(c) the symmetry $p_t(\vec{x}) = p_t(-\vec{x})$, for $t > o$, there exists a D such that

$$p_t(\vec{x}) = \left(\frac{1}{4\pi D t}\right)^{d/2} \exp\left(-\vec{x}^2/4Dt\right) \quad , \tag{I-9}$$

and

$$\frac{\partial p_t(\vec{x})}{\partial t} = D \nabla^2 p_t(\vec{x}) \quad , \quad t \geqslant o \quad .$$

Probably the most important property of a Wiener process is that the trajectories are continuous but almost nowhere differentiable - the velocities are not defined. In fact the trajectories may not be rectifiable. In modern parlance, these are fractals.

The Wiener process also induces a Wiener measure in the sample space of paths. The latter may be obtained as an n-fold convolution of $p(t,\vec{x})$ in the limit $n \to \infty$:

$$\lim_{\substack{N \to \infty \\ \Delta t \to o \\ N\Delta t = t}} \left(\frac{1}{4\pi D \Delta t}\right)^{\frac{dN}{2}} \exp\left\{-\frac{\Delta t}{4D} \sum_{n=o}^{N} \frac{(\vec{x}_{n+1} - \vec{x}_n)^2}{(\Delta t)^2} \, d\vec{x}_1 \ldots\ldots d\vec{x}_N\right\}$$

$$\longrightarrow e^{-\frac{1}{4D} \int_{\vec{x}(o)}^{\vec{x}(t)} \left(\frac{d\vec{x}}{dt}\right)^2 dt} \mathcal{D}[\vec{x}(t)] \tag{I-10}$$

where $\mathcal{D}[\vec{x}(t)]$ denotes the element of volume in the function space, and we have a functional integration. The connection with the path integral formulation of quantum mechanics is obvious - after all diffusion equation is essentially the Schrödinger equation with imaginary time. This indeed forms the basis of the 'Feynman-Kac' formula that enables one to evaluate the average of the path-dependent quantities [8] :

$$\left\langle \exp\left(-\int_o^t V(\vec{x}(t'))\, dt'\right)\right\rangle = \int K_\infty^V (\vec{x},t;o,o)\, dx \tag{I-11}$$

where K_∞^V is the propagator for the diffusion equation in a potential, i.e. for

$$\frac{\partial n}{\partial t} = D \nabla^2 n + V(\vec{x})\, n \tag{I-12}$$

A rather interesting example would be to evaluate [9]

$$\langle \exp(-\nu \mid C_t^a(\vec{x}(.))\mid)\rangle , \quad \text{where} \quad C_t^a(\vec{x}(.))$$

is the volume of the Wiener Sausage, i.e.

$$C_t^a(\vec{x}(.)) = \cup\, S(\vec{y},a) \quad , \quad \vec{y} \in C_t(\vec{x}(.)) \; , \qquad \text{(I-13)}$$

where $S(\vec{y},a)$ is a sphere of radius a centered at point $y \in$ the image set $C_t(x(.))$ of the path $\vec{x}(s)$, $0 \leqslant s \leqslant t$, starting at $\vec{x}(o) = 0$. These spheres may overlap and that is the problem. It turns out that the above can be evaluated to give the remarkable result [9]

$$\lim_{r\to\infty} \frac{1}{t^{d/d+2}} \ln \langle \exp(-\nu \mid C_t^a(x(.))\mid)\rangle = -\left(\frac{d+2}{2}\right)\left(\frac{2\gamma_d}{d}\right)^{\frac{d}{d+2}} \nu^{\frac{2}{d+2}} \qquad \text{(I-14)}$$

where γ_d is the lowest eigenvalue of $-\frac{1}{2}\nabla_d^2$, with the solution vanishing at the surface of a unit sphere. Physically, the quantity evaluated above may give the survival probability of a particle diffusing in a medium with a uniform distribution of a reactant, a being the 'lethal radius' or the radius of the sphere of influence. Relevance to the Glarum defect - diffusion - relaxation model is obvious.

I.3 Phase-space Markoff - Process and Ornstein-Uhlenbeck Brownian Motion [1,7] :

Brownian motion as described by the Wiener process is a position Markoff process that has a purely stochastic kinematics without any dynamical content. The corresponding diffusion equation is nothing but a rate equation (a master equation). The process is unphysical in that the paths are almost always non-differentiable and hence the velocity undefined. In particular $\langle x^2(t)\rangle \propto t$ on all time scales. For a physical system, the motion must be free (ballistic) and thus $\langle x^2(t)\rangle \propto t^2$ on a sufficiently short time scales. This would make the path differentiable and the velocity well defined. This is demanded by the equipartition law too. Furthermore, the Newtonian dynamics $\vec{F} = m\,\vec{a}$ must be built into it. Thus, one is tempted to replace (I-6) by a phase-space Markoff process

$$d\,\vec{x}(t) = \vec{v}(t)\,dt$$

$$d\,\vec{v}(t) = -\frac{1}{M}\vec{\nabla}\, V\, dt + d\,\vec{B}(t) \qquad \text{(I-15)}$$

with $\vec{B}(t)$ a Wiener process and $d\,\vec{B}(t)$ independent of $\vec{x}(s)$ and $\vec{v}(s)$ for $s \leqslant t$. This is, however, not quite all right yet. It supposes the stochastic process characterizing the medium (the 'bath') given independently of the test particle, and thus neglects the reaction of the test particle back on the medium. The medium acts but cannot be acted upon ! As can be shows quite trivially (see Lecture II), this would lead to an indefinite acceleration of the particle. Such a stochastic acceleration is, of course, possible in some cases The Fermi mechanism of stochastic acceleration of cosmic ray particles by 'collisions' with the inhomogeneous magnetic fields frozen - into and co-moving with the material clouds moving at random, is an example in point. Here the particle is hardly expected to affect the motion of these clouds. In general, however the particle will act back on the medium and this reaction will be fed back to the particle with some delay. This would imply a memory for the process B(t) making it non-Markoffian and very hard to calculate. However, for a Brownian particle of large inertia diffusing through a dense medium of lighter (mass m) particles, the situation is not quite so bad. For instance, there must be a Doppler Friction inasmuch as there are more collisions head on (a priori !) than from behind (a posteriori !). The reaction may be simulated by an instantaneous dynamical viscous drag $-\gamma\vec{v}$, treating $d\vec{B}(t)$ as the <u>residual</u> 'bath' force. Thus one gets [7]

$$\begin{aligned} d\,\vec{x}(t) &= \vec{v}(t)\,dt \\ d\,\vec{v}(t) &= -\gamma\vec{v}(t)\,dt - \frac{1}{M}\,\vec{\nabla}\,V + d\,\vec{B}(t)\ , \end{aligned} \tag{I-16}$$

where the Wiener process $\vec{B}(t)$ has now $\langle dB(t)^2\rangle = 2d(\gamma k_B T/M)\,dt$ to ensure eventually (as $t \to \infty$) equipartition law (thermalization). This is the stochastic Langevin equation describing the Ornstein-Uhlenbeck process. It gives $\langle x^2(t)\rangle \propto 2dDt$ for $t \gg \gamma^{-1}$ but $\langle x^2(t)\rangle \propto t^2$ for $t \ll \gamma^{-1}$. Here $D = (k_B T)/m\gamma$. Modern theory of Brownian motion is based on this. It is readily seen, that in the limit of large friction γ , one recovers the position Markoff process discussed earlier.

It must be re-emphasized that when the inertia of the Brownian particle (M) is comparable to that of the bath particles (m) (e.g. in the problem of self-diffusion) the memory effects are important. In fact the test particle sets up a vortex - like back flow so that the momentum imparted to the medium head on is returned to the test particle from the rear in the 'antiperistaltic' fashion. This, for instance, makes the velocity-velocity correlation function $\langle v(o)\ v(t)\rangle$ start out analytically with zero slope at t=o and fall off asymptotically slower than the exponential - indeed algebraically as $t^{-d/2}$. Such long Non-Debye tails have been seen in molecular dynamical calculations [10] . The Ornstein-Uhlenbeck process on the other hand predicts an exponential correlation $e^{-\gamma|t|}$ which is non-analytic at t=o. There is in fact an inconsistency [5] due to the assumption that dB(t) is independent of $\vec{x}(s)$ and $\vec{v}(s)$ for $s \leqslant t$, which must break down as $s \to t$. In order to incorporate the Non-Markoffian (memory) effects, one must consider a fully dynamical treatment such as the one due to Zwanzig [11] and Mori [12] . One can recover [13] the Markoffian Langevin equation perturbatively as $(m/M) \to 0$. The conditions under which the system can act as its own bath were examined by Ford, Kac and Mazur (FKM) [14] . Non-Markoffian effects are particularly important for thermally activated escape over barriers [15] .

The Ornstein-Uhlenbeck process will form the basis for our quantum treatment in Lecture II. By quantization what is implied is treating the deterministic part of the Hamiltonian including frictional term quantum mechanically. The stochastic driving process is still to be treated classically.

A technical remark now. The various stochastic processes discussed above and the corresponding equations for the probability densities $p(\vec{x},t)$ generated by them are related very generally as follows [6] . For a system of stochastic differential equations

$$\dot{x}_\nu = f_\nu (\vec{x}, t; \omega), \quad \nu = 1,2,\ldots\ldots.n, \tag{I-17}$$

with ω the random variable or function, one can write down the Liouville equation for the phase-space density

$$\frac{\partial \rho}{\partial t} = -\sum_\nu \frac{\partial}{\partial x_\nu} (\rho \, f_\nu) \tag{I-18}$$

Then it is readily shown that

$$p(\vec{x},t) = \langle \rho(\vec{x},t) \rangle_\omega \tag{I-19}$$

While performing ensemble averages $\langle \cdot\,\cdot\,\cdot\,\cdot\,\cdot \rangle_\omega$ of functionals $F[\omega(t)]$ over the gaussian random variables $\omega(t)$, the following identity due to Novikov [16] comes in very handy

$$\langle \omega_i(t) \, F[\vec{\omega}(t)] \rangle_\omega = \sum_j \int dt' \, \langle \omega_i(t) \, \omega_j(t') \rangle_\omega \cdot \cdot \left\langle \frac{\delta F[\vec{\omega}(t)]}{\delta \omega_j(t') \, \delta t'} \right\rangle_\omega \tag{I-20}$$

When the random variable is delta-correlated, i.e. it has white noise,it enables one to close the hierarchy of coupled equations. Thus (I-18) leads to the associated Fokker-Planck equation in general.

And finally a brief excursion into stochastic quantization of Nelson [7] . The original physical motivation namely that of providing a 'hidden variable' classical substratum to quantum mechanics is of no concern to us here. We are only concerned with the mathematical aspect. The essential point is that for all measurements reducible to position measurements, any state of the system described by the Schrödinger equation

$$i \hbar \frac{\partial \phi}{\partial t} = -\frac{\hbar^2}{2M} \nabla^2 \phi + V(\vec{x},t)\phi \tag{I-21}$$

is equivalent to a position Markoff process

$$d\,\vec{x}(t) = -\vec{b}\,(\vec{x}(t), t)\, dt + d\,\vec{W}(t) \quad , \tag{I-22}$$

with $\vec{W}(t)$ a Wiener process ($\vec{W}(t) - \vec{W}(s)$ independent of $\vec{W}(r)$ for $r \leqslant s \leqslant t$) having a diffusion coefficient $= \hbar/2m$. The drift (forward) velocity $\vec{b}\,(\vec{x}(t), t)$ is given by $\vec{b} = \vec{v}$ (current velocity) $+ \vec{u}$ (osmotic velocity), where

$$\frac{\partial \vec{u}}{\partial t} = -\frac{\hbar}{2M} \vec{\nabla} (\vec{\nabla} \cdot \vec{v}) - \vec{\nabla}(\vec{u} \cdot \vec{v})$$

$$\frac{\partial \vec{v}}{\partial t} = -\frac{1}{M} \vec{\nabla} V - (\vec{v}.\vec{\nabla}) \vec{v} + (\vec{u} .\vec{\nabla}) + \frac{h}{2M} \vec{\nabla}^2 \vec{u} \qquad \text{(I-23)}$$

Thus the velocity-field equations are to be solved subject to the given initial conditions (Cauchy data) $\vec{u}_o(x)$ and $\vec{v}_o(x)$ that can be related to the given initial state through

$$\phi = e^{R+iS} , \quad \frac{m\vec{v}}{\hbar} = \vec{\nabla} S , \quad \frac{m}{\hbar} \vec{u} = \vec{\nabla} R \qquad \text{(I-24)}$$

The probability density $\rho = | \phi |^2$ satisfies the (forward) Fokker-Planck equation

$$\frac{\partial \rho}{\partial t} = - \vec{\nabla}.(\rho \vec{b}) + \nu \nabla^2 \rho , \quad \nu = \frac{\hbar}{2M} \qquad \text{(I-25)}$$

generated by (I-22).

In this description one combines the kinematics of Wiener process with the dynamics (force-acceleration relation) of the Ornstein-Uhlenbeck process sans dissipation. All I want to emphasize now is that for $V(\vec{x},t)$ a random variable, $\vec{b}(\vec{x}(t), t)$ will be a stochastic process and hence equation (I-25) should lead to a renormalized diffusion constant as discussed earlier [3] . Thus the problem of a quantum particle moving in a time-random potential reduces to that of a classical diffusion in a turbulent medium. In point of fact stochastic quantization may enable one to treat at once intrinsic quantum mechanical probability along with the classical randomness (of thermal origin).

We will end with a general remark. Ultimately, diffusion refers to time dependent randomness, i.e. randomness may be taken to reside in the particle itself (like the Mexican jumping beans). This must be distinguished from the case where the randomness lies in the medium (spatial). The latter is a much more difficult (percolation) problem and can lead to localization. Ultimately, this difficulty may be attributed to our inability to define meaningfully a Markoff process for multi-dimensional 'time'.

II. Quantum Diffusion

The Ornstein-Uhlenbeck process is perhaps the simplest stochastic process that is a physically admissible idealization and has a non-trivial domain of validity. It has intuitive appeal and forms the basis of modern theory of Brownian motion. It approximates the time-dependent statistical mechanics of a many-body system where one identifies the few, relatively slow hydrodynamical degrees of freedom of interest, widely separated on time-scale from the almostly infinitely many, fast microscopic degrees of freedom, the details of which are of no interest to us, and that collectively form the thermal 'bath'. It thus provides a generalized hydrodynamics in which the hydrodynomical degrees of freedom with long auto-correlation times obey macroscopic equations of motion driven parametrically by the 'noise' terms having short auto-correlation time, representing the influence of the 'bath' forces, whose statistical properties are assumed simple and given. The reaction of the hydrodynamical degrees of freedom back on the bath is thus ignored except in some average sense, e.g., through the dynamical friction ($-\beta\vec{v}$) term in the Langevin equation. I must hasten to add that while such a 'back action evasion' may be justified by intuitive appeal to large inertia, weak coupling to bath and to the large number of 'bath' degrees of freedom, the intuition is at times belied by explicit microscopic calculations possible for some simple nonlinear systems.

With these remarks, we now turn to the problem of quantization of the 'Brownian motion a la Langevin'. We have in mind a particle (electron, say) weakly coupled to the 'bath' (thermal phonons). At not too low a temperature, the auto-correlation time for the phonon displacement field is sufficiently short compared to the electron velocity auto-correlation time to justify the assumptions of Langevin. By quantization here we mean treating the electron motion quantum mechanically, but treating the bath force still as a given stochastic process. It is tempting therefore to consider as the quantum analogue a particle moving in a time-random potential, as indeed has been done by several workers [17-21]. The existing theoretical treatments are based almost entirely on the lattice (L) Hamiltonian, namely the tight-binding one-band Hamiltonian [19]

$$H_L = \sum_i E_i(t) \mid i \rangle\langle i \mid + \sum_{i\neq j} V_{ij}(t) \mid i \rangle\langle j \mid \quad , \tag{II-1}$$

in obvious notation. The dynamical disorder is introduced by treating the potentials, i.e., the site-diagonal and the off-diagonal matrix elements, as random c-number variables, evolving stochastically in time. Such a time dependence is known to arise from the random modulation of the potential by the incoherent lattice vibrations of thermal origin. It must be emphasized here that in all the treatments referred to above, as also in the treatment to follow, this time dependence is taken to be parametric in that the potential is supposed to introduce no additional dynamical degrees of freedom in the problem. For the Gaussian choice of randomness having a white-noise spectrum, i.e., δ-correlated in time but arbitrarily correlated in space, the problem has been solved exactly by several workers [17-21]. In all cases one obtains a classical diffusive behavior in that the mean square displacement $\langle x^2(t)\rangle \sim t$, for $t\to\infty$, implying a well defined diffusion constant and hence mobility. This common result, however plausible and expected from the physical point of view, is surprising when analyzed more carefully. Indeed, as the following exact treatment [22] reveals, for the corresponding continuum problem we have $\langle x^2(t)\rangle \sim t^3$ asymptotically, implying nondiffusive motion. Thus, the diffusive behavior obtained by the other workers is due presumably to the specific nature of the one-band lattice Hamiltonian.

In order to appreciate this point fully it is expedient to consider first the related problem of classical diffusion a la Langevin equation in a spatial one-dimensional continuum :

$$m \; du/dt = - \Gamma u + f(t) \; , \qquad \text{(II-2)}$$

where the fluctuating random force f(t) and the concomitant dissipation represented by the frictional coefficient are related by the fluctuation-dissipation theorem, i.e.,

$$\langle f(t) \; f(t') \rangle = 2k_B T \Gamma \; \delta(t-t') \equiv \Delta^2 \; \delta(t-t') \; .$$

As is well known this gives a mean square displacement $\langle x^2(t) \rangle \sim 2Dt$, for $t \gg m\Gamma^{-1}$, defining the diffusion constant $D = k_B T/\Gamma$. If, however, we omit the dissipative term ($-\Gamma u$) from Eq.(II-2), i.e., we set

$$m \; du/dt = f(t) \quad , \qquad \text{(II-3)}$$

we can readily show that

$$\langle x^2(t) \rangle \sim (\Delta^2/4m^2)t^3, \quad \text{for } t \to \infty \quad . \qquad \text{(II-4)}$$

This implies a nondiffusive random motion. Here the particle continues to absorb energy from the fluctuating force and accelerates indefinitely. In short the particle 'heats up' to an infinite temperature. Now, the quantum mechanical treatment based on the Hamiltonian H_L in (II-1) corresponds precisely to this nondissipative classical system in that no dissipation is incorporated explicitly in H_L. And yet the mean - square displacement calculated exactly from (II-1) shows a diffusive behavior as noted above. In the following, we address ourselves to this paradoxical situation. More specifically, we will first show that the exactly solvable continuum analog of (II-1) also reproduces t^3 behavior as in (II-4). We then argue that the diffusive behavior obtained by other workers referred to above is entirely due to the one-band lattice nature of the Hamiltonian given in (II-1). And finally we will present a proper treatment of quantum diffusion including friction.

To this end we will now obtain an exact solution of the quantum problem in a continuum. For simplicity we shall treat the case of one space dimension. Generalization to arbitrary dimension is straightforward. The quantum evolution is now given by the time-dependent Schrodinger equation,

$$i\hbar \frac{\partial \psi}{\partial t} = - \frac{\hbar^2}{2M} \frac{\partial^2 \psi}{\partial x^2} + V(\vec{x},t)\psi \quad , \qquad \text{(II-5)}$$

where $V(\vec{x},t)$ is the stochastic potential assumed to be Gaussian, with space-time correlation

$$\langle V(x,t) \; V(x',t') \rangle = V_o^2 \; \delta(t-t') \quad g(x-x') \qquad \text{(II-6)}$$

The physical quantities of interest can be conveniently expressed in terms of the reduced density matrix $\rho(x',x,t)$ where

$$\rho(x',x,t) \equiv \psi^*(x',t) \; \psi(x,t) \quad , \qquad \text{(II-7)}$$

and the angular brackets denote the average over the stochastic potential. Clearly

$\rho(x', x,t)$ is a functional of the Gaussian random variable $V(x,t)$ and hence the Novikov theorem applies. Following essentially the earlier treatments, we get the equation of motion

$$\frac{\partial}{\partial t}\langle \rho(x',x,t)\rangle = -\frac{i\hbar}{2M}\left(\frac{\partial^2}{\partial x^2} - \frac{\partial^2}{\partial x'^2}\right)\langle \rho(x', x,t)\rangle - \frac{V_o^2}{\hbar^2}\left[g(0) - g(x-x')\right]\langle \rho(x,x',t)\rangle \qquad \text{(II-8)}$$

This has to be solved subject to the initial condition that the particle was prepared initially in a wave packet centered at the origin, x=o. We shall take conveniently

$$\rho(\vec{x}', x,t=o) = \psi^*(x',t=o)\ \psi(x,t=o),$$

where

$$\psi(x,t=o) = \left((2\pi)^{\frac{1}{4}}\ \sigma^{\frac{1}{2}}\right)^{-1} \exp\left(-x^2/4\sigma^2\right) \qquad \text{(II-9)}$$

This ensures correct normalization. Here σ denotes the spatial spread of the initial wave packet. Because of the unbounded nature of the kinetic energy operator in the continuum limit, it is necessary to choose a wave packet with $\sigma > 0$. The asymptotic ($t \to \infty$) behavior is, of course, independent of the precise form of the wave packet. This problem does not arise in the case of the lattice Hamiltonian H_L which is bounded. Equation (II-8) can be solved by first taking the time Laplace transform and then considering the resulting hyperbolic equation in the two independent variables x and x'. We get

$$\frac{2i\hbar}{M}\frac{\partial^2}{\partial x\,\partial y}\tilde{R}(X,Y,s) + \left(s + \frac{V_o^2}{\hbar^2}g(o) - \frac{V_o^2}{\hbar^2}g(y)\right)\tilde{R}(X,Y,s) = R(X,Y,t=o), \qquad \text{(II-10)}$$

where we have introduced the characteristic coordinates $X = x + x'$, $Y = x - x'$. Here s is the Laplace transform variable. We have defined

$$R(X,Y,t) \equiv \rho(x',x,t)\ ;\ \tilde{R}(X,Y,s) = \int_0^\infty R(X,Y,t)\ e^{-st}\,dt \qquad \text{(II-11)}$$

This mean square displacement can be expressed as

$$\langle x^2(t)\rangle = \frac{1}{8}\ \frac{\partial^2}{\partial K^2}\ \bar{R}(K,y=o,t)\ \Big|_{K=o}\ , \qquad \text{(II-12)}$$

with

$$\bar{R}(K,y,t) = \int_{-\infty}^{\infty} R(X,Y,t)\ e^{iKx}\,dx\ . \qquad \text{(II-13)}$$

Here an overbar denotes the spatial Fourier transform while a tilde denotes the time Laplace transform. Equation (II-12) holds provided $\bar{R}(K,Y=0,t)$ is analytic in K at K=0.

Equation (II-10) can be converted into an ordinary differential equation in Y by taking Fourier transform with respect to X, which can then be solved readily subject to the initial condition to give

$$\tilde{\bar{R}}\ (K,Y=0,s) = \int_0^{\infty} 2 \exp\left[-\left(2\sigma^2 + \frac{\hbar^2 y^2}{2M^2\sigma^2}\right) K^2\right] e^{-sy}$$

$$\exp\left\{-\frac{V_0^2}{\hbar^2}\left[g(0)\,y - \int_0^y g\left(\frac{2\hbar\,|K|\,'}{M}\right) dy'\right]\right\} dy \qquad \text{(II-14)}$$

The right-hand side of this equation is already in the form of a Laplace transform. Hence we get at once on inversion

$$\bar{R}\ (K,Y=0,t) = 2 \exp\left[-\left(2\sigma^2 + \frac{\hbar^2 t^2}{2M^2\sigma^2}\right) K^2\right]$$

$$\exp\left\{-\frac{V_0^2}{\hbar^2}\left[g(0)\,t - \int_0^t g\left(\frac{2\hbar\,|K|\,y'}{M}\right) dy'\right]\right\} \qquad \text{(II-15)}$$

One can confirm that (II-15) fulfills the normalization and the initial condition. We now choose an explicit form for the function $g(Y)$. For simplicity we take it to be Gaussian, i.e.,

$$g(Y) = \left[(2\pi)^{\frac{1}{2}}\alpha\right]^{-1} \exp\left(-Y^2/2\alpha^2\right) \qquad \text{(II-16)}$$

With this choice, $\vec{R}(K,Y=0,t)$ can be seen to be analytic in K at K=0. Thus from (II-12), (II-15) we get for the mean-square displacement

$$\langle x^2(t)\rangle = \sigma^2 + \frac{\hbar^2}{4M^2\sigma^2}\,t^2 + \frac{1}{3\sqrt{2\pi}}\,\frac{V_0^2}{M^2\alpha^3}\,t^3 \qquad \text{(II-17)}$$

This is an exact result. It shows clearly that the particle motion is nondiffusive on any time scale. In fact, the above result is quite general and depends only on the fact that $g(Y)$ is an even function of Y and is analytic in Y at Y=0. The special case $g(Y) \sim e^{-\alpha|y|}$ which is not analytic at Y=0 calls for a somewhat more detailed evaluation by quadrature. Thus, we confirm that the quantum motion in a fluctuating continuum gives nondiffusive motion. The result is essentially identical to that for the classical motion in a fluctuating medium as in (II-3) and (II-4). In point of fact one may choose $V(x,t) = xf(t)$ such that the random force obtained as the gradient of potential $V(x,t)$ is actually $f(t)$ as in (II-3). One can readily verify that the asymptotic time behavior remains cubic as obtained above.

The fact that the exact quantum treatment on the lattice gives diffusive behavior has, therefore, to do with the specific nature of the lattice Hamiltonian H_L. The question is how to understand this difference. The point is that a one-band lattice Hamiltonian has a momentum cutoff inherent in it. This limiting momentum is related to the Bragg reflection at the Brillouin zone boundary or, what is essentially the same, one has the Umklapp process. The lattice acts as an infinite momentum sink and prevents indefinite acceleration of the particle. More transparently, as the particle quasimomentum increases towards the limiting value, the group velocity decreases and even reverses sign. Since it is the group velocity that leads to physical displacement the above results are understandable.

In order to see more clearly how such a limiting momentum can lead to diffusive motion, it is very revealing to consider again the classical motion in a fluctuating medium described by (II-3), with the proviso that the physical velocity u be defined modulo some limiting velocity u_o, say. This mathematically simulates the umklapp process. For instance, we could redefine physical velocity as

$$v = u_o \sin (2\pi\, u/u_o) \tag{II-18}$$

and calculate the mean square displacement, with v as the physical velocity. We get

$$\langle x^2(t)\rangle = \langle \{ \int_0^t u_o \sin [\, 2\pi \int_0^{t'} f(t'')\, dt''] \, dt' \}^2 \rangle \ ,$$

$$= u_o^2\, a\, (t + 2a\, e^{-t/a} - \frac{1}{2}\, a\, e^{-2t/a} - \frac{3}{2}\, a) \ , \tag{II-19}$$

where $a = u_o^2 / 2\pi^2 \Delta^2 M^2$. It is clear that

$$\langle x^2 (t)\rangle \sim u_o^2\, a\, t \qquad \text{as } t \to \infty \ , \tag{II-20}$$

which is again diffusive. In deriving this we have used the well-known result $\langle \exp [\, i \int_0^t f(t')dt'] \rangle = \exp [\, - (\frac{1}{2}\, \Delta^2) t]$. It seems clear, therefore, that the quantum treatment based on the lattice Hamiltonian is <u>per se</u> not a quantum analog of the Brownian motion. The diffusive behavior with H_L results entirely from the momentum absorption by the lattice via Bragg reflections. The latter is absent in the case of the continuum, and hence the nondiffusive behavior.

Finally, we must clarify that for a quantum particle in a real fluctuating continuum, we do expect a diffusive behavior. Here the effect of the interaction of the test particle with the dynamical degrees of freedom of the background fluctuating medium, however, cannot be represented entirely by a stochastic potential V(x,t) having a parametric time dependence. We must necessarily incorporate the analog of a dissipative term as well. In the following, quantum diffusion in dynamically disordered medium is reformulated treating explicitly the friction concomitant with the randomly fluctuating potential. The correct diffusive behaviour, namely $\langle x^2(t)\rangle \sim t$ is obtained both in the lattice as well in the continuum limit. This cures the unphysical $\langle x^2(t)\rangle \sim t^3$ behaviour inherent in the existing treatments, resulting from the neglect of friction as noted above. I will take for simplicity the spatial dimensionality to be unity, but as can been seen readily from the treatment, there is no loss of generality on this count.

As the neglect of frictional effects shows up rather dramatically in the continuum case, we shall consider this case first. Since quantum mechanics subsumes a classical Lagrangian, we are led to considering the classical limit first. Here the situation is familiar. The motion of the particle acted upon by the stochastic bath forces of thermal orgin is known to be an Ornstein-Uhlenbeck process described well by the Langevin equation

$$\frac{d^2x}{dt^2} = -\gamma\frac{dx}{dt} - \frac{1}{M}\frac{\partial V}{\partial X} + \frac{1}{M} f(t) \quad , \tag{II-21a}$$

with

$$\langle f(t)\, f(t')\rangle = 2\gamma M\, k_B T\, f_0^2\, \delta(t-t') \tag{II-21b}$$

Equation (II-21b) related explicitly the friction coefficient γ to the random force f(t), assumed to be Gaussian and delta-correlated. Here V(x) is a given static potential. The dissipative Lagrangian L_D associated with (II-21) is now readily constructed following the approach due originally to Helmholtz [23] . For example, if

$$G = \ddot{q} + g\,(q,\dot{q},t) = 0 \tag{II-22}$$

is the given equation, then there exists an 'integrating factor' $f(q,\dot{q},t)$ and an effective Lagrangian $L^*(q,\dot{q},t)$ such that

$$f\,G = \frac{d}{dt}\frac{\partial L^*}{\partial \dot{q}} - \frac{\partial L^*}{\partial q} , \tag{II-23}$$

where f is given by the solution of

$$g\,\frac{\partial\, \ell n f}{\partial \dot{q}} + \frac{\partial g}{\partial \dot{q}} = \frac{\partial \ell n f}{\partial q}\,\dot{q} + \frac{\partial f}{\partial t} \tag{II-24}$$

Thus one can solve for L^*. Since fG = 0 implies G = 0 (for $f \neq 0$) L^* is the Lagrangian sought for. In the present case one readily obtains

$$L_D^* = e^{\gamma t}\left[\frac{1}{2}M\dot{x}^2 - V(x) + x f(t)\right] \tag{II-25}$$

Defining the canonical momentum $p = \partial L_D^* / \partial \dot{x} = m\dot{x}\, e^{\gamma t}$, one gets the associated Hamiltonian

$$H_D^* = \dot{x}\,p - L = e^{-\gamma t}\frac{1}{2M}p^2 - e^{\gamma t} f(t)\,x + e^{\gamma t} V(x) \tag{II-26}$$

This is the well known frictional Hamiltonian and has been used extensively in other contexts, e.g. frictional effects in nuclear reactions [23-25] . In the following we will drop superscript star.

Now, for the quantum treatment we can proceed via the path-integral formulation using the Lagrangian L_D (as we will do in the continuum case), or via the canonical quantization of H_D (as we will do for the lattice case). Thus, we will be illustrating both approaches without being repetitive.

In order to calculate the mean-squared displacement $\langle x^2(t)\rangle$, let the particle be prepared initially in a gaussian wave-packet centered at origin i.e.,

$$\psi(x,t{=}o) = \left(\frac{1}{\sqrt{2\pi}\,\sigma}\right)^{\frac{1}{2}} \exp\left(-\, x^2/4\sigma^2\right) \tag{II.27}$$

We then have

$$\langle x^2(t)\rangle = \left\langle \int_{-\infty}^{\infty} x^2\, \psi^*(x,t)\, \psi(x,t)\, dx \right\rangle \quad , \tag{II-28}$$

where the time developed $\psi(x,t)$ can be related to the initial state $\psi(x,t{=}o)$ by the propagator $K_D(x,t', x',t')$ as

$$\psi(x,t) = \int_{-\infty}^{\infty} K_D(x,t, x',o)\, \psi(x',o)\, dx' \qquad t \geqslant o \tag{II-29}$$

Here the angular bracket denotes average over f(t).

Finding $K_D(k,t;\ x',t')$ corresponding to L_D (with $V(x)=0$ for the present case) is highly non-trivial because of the exponential $(e^{-\gamma t})$ prefactor that complicates the associated measure, or the normalization. Fortunately, this problem has been solved for the arbitrary bilinear Lagrangian by Papadopoulos [26] . Specializing his results to the present case, we get after some straightforward quadrature

$$\langle x^2(t)\rangle = \sigma^2 + \frac{\hbar^2}{M^2\sigma^2\gamma^2} \sinh^2\left(\frac{\gamma t}{2}\right) e^{-\gamma t} +$$

$$+ \frac{2k_BT}{M} e^{-\gamma t}\left[t\, e^{\gamma t} + e^{-\gamma t}\left(e^{2\gamma t} - 1\right)\frac{1}{2\gamma} - \frac{2}{\gamma}\left(e^{\gamma t} - 1\right)\right]$$

$$\simeq \left(\frac{2k_BT}{M\gamma}\right) t \ , \qquad \text{for} \qquad t \gg \frac{1}{\gamma} \ . \tag{II-30}$$

This implies diffusive behaviour. One readily verifies, that neglecting friction,i.e., setting $\gamma \simeq 0$, recovers the unphysical behaviour

$$\langle x^2(t)\rangle \sim t^3 \ , \quad \text{for} \quad \frac{M\sigma^2}{\hbar} \ll t \ll \frac{1}{\gamma} \tag{II-31}$$

Identical results are obtained if we calculated using the frictional Hamiltonian H_D instead.

We will now consider the case of the lattice Hamiltonian in the one-band tight-binding limit, considered in most existing treatments sans friction. Taking cue from the form of H_D in (II-26), one is tempted to consider a lattice Hamiltonian with the off-diagonal (kinetic) matrix elements having the factor $e^{-\gamma t}$, and the diagonal (potential) matrix elements having the factor $e^{+\gamma t}$, apart from being random. As we shall presently see this turns out to be incorrect.

In the case of the lattice, the potential $V(x)=V(x+a)$, where 'a' is the period. The quantum Hamiltonian H_D must be re-written in the representation in which the basis functions are the complete orthonormal set of Wannier functions $\chi_m^\alpha(x)$ corresponding to the non-random Hamiltonian $\hat{p}^2/2m + V(x)$:

$$\sum_{\alpha,m} \chi_m^{\alpha *}(x)\ \chi_m^{\alpha}(x') = \delta(x-x') \ ,$$

$$\int_{-\infty}^{\infty} \chi_m^{\alpha *}(x)\ \chi_n^{\beta}(x)\, dx = \delta_{\alpha\beta}\delta_{mn} \ , \tag{II-32}$$

where α is the band index and m labels the sites. Now, to project out the one-band Hamiltonian one has only to retain the matrix elements diagonal in the band index. We then have

$$H_{DL} = \sum_{m,\Delta} e^{\gamma t} V_o \, | m \rangle\langle m + \Delta | + f(t)\, a\, e^{\gamma t} \sum m \, | m \rangle \langle m |, \quad \text{(II-33)}$$

where Δ spans nearest neighbours. Here we have kept only the dominant matrix elements, e.g. $e^{\gamma t} + e^{-\gamma t} \simeq e^{\gamma t}$ since we will be interested in the limit $\gamma t \gg 1$. I must hasten to add that none of these simplifying approximations are essential to what follows. H_{DL} is the simplest lattice Hamiltonian incorporating dissipation. Both diagonal as well as off-diagonal terms carry the factor $e^{\gamma t}$. It is not at all obvious that the exponentially growing non-random off-diagonal terms will be off-set by the exponentially growing random diagonal terms to give a diffusive behaviour for long times. But this is precisely what happens as will be shown in the following.

We will now calculate $\langle x^2(t) \rangle$. Let the particle be prepared at the original site m = o at t = o. It is convenient to introduce a reduced density matrix $\langle \rho_{mn}(t) \rangle = \langle a^*_m(t)\, a_n(t) \rangle$, where a_m is the wave amplitude at site m. The initial condition is $\langle \rho_{mn}(t{=}o) \rangle = \delta_{m,o}\, \delta_{n,o}$. The equation of motion for $\langle \rho_{mn}(t) \rangle$ is readily found to be

$$i\hbar \frac{\partial \tilde{\rho}_{p,q}(\tau)}{\partial \tau} = \frac{4V_o}{(2\gamma\tau)^{\frac{1}{2}}} \cdot \tilde{\rho}_{p,q}(\tau) \sin p \sin (q/2) - \left(\frac{f_o^2\, a^2}{2\, i\, \hbar}\right) \cdot \frac{\partial^2 \tilde{\rho}_{p,q}(\tau)}{\partial p^2}, \quad \text{(II-34)}$$

where we have introduced the lattice Fourier transform

$$\langle \tilde{\rho}_{p,q} \rangle = \sum_{m,n} \langle \rho_{mn}(t) \rangle \, e^{ikm - ik'n} \quad \text{(II-35)}$$

with

$$p + \frac{1}{2} q = k, \quad p - \frac{1}{2} q = k', \text{ and } e^{2\gamma t} = \tau .$$

Thus all summations over p,q span first Brillouin zone. As before, we have used the Novikov identity for functionals of gaussian random variables to close the hierarchy of equations. As we are interested only in some moments, e.g., $\langle x^2(t) \rangle$, it is not necessary to solve (II-34) for $\tilde{\rho}_{p,q}(\tau)$. We simply have to expand both sides of (II-34) in powers of q and equate coefficients of like powers. The first three equations are

$$\dot{\tilde{\rho}}_{p,o}(\tau) = D_o \frac{\partial^2 \tilde{\rho}_{p,o}(\tau)}{\partial p^2}, \quad \text{(II-36)}$$

$$\dot{\tilde{\rho}}'_{p,o}(\tau) = \left(\frac{2\gamma}{i\hbar\sqrt{2\gamma}}\right) \frac{1}{\tau^{\frac{1}{2}}} \tilde{\rho}_{p,o}(\tau) \sin p + D_o \frac{\partial^2 \tilde{\rho}'_{p,o}(\tau)}{\partial p^2} \quad \text{(II-37)}$$

$$\dot{\tilde{\rho}}''_{p,o}(\tau) = \left(\frac{4\gamma}{i\hbar\sqrt{2}\gamma}\right)\frac{1}{\tau^{\frac{1}{2}}}\,\tilde{\rho}'_{p,o}(\tau)\sin p + D_o\frac{\partial^2\tilde{\rho}''_{p,o}(\tau)}{\partial p^2}, \tag{II-38}$$

with $D_o = a^2 f^2/2\hbar^2$, prime denoting $\partial/\partial q$, and overhead dot denoting $\partial/\partial\tau$. The mean-squared displacement, or rather the diffusion constant D, if it exists, can be expressed as

$$D = \lim_{t\to\infty}\frac{1}{2}\frac{d}{dt}\langle x^2(t)\rangle = -\lim_{\tau\to\infty}\left(\frac{\gamma a^2\tau}{\pi}\right)\int_{-\pi}^{\pi}\dot{\tilde{\rho}}''_{p,o}(\tau)\,dp$$

$$= \lim_{\tau\to\infty} - \left(\frac{2V_o}{i\hbar\pi}\right)(2\gamma\tau)^{\frac{1}{2}}\int_{-\pi}^{\pi}\tilde{\rho}'_{p,o}(\tau)\sin p\,dp \tag{II-39}$$

In getting the last equality, we have made use of (II-38) and the periodic boundary condition in p so that the contribution of the last term in (II-38) vanishes identically. All we have to do now as to solve (II-36) for $\tilde{\rho}_{p,o}(\tau)$, substitute the solution in (II-37) as the inhomogeneous term and solve for $\tilde{\rho}'_{p,o}(\tau)$, and finaly substitute this solution in (II-39). Since the equations are of the parabolic type, one simply needs the Green function $G(p,\tau;p',\tau')$ subject to the periodic boundary condition. This is readily found to be

$$G(p,\tau;p',\tau') = \sum_{n=-\infty}^{\infty}\frac{1}{[4\pi D_o(\tau-\tau')]^{\frac{1}{2}}}\,e^{-\frac{((p-p')-2\pi n)^2}{4D_o(\tau-\tau')}} \tag{II-40}$$

Thus, the diffusion constant turns out to be

$$D = \lim_{\tau\to\infty}\left(\frac{8V_o^2a^2}{4\pi\hbar^2}\right)\int^{\tau} e^{-D_o(\tau-\tau')}\left(\frac{\tau}{\tau'}\right)^{\frac{1}{2}}d\tau'$$

$$= \frac{2V_o^2}{\pi M\gamma k_BT} \tag{II-41}$$

This would correspond to a temperature - dependent mobility. This expression is similar to but not quite the same as the one we would obtain in the conventional treatment of dynamic disorder for a lattice Hamiltonian with site-diagonal disorder.

Let me now put this work in the proper perspective. We considered the quantum motion of a particle (subsystem) under the influence of its interaction with the rest of the system (bath). If the bath is taken to act on the particle parametrically as a c - number stochastic force, we get non-diffusive motion i.e. indefinite acceleration. (The fact that for a one-band lattice Hamiltonian one got diffusive behaviour is due to the momentum cut off inherent in such a model). We attributed this unphysical motion to the neglect of friction concomitant with the fluctuating stochastic force. We then introduced friction through the phenomenological frictional Hamiltonian and recovered the physical, diffusive motion. But, now we must ask if the frictional Hamiltonian (II-26) is consistent with quantum mechanics, vis - a - vis the fact that $[\hat{x}, \hat{p}_{kin}] = i\hbar e^{-\gamma t} \longrightarrow 0$ as $t \longrightarrow \infty$, implying violation of the uncertainty principle. This is concerned with the basic problem of treating dissipation in quantum mechanics that has a long history of non-success starting from the introduction of the frictional Hamiltonian almost half a century ago by Bateman [27] and Kanai [28] through the classic work of Ford-Kac-Mazur (FKM) [14]. Comprehensive reviews have appeared since then [24,25]. Very briefly, just that the Lagrangian (II-25) gives the correct classical equation of motion (i.e. the Langevin equation (II-21)) does not imply that it also gives a unique and correct quantum mechanics. Indeed, it has been claimed that only a coordinate dependent dissipative potential is consistent with quantum mechanics [29]. One must note, however, that $[\hat{x}, \hat{p}_{kin}] = i\hbar e^{-\gamma t}$ by itself may not imply violation of $\Delta x \, \Delta p_{kin} \geqslant \hbar/2$. We must evaluate the uncertainty product in the time-evolved wavefunction. For the case of damped harmonic oscillator, and most studies are limited this case, it has been possible in some sense to preserve the uncertainty principle [24]. More recently, however, the violation for certain initial states and certain specific time instants has been reaffirmed [30]. In fact, it has been shown that the Schrödinger and the Heisenberg pictures of the evolution consistent with equilibration as $t \longrightarrow \infty$ do not exist for the frictional Hamiltonian. It seems that the effect of the bath dynamical degrees of freedom may not be representable entirely by a c - number white noise stochastic force [31]. In point of fact the FKM treatment of the quantum Langevin equation requires a quantized gaussian noise, with non-zero correlation time $\hbar/k_B T$ for the case of coupled Harmonic oscillators. Thus, the status of the frictional Hamiltonian is far from secure. We must add, however, that in the present case of a freely diffusive particle, one can explicitly show that the uncertainty product $\Delta x \, \Delta p_{kin} \geqslant \hbar/2$ for all times (and the quantum statistical expectation of the frictional Hamiltonian goes over to $1/2 \, k_B T$ as $t \rightarrow \infty$). But, of course, one must show that the inequality holds with probability unity and not just on the average. This is under investigation.

References

1. S. Chandrasekhar, Rev. Mod. Phys. 15, 1 (1943). See also the reprint collection 'Noise and Stochastic Processes' ed. Nelson Wax (Dover - New York, 1954).
2. R. Kubo, J. Phys. Soc. Japan 12, 570 (1957) also Rep. Prog. Phys. 29, 255 (1966).
3. Y. Pomeau and P.Résibois, Phys. Rep. 19, 63 (1975).
4. R.F. Fox, Phys. Rep. 48, 179 (1978).
5. G. Wyllie, Phys. Rep. 61, 327 (1970).
6. N.G. van Kampen, Phys. Rep. 24 C, 172 (1976).
7. Edward Nelson, Dynamical Theories of Brownian Motion (Mathematical Notes, Princeton University Press, 1967).
8. "New Stochastic Methods in Physics", ed. C.De Witt - Morette and K.D. Elworthy, Phys. Rep. 77 (1981).
9, M.D. Donsker and S.R.S. Varadhan, Comm. Pure Appl. Math. 28, 525 (1975).
10. B.J. Alder and T.E. Wainwright, Phys. Rev. A1, 18 (1970).

11. R. Zwanzig, Ann. Rev. Phys. Chem. 16, 67 (1955).
12. H. Mori, Prog. Theor. Phys. 33, 423 (1965), also 34, 399 (1965).
13. P. Résibois and J. Lebowitz, Phys. Rev. 139, 1101 (1965).
14. G.W. Ford, M. Kac and P. Mazur, J. Math. Phys. 6, 504 (1965).
15. Benny Carmeli and Abraham Nitzan, Phys. Rev. Lett. 49, 423 (1982).
16. E.A.Novikov, Zh. Eksp. Teor. Fiz. 47, 1919 (1964) [Sov. Phys. JETP 20, 1990 (1965)].
17. H. Haken and P. Reinecker, Z. Phys. 249, 253 (1972).
18. A.A. Ovchinnikov and N.S. Erikhman, Zh. Eksp. Teor. Fiz. 69, 1474 (1974) [Sov. Phys. JETP 40, 733 (1974)].
19. A. Madhukar and W. Post, Phys. Rev. Lett. 39, 1424 (1977).
20. H. Metiu, K. Kitahara, R. Silbey and J. Ross, Chem. Phys. Lett. 43, 189 (1976).
21. Y. Inaba, J. Phys. Soc. Japan, 50, 2473 (1981).
22. A.M. Jayannavar and N. Kumar, Phys. Rev. Lett. 48, 553 (1982).
23. See, e.g. P. Havas, Nuovo Cimento Suppl. 5, 363 (1967).
24. H. Dekker, Phys. Rep. 80, 1 (1981).
25. R.W. Hasse, Rep. Prog. Phys. 41, 1027 (1978).
26. G.J. Papadopoulos, Phys. Rev. D11, 2870 (1976).
27. H. Bateman, Phys. Rev. 38, 815 (1931).
28. E. Kanai, Prog. Theor. Phys. 3, 440 (1948).
29. Wesley, E. Brittin. Phys. Rev. 77, 396 (1950).
30. R. Alicki and J. Messer, J. Phys. A 15, 3543 (1982).
31. G. Iche and P. Nozíeres, Physica, 91A, 485 (1978).

RELAXATION OF SINGLE DOMAIN MAGNETIC PARTICLES

Deepak Kumar
Department of Physics
University of Roorkee, Roorkee 247667

In this article, we intend to describe a particularly illustrative application of the general theory of stochastic processes which has been discussed in great detail by various authors in this volume. The example is taken from the field of magnetism and is based on the work of Brown [1] . Brown considered the relaxation of single domain particles. This mechanism of relaxation has been invoked by a number of workers to understand relaxation and response behaviour of spin-glasses in recent years.

1. Description of the system

Let us begin the discussion by explaining what is meant by a 'single domain particle'. A body made up of magnetic material, say iron, has no magnetic moment under usual conditions even if it is well below its Curie temperature. This is so because the body gets divided into various domains in which the spontaneous magnetisation points in different directions. However, if the size of the body is reduced, there comes a point beyond which the body has just one domain. For example, iron particles having radius below 150 Å stay in a single domain. At this point, the magnetostatic energy, and the energy of forming domain wall compare in such a way that single domain state becomes preferable. Here we shall consider an assembly of such single domain particle which are assumed to be non-interacting [2] . The magnetic state of such a particle is characterised by its magnetisation vector $\vec{M}$

$$\vec{M} = vM_s\hat{n} \tag{1.1}$$

where v is the volume of the particle, M_s the saturation magnetisation, and $\hat{n}$ a unit vector. There are generally two contributions to the magnetic energy. The first one is due to the external field $\vec{H}$, given by $-\vec{M}.\vec{H}$, the second due to the magnetic anisotropy energy which arises from the crystalline structure of the material. The typical forms for anisotropy energy are

(i) Cubic Anisotropy (Example : iron)

$$E = -vA\left[n_x^4 + n_y^4 + n_z^4\right] \tag{1.2}$$

(ii) Uniaxial Anisotropy (Example : Cobalt)

$$E = +vK\left[1 - n_z^2\right] \tag{1.3}$$

where A and K are constants, and n_j $(j = x,y,z)$ are the components of $\hat{n}$. Since the purpose of the article is illustrative, we shall consider henceforth particles of uniaxial anisotropy. Choosing the z-axis along the anisotropy axis, and denoting by θ the angle between the z-axis and $\hat{n}$, the anisotropy energy may be written as

$$E(\theta) = vK\ \mathrm{Sin}^2\theta \qquad (1.4)$$

2. Relaxation : a simple description

We now define a distribution function $f(\Omega)$, such that $f(\Omega)\,d\Omega$ denote the fraction of particles whose magnetisation vector $\hat{n}$ lies between solid angles Ω and $\Omega + d\Omega$. If the assembly is in thermal equilibrium

$$f_{eq}(\Omega)\,d\Omega \propto e^{-(vK/kT)\,\mathrm{Sin}^2\theta}\ \mathrm{Sin}\,\theta\,d\theta\,d\phi \qquad (2.1)$$

If at time t=o, one starts with an arbitrary distribution $f(\Omega)$, the system eventually relaxes to the equilibrium distribution given in (2.1). However if $vK/kT \ll 1$, the equilibrium distribution is more or less uniform, and the relaxation occurs on a microscopic time scale. On the other hand, if $vK/kT \gg 1$, most of the particles in equilibrium are either near $\theta \simeq 0$ or $\theta \simeq \pi$. The relaxation now occurs in two stages as illustrated in Fig.1. In the first stage the particles in the region $0 \leqslant \theta < \frac{\pi}{2}$ quickly slide to the potential valley $\theta \simeq 0$, while those in the region $\frac{\pi}{2} < \theta \leqslant \pi$ slide into the potential valley $\theta \simeq \pi$. This process occurs on the microscopic time scale, but leads to

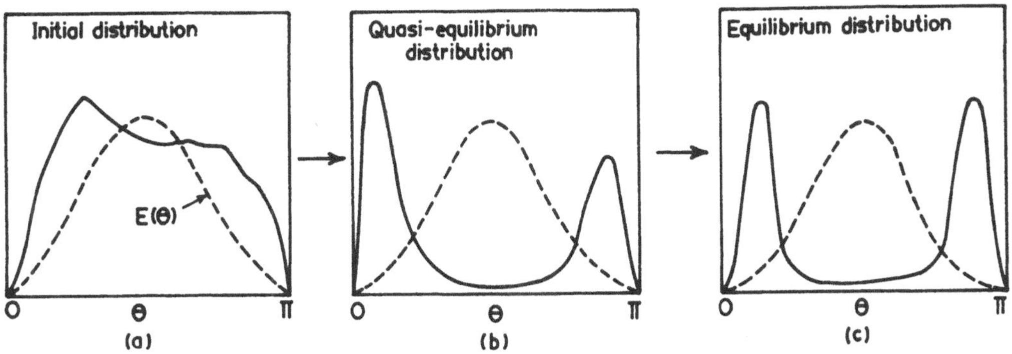

Fig.1 This series of figures depicts the evolution of the distribution function towards thermal equilibrium. Fig. a shows the initial distribution. The first stage of evolution leading to quasi-equilibrium distribution (Fig.b) occurs on microscopic time scale $\tau_\perp$. The second stage of evolution from quasi-equilibrium state occurs on much larger time scale of $\tau_\perp e^{\beta K v}$, as it involves thermally activated tunnelling across the barrier. Fig.c shows the equilibrium distribution.

only a quasi-equilibrium (or metastable) state, because though the equilibrium is achieved within each valley of the potential, the overall equilibrium between the two valleys is not reached (see Fig.1). In order that the global equilibrium may be established, the magnetic vector of some particles must go from one valley to the other. This requires a thermally activated tunnelling across a barrier, which is a far slower process. This sort of situation was considered by Neél long ago who described the evolution from quasi-equilibrium to global equilibrium in terms of simple rate equations [3] . Let us define the fractional populations in the two valleys n_1 and n_2 as follows

$$n_1(t) = \int_0^{\theta_1} f(\theta)\ \mathrm{Sin}\theta\ d\theta; \quad n_2(t) = \int_{\theta_2}^{\pi} f(\theta)\ \mathrm{Sin}\theta\ d\theta \qquad (2.2)$$

where θ_1 and θ_2 are two arbitrary angles such that $\theta_1 \sim 0$ and $\theta \sim \pi$(See Fig.1). Since the number of particles in other angular regions is negligible ($vK/kT \gg 1$), the thermal evolution can be well described in terms of merely $n_1(t)$ and $n_2(t)$. One can easily write down the rate equations for these two as

$$\frac{dn_1(t)}{dt} = -\frac{dn_2(t)}{dt} = -\nu_{12}\, n_1 + \nu_{21}\, n_2 \qquad (2.3)$$

where ν_{12} and ν_{21} are the rates at which the particles reorient from one region to the other, and Neél assumed the usual Arrhenius form for these

$$\nu_{12} = \nu_{21} = \nu_o\ e^{-vK/kT} \qquad (2.4)$$

where ν_o is some kind of an attempt frequency.

3. A First Principles Description of Relaxation : The Fokker Planck Equation

While the above approach is intuitively appealing, it needs justification from a more basic point of view. The basic theory should be able to describe the thermal evolution generally, i.e. in situations in which the restriction of large barrier (i.e. $vK/kT \gg 1$) can be removed. Moreover, in a basic theory, one should be able to derive the rate equations and find explicit expressions for the rates ν_{12} and ν_{21}, which incorporate precessional motion of spins, details of the potential barrier, etc. Brown essentially provided such a theory.

Brown began his considerations by first constructing an appropriate Langevin equation for the precession of the magnetisation vector $\vec{M}$ [4,5] . To construct this equation let us first write the deterministic equation of motion for $\vec{M}$, which reads

$$\frac{d\vec{M}}{dt} = \gamma \vec{M} \times (\vec{H}_e - \frac{\partial E(\vec{M})}{\partial \vec{M}}) = \gamma \vec{M} \times \vec{H} \qquad (3.1)$$

where γ is the gyromagnetic ratio and $\vec{H}_e$ is the external field. This equation describes the precession of $\vec{M}$ around the total effective field $\vec{H}$, as shown in Fig.2. If we now let the particle interact with a heat bath, two new terms should be added to Eq. (3.1). First, there is the frictional term, which causes the precession-moment to spiral towards the axis of the effective field, so that it reaches the lowest energy state. Second, the bath exerts a random force which essentially provides thermal energy to the moment. These physical requirements, in addition to the requirement that the magnitude $\vec{M}$ is a fixed quantity, permit us to introduce the additional terms in the following way

$$\frac{d\vec{M}}{dt} = \gamma \vec{M} \times [\ \vec{H} - \eta(\vec{M}\times\vec{H}) + \vec{h}'(t)] \qquad (3.2)$$

where η is the co-efficient of viscosity and $\vec{h}'(t)$ is a random field, which is taken

taken to be a stationary Gaussian random process. Its correlations are given by

$$\langle h'_\alpha (t_1)\, h'_\beta (t_2) \rangle = D_{\alpha\beta}\, \delta(t_1-t_2) \tag{3.3}$$

where we take

$$D_{xx} = D_{yy} = \frac{1}{\tau_\perp}\ ; \quad D_{zz} = \frac{1}{\tau_\parallel}\ ; \quad D_{xy} = D_{yz} = 0 \text{ etc.} \tag{3.4}$$

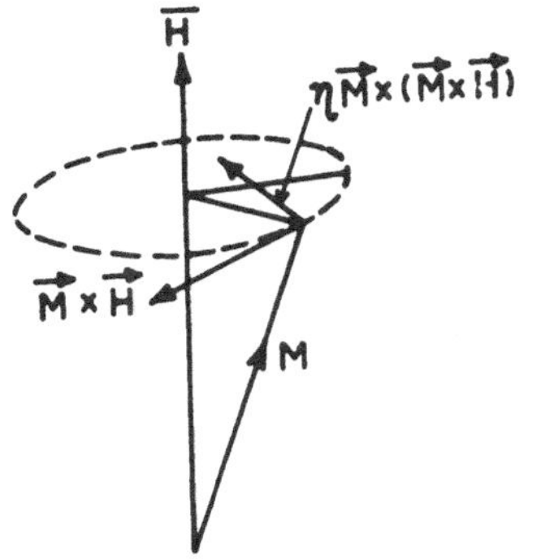

Fig. 2 Torques on the magnetisation vector according to Langevin equation.

Using the Langevin equation (3.2) and properties given in Eqs.(3.3) and (3.4), the Fokker-Planck equation can be derived following the methods described in this volume. In the present case the Fokker-Planck equation turns out to be

$$\frac{\partial}{\partial t} (\vec{M},t) = +\gamma[\vec{L}.\vec{H}\, f - \eta\vec{M} \times \vec{L}.\vec{H}\, f + 1/2\, L_\alpha D_{\alpha\beta} L_\beta f\,] \tag{3.5}$$

where we have defined the operator

$$\vec{L} = \vec{M} \times \frac{\partial}{\partial \vec{M}} \quad . \tag{3.6}$$

We shall now effect a simplification by considering an axially symmetric situation. We take the anisotropy energy E and the initial distribution $f(\theta,\phi,0)$ to be independent of ϕ, where θ,ϕ denote the direction of magnetisation in polar coordinates. Under these conditions $f(\theta,\phi,t)$ remains independent of ϕ for all t and Eq.(3.5) may be written as

$$\frac{\partial f(\theta,t)}{\partial t} = \frac{1}{2\tau_\perp}\, \frac{1}{\sin\theta}\, \frac{\partial}{\partial\theta} \left[\sin\theta \left\{ \frac{1}{kT}\, \frac{\partial E(\theta)}{\partial\theta} f + \frac{\partial f}{\partial\theta} \right\} \right] \tag{3.7}$$

where we have used the Einstein relation

$$2\tau_\perp \eta \gamma = \frac{1}{kT} \tag{3.8}$$

to ensure that $f(\theta,t)$ approaches the equilibrium distribution in the limit $t\to\infty$. As a first consequence of Eq. (3.5) let us consider the time evolution of the average component of magnetisation. In the case of zero anisotropy and an external field H_o in z-direction [7] one finds

$$\frac{d\langle M_z\rangle}{dt} = - \frac{\langle M_z\rangle}{T_1} + \frac{H_o}{kT}\, \frac{\langle M_x^2 + M_y^2\rangle}{2T_1} \tag{3.9}$$

$$\frac{d\langle M_x\rangle}{dt} = - \gamma H_o \langle M_y\rangle - \frac{\langle M_x\rangle}{T_2} - \frac{\eta H_o}{2kT}\, \frac{\langle M_z M_x\rangle}{T_1} \tag{3.10}$$

$$\frac{d\langle M_y\rangle}{dt} = \gamma H_o \langle M_x\rangle - \frac{\langle M_y\rangle}{T_2} - \frac{\eta H_o}{2kT} \frac{\langle M_z M_y\rangle}{T_1} \tag{3.11}$$

where $T_1 = \tau_{\parallel}$ and $T_2^{-1} = 1/2(\tau_{\parallel}^{-1} + \tau_{\perp}^{-1})$. These equations reduce to the well known Bloch equations in the limit $M_s H_o/kT \ll 1$. The Fokker-Planck equation discussed above can, in principle, describe relaxation in arbitrary situations. However, analytic results can be obtained only in certain simple cases. In the next section, we shall study the solution of the axially symmetric equation (3.7) in the large barrier limit.

4. Variational Solution in Large Barrier Limit

We can write Eq. (3.7) schematically as

$$\frac{\partial f}{\partial t'} = Df \tag{4.1}$$

where $t' = t/2\tau_{\perp}$ and

$$D = \frac{1}{\sin\theta} \frac{\partial}{\partial\theta} \left[\sin\theta \left\{ \frac{1}{kT} \frac{\partial E(\theta)}{\partial\theta} + \frac{\partial}{\partial\theta} \right\} \right]. \tag{4.2}$$

Noting that $e^{\beta E(\theta)} D$ is a Hermitian operator ($\beta = 1/kT$) and $D\, e^{-\beta E(\theta)} = 0$, we can write down the time dependent solution of Eq. (4.1) as

$$f(\theta,t) = A_o\, e^{-\beta E(\theta)} + \sum_n A_n\, e^{-p_n t'} F_n(\theta) \tag{4.3}$$

where $F_n(\theta)$'s are eigenfunctions of D, satisfying

$$D\, F_n(\theta) = -\, p_n\, F_n(\theta) \quad . \tag{4.4}$$

F_n's form a complete orthonormal set, with the following definition of the scalar product

$$(F_n, F_m) = \int_0^{\pi} F_n(\theta)\, F_m(\theta)\, e^{\beta E(\theta)} \sin\theta\, d\theta = \delta_{n,m} \quad . \tag{4.5}$$

Following physical ideas of Kramers [5], Brown [1] argued that only the smallest eigenvalue p_1 (other than zero, of course) describes the long time evolution from the quasi-equilibrium state. The higher eigenvalues are important only in stablizing quasi-equilibrium in the early stages of relaxation. More precisely, we show in the appendix that $p_1 \sim O(e^{-\beta K})$ while $p_i \sim O(1)$ for $i \geqslant 2$.

To evaluate p_1 and F_1, Brown gave a variational treatment. Writing $F_1(\theta) = e^{-\beta E(\theta)} \Phi(\theta)$, one can easily show that the above eigenvalue problem is equivalent to the minimisation of the functional

$$D[\Phi] = \int_0^{\pi} e^{-\beta E} \left(\frac{d\Phi}{d\theta}\right)^2 \sin\theta\, d\theta \tag{4.6}$$

under the constraints

$$H[\Phi] = \int_0^{\pi} e^{-\beta E}\, \Phi^2 \sin\theta\, d\theta = 1 \tag{4.7}$$

and

$$(F_1, e^{-\beta E}) = \int_0^{\pi} e^{-\beta E} \Phi \operatorname{Sin}\theta \, d\theta = 0 \tag{4.8}$$

The constraint (4.7) corresponds to the normalisation of eigenfunction, while (4.8) to its orthogonality to the lowest eigenfunction.

The choice of Φ is now dictated by the following requirements. (i) $\Phi(\theta)$ must change sign within the interval $0 \leqslant \theta \leqslant \pi$ to satisfy Eq. (4.8). (ii) To keep $D[\Phi]$ small, one should concentrate large values of $|d\Phi/d\theta|$ near the minima of $e^{-\beta K \operatorname{Sin}^2\theta}$, i.e., $\theta \approx \pi/2$. Thus Brown allowed Φ to take constant values ϕ_1 in $(0,\theta_1)$ and ϕ_2 in (θ_2,π) and determined the derivative $d\Phi/d\theta$ in the region (θ_1,θ_2) from the minimisation of $D[\Phi]$. The form of the solution is shown schematically in Fig. 3. Under the condition $\beta K \gg 1$, the constraints (4.7) and (4.8) can be evaluated by neglecting the constribution to the integrals from the region (θ_1,θ_2) where $e^{-\beta E}$ is very small. Thus the two constraints require

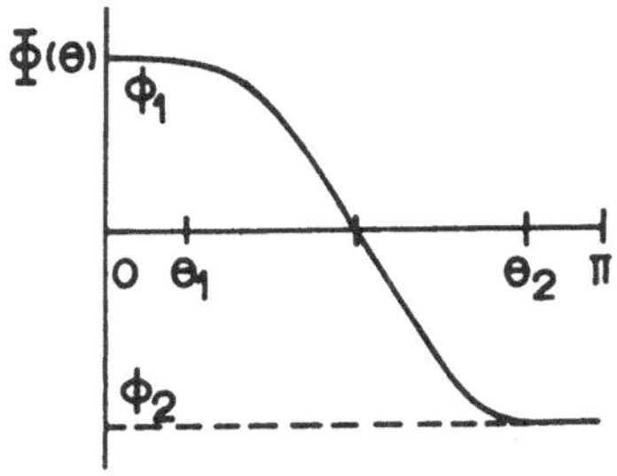

Fig. 3 Variational choice of $\Phi(\theta)$.

$$I_1\phi_1^2 + I_2\phi_2^2 = 1 \quad ; \quad I_1 \phi_1 + I_2 \phi_2 = 0 \tag{4.9}$$

where

$$I_{1,2} = \int_{R_1, R_2} e^{-\beta E(\theta)} \operatorname{Sin}\theta \, d\theta \tag{4.10}$$

where R_1, R_2 denote respectively the integration regions $(0,\theta_1)$ and (θ_2,π). Now we minimise $D[\Phi]$ in (θ_1,θ_2) with the condition that $\Phi(\theta_1) = \phi_1$ and $\Phi(\theta_2) = \phi_2$. This yields

$$\frac{d}{d\theta}\left[e^{-\beta E} \frac{d\Phi}{d\theta} \operatorname{Sin}\theta \right] = 0 \tag{4.11}$$

or

$$\frac{d\Phi}{d\theta} = \frac{A \; e^{\beta E}}{\operatorname{Sin}\theta} \tag{4.12}$$

where A is an undetermined constant. Integration of Eq. (4.12) yields

$$\phi_1 - \phi_2 = A \int_{\theta_1}^{\theta_2} \frac{e^{\beta E}}{\operatorname{Sin}\theta} d\theta = A \, I_m \tag{4.13}$$

where

$$I_m \simeq \sqrt{\frac{2\pi}{\beta E''(\pi/2)}} \; e^{\beta E(\pi/2)} \tag{4.14}$$

Equations (4.9), (4.10), (4.13) and (4.14) determine the solution. Using Eqs. (4.9), we get

$$\phi_1 = \frac{C}{I_1} \; ; \; \phi_2 = -\frac{C}{I_2} \quad ; \quad C = \sqrt{\frac{I_1 \, I_2}{I_1+I_2}} \tag{4.15}$$

Substituting Eq. (4.15) in (4.13), one can find A as well. Finally, the eigenvalue p_1

is given as

$$p_1 = D[\Phi]_{min} = \frac{1}{I_m}\left(\frac{1}{I_1} + \frac{1}{I_2}\right) \tag{4.16}$$

$$= \frac{4(\beta K v)^{3/2}}{\pi^{1/2}} e^{-\beta K v} \tag{4.17}$$

where the second line follows by substituting the approximate values for the integrals when $VK \gg kT$. The variational solution for long times is

$$f(\theta,t) = e^{-\beta E(\theta)} \left[A_o + A_1 \Phi(\theta) e^{-P_1 t'} \right] \tag{4.18}$$

From this solution, it is easy to derive the Neél picture. Let us calculate n_1 and n_2 as defined in (2.2). Using (4.18), one finds

$$n_{1,2} = I_{1,2} (A_o + A_1 \phi_{1,2} e^{-p_1 t'}) \tag{4.19}$$

The constants A_o and A_1 are determined from the normalisation of the distribution function and the initial condition. However, here we shall eliminate them in favour of n_1, n_2 and their time derivatives. Using Eq.(4.10), one gets

$$n_1 + n_2 = A_o (I_1 + I_2) \tag{4.20}$$

which shows that the total number of particles in the two potential valleys is a constant within this approximation. Further

$$\frac{dn_1}{dt} = - I_1 \phi_1 \frac{p_1}{2\tau_\perp} A_1 e^{-p_1 t'} = - \frac{dn_2}{dt} \tag{4.21}$$

We can now use Eqn. (4.19) to eliminate $A_1 e^{-p_1 t'}$ in favour of n_1 and n_2. This yields

$$\frac{dn_1}{dt} = - \frac{dn_2}{dt} = - \nu_{12} n_1 + \nu_{21} n_2 \tag{4.22}$$

with

$$\nu_{12} = \frac{1}{2\tau_\perp} \frac{p_1 \phi_1}{\phi_1 - \phi_2} = \frac{1}{2\tau_\perp} \frac{1}{I_1} \frac{1}{I_m} \simeq \frac{(\beta K v)^{3/2}}{\tau_\perp \pi^{1/2}} e^{-\beta K v} \tag{4.23}$$

$$= \nu_{21} \tag{4.24}$$

In the next section, we discuss the application of (4.22) for calculating the magnetic response of an assembly of single domain particles.

5. Non-equilibrium Response

In the large barrier limit, the magnetisation of the assembly is

$$M = vM_s [n_1(t) - n_2(t)] \tag{5.1}$$

Using rate equations, it is a simple matter to see that

$$\frac{dM}{dt} = - \nu M \tag{5.2}$$

where $\nu = (\nu_{12} + \nu_{21})$. Now suppose at time $t = 0$ we apply a field H along the anisotropy axis of the particles, assuming that anisotropy axes of all the particles are aligned. The rates in the presence of the field can again be calculated using the formalism of Sec.3. In the rate-equation description, however, the effect of the field can be easily determined using Arrhenius-like arguments. The energy of the n_1 particles is lowered by $+vM_sH$, so that the rate ν_{12} becomes ($M_sH \ll K$)

$$\nu_{12} = \frac{(v\beta K)^{3/2}}{\tau_\perp \pi^{1/2}} e^{-\beta v(K+M_sH)} \tag{5.3}$$

Similarly the energy of the n_2 particles is raised by vM_sH, and corresponding rate ν_{21} becomes

$$\nu_{21} = \frac{(v\beta K)^{3/2}}{\tau_\perp \pi^{1/2}} e^{-\beta v(K-M_sH)} \tag{5.4}$$

Substituting these rates in Eq.(4.22), it is an easy matter to show that

$$M(t) = M(o)\, e^{-\nu t} + vM_s(1-e^{-\nu t}) \tanh \frac{vM_sH}{kT} \tag{5.5}$$

where

$$\nu = \nu_{12} + \nu_{21} = 2 \frac{(v\beta K)^{3/2}}{\tau_\perp \pi^{1/2}} e^{-\beta Kv} \operatorname{Cosh} \beta vM_sH \tag{5.6}$$

The susceptibility $\chi_\parallel$ along the easy axis is then

$$\chi_\parallel = \frac{(vM_s)^2}{kT} (1-e^{-\nu t}) \tag{5.7}$$

The situation considered above is somewhat artificial, because in most physical situations, the anisotropy axes of different particles are in different directions. So one really needs the response of the particles to a field in an arbitrary direction [9]. This response can be expressed in terms of the parallel response determined above and the response to a field perpendicular to the easy axis. If α is the angle between the easy axis and the field,

$$M = M_\parallel \operatorname{Cos}\alpha + M_\perp \operatorname{Sin}\alpha = (\chi_\parallel \operatorname{Cos}^2\alpha + \chi_\perp \operatorname{Sin}^2\alpha)\, H \tag{5.8}$$

Averaging over α then yields for susceptibility χ

$$\chi = 1/3\, [2\chi_\parallel + \chi_\perp] \tag{5.9}$$

The time dependent perpendicular susceptibility χ can be obtained by noting that the perpendicular field changes the energy of the particles according to the equation

$$E(\theta) = v\,[K\operatorname{Sin}^2\theta - M_sH \operatorname{Sin}\theta] \tag{5.10}$$

which shifts the energy minima to θ_1 and θ_2 given by

$$\theta_1 = \operatorname{Sin}^{-1} \frac{M_sH}{2K} \approx \frac{M_sH}{2K} = \pi - \theta_2 \ . \tag{5.11}$$

When the field is switched on at t=o, the response occurs in the following way. On time scales of order $\tau_\perp$, the quasi-equilibrium state is established, as the particles in the two valleys simply adjust to the new potential. This occurs quickly since no tunnelling across the barrier is involved. The contribution to $M_\perp$ is then

$$M_\perp = (n_1 \mathrm{Sin}\theta_1 + n_2 \mathrm{Sin}\theta_2)\, vM_s = \frac{vM_s^2}{2K} nH \tag{5.12}$$

Since $n = n_1 + n_2$ is a constant of motion in this approximation, the perpendicular response has only quick component. If we had considered an asymmetric barrier, in which case $\mathrm{Sin}\theta_1 \neq \mathrm{Sin}\theta_2$, the perpendicular response would also have a slow component. On time scales of interest here, we can put together all these results to write for susceptibility

$$\chi = \frac{N}{3}\left[2\frac{(vM_s)^2}{kT}(1-e^{-\nu t}) + \frac{vM_s^2}{2K}\theta(t)\right] \tag{5.13}$$

where $\theta(t)$ is the step function. From this, relaxational response it is easy to derive the equilibrium frequency dependent response via the use of the fluctuation-dissipation theorem [9]

$$\chi(\omega) = \frac{N}{3}\left[\frac{2(vM_s)^2}{kT}\frac{\nu}{\nu+i\omega} + \frac{vM_s^2}{2K}\right. \tag{5.14}$$

Appendix

In this appendix we show that in the large barrier limit only one eigenvalue of the Fokker-Planck equation (3.10) is of order $e^{-\beta K}$, while all others are of O(1). Thus only one eigenvalue is needed to describe the slow relaxation from the quasi-equilibrium state. A transparent proof of the statement can be made by first transforming Eq. (3.10) into a Schrödinger like equation (10). The transformation is

$$f(\theta,t) = e^{-\beta E(\theta/2}\, g(\theta,t) \tag{A.1}$$

Substitution in Eq. (3.10) leads to the following equation for $g(\theta,t)$

$$-\frac{\partial g}{\partial t'} = \left[-\frac{1}{\mathrm{Sin}\theta}\frac{\partial}{\partial\theta}\left(\mathrm{Sin}\theta\frac{\partial g}{\partial\theta}\right) + U(\theta)g\right] \tag{A.2}$$

where

$$U(\theta) = -\frac{1}{2}\frac{1}{\mathrm{Sin}\theta}\frac{\partial}{\partial\theta}\left(\mathrm{Sin}\theta\frac{\partial E}{\partial\theta}\right) + \frac{\beta^2}{4}\left(\frac{\partial E}{\partial\theta}\right)^2 \tag{A.3}$$

Substituting for E from Eq. (1.4), gives

$$U(\theta) = (\beta vK)^2 \mathrm{Sin}^2\theta\, \mathrm{Cos}^2\theta - \beta vK(2\mathrm{Cos}^2\theta - \mathrm{Sin}^2\theta) \tag{A.4}$$

Writing $g(\theta,t) = e^{-pt}G(\theta)$, (A.2) can be reduced to the standard quantum mechanical eigenvalue pronlem

$$pG(\theta) = -\frac{1}{\mathrm{Sin}\theta}\frac{\partial}{\partial\theta}\left(\mathrm{Sin}\theta\frac{\partial G}{\partial\theta}\right) + U(\theta)G. \tag{A.5}$$

The potential $U(\theta)$ is plotted in Fig.4. As expected, the minimum potential regions are $\theta \approx 0$ and $\theta \approx \pi$. In the large barrier limit, as a first approximation the tunnelling between the two potential wells can be neglected, and one finds that the small eigenvalues p_i^o's are doubly degenerate, with their wavefunctions confined to either

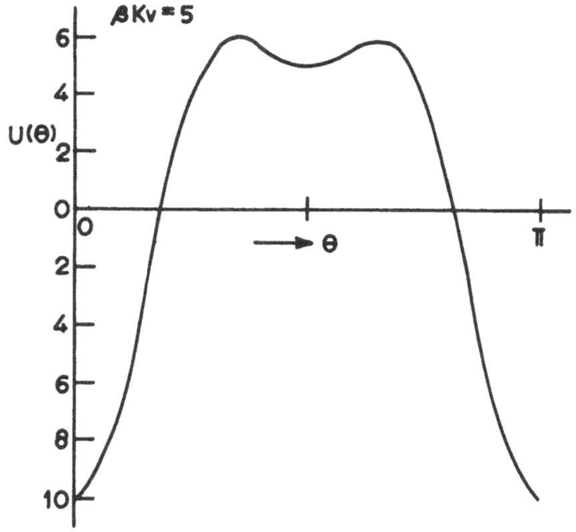

Fig.4 Plot of the effective potential $U(\theta)$

of the two potential wells. When the tunnelling is included the degeneracy of the levels is lifted and

$$p_i = p_i^o \pm t_i$$

where t_i's are the tunnelling amplitudes, which can be easily estimated in the WKB approximation to be

$$t_1 \approx e^{-\beta Kv}$$

Thus the first two levels are split by an amount $e^{-\beta Kv}$. But we know the lowest one to be zero, and thus $p_1 \approx e^{-\beta Kv}$. The higher eigenvalues are of order unity.

References

1. W.F. Brown, Phys. Rev. 130, 1677 (1963)
2. C.P. Bean and J.D. Livingston, J. App. Phys. 30, 120 (1959)
3. L. Neél , Ann. Geophys. 5, 99 (1949); Adv. Phys. 4, 191 (1955)
4. T.L. Gilbert, Phys. Rev. 100, 1243 (1955)
5. L. Landau and E. Lifshitz, Phys. Z. Sowjetunion 8, 153 (1935)
6. See also, S. Chandrasekhar, Rev. Mod. Phys. 15, 1 (1943)

7. R. Kubo and N. Hashitsume, Prog. Theoret. Phys. Supp. No.46, 210 (1970)
8. H.A. Kramers Physica 7, 284 (1940), see also articles by S. Dattagupta and V. Srinivasan in this volume
9. D. Kumar and S. Dattagupta, to be published in J. Phys. C (1983)
10. N.G. Van Kampen, J. Stat. Phys. 17, 71 (1977)

LANGEVIN EQUATION - APPLICATION TO LIQUID STATE DYNAMICS

K. N. Pathak
Department of Physics
Panjab University
Chandigarh 160014, India

1. Introduction

Matter can exist either as gas, liquid or solid. Sometimes one also speaks of a fourth phase i.e. a 'plasma'. The dynamics of a phase depends strongly in which state it exists. For a gas with low density the atoms move freely with binary collisions where only two atoms are involved at a time. The collision provides a mechanism for the transfer of energy and momentum which brings the system towards equilibrium. The dynamics of such a system can be formally worked out using Boltzmann transport equation.

For solids, particularly crystalline ones, there is a well known description called as Born-von-Karman theory. In this case the atoms are localized and make small excursions from the equilibrium positions. This and the translational symmetry of the system facilitates the solution of the problems in terms of the phonons.

The interaction among the phonons is normally weak below about one-third of the melting temperature and can be treated in usual fashion using the perturbation theory [1,2]. Near melting temperature and for quantum crystals above description is not applicable and the method of self-consistent approximations is to be employed [3].

The problem of atomic motion in liquids is considered more complicated and still we do not have any fully satisfactory theory. In comparison with the crystals, the regular order is lost and for the liquids [4] the atomic displacements are no longer small. This, as well as the fact that liquid densities are not low, make the problem more difficult but interesting.

The main problems of the liquid state [4,5] are the thermodynamic properties - such as equation of state, specific heat; transport properties - diffusion and heat transport; phase transition - condensation and crystallization; space-time correlation function in fluids and collective excitations.

In the set of two lectures I shall deal with only two problems, (1) motion of a tagged particle in a fluid i.e. diffusion and (2) collective excitations using the generalized Langevin equation [6,7].

2. Basic Definitions and Relevant Correlation Functions

We consider a fluid in which there are n_o number of tagged particles (may be a radioisotope of the host fluid). These particles will diffuse through the host fluid as the time progresses and one observes the concentration $n_o(\vec{r},t)$ of tagged particles which is governed by Fick's law

$$\vec{J}_0(\vec{r},t) = -D\,\vec{\nabla} n_0(\vec{r},t) \tag{2.1}$$

where $\vec{J}_0$ is the tagged particle current density and D is the self-diffusion coefficient. The concentration of the tagged particles is given by usual continuity equation

$$\frac{\partial n_0(\vec{r},t)}{\partial t} + \vec{\nabla}\cdot\vec{J}_0(\vec{r},t) = 0 \tag{2.2}$$

Combining eqs. (2.1) and (2.2) one finds that the tagged particle concentration $n_0(\vec{r},t)$ obeys the well known diffusion equation

$$\frac{\partial n_0(\vec{r},t)}{\partial t} = D\nabla^2 n_0(\vec{r},t) \tag{2.3}$$

We also introduce Van Hove self-correlation function $G_s(\vec{r},t)$ which describes the probability of finding the tagged particle at space time point $\vec{r},t$ if it was initially located at $\vec{r}$=o, t=o. It is clear that equation (2.3) should also be obeyed by $G_s(\vec{r},t)$ for times larger than the collision time (i.e. $t \gg \tau_c$) :

$$\frac{\partial G_s(\vec{r},t)}{\partial t} = D\nabla^2 G_s(\vec{r},t) \tag{2.4}$$

with the boundary condition

$$G_s(\vec{r},t) = \delta(r) \,, \tag{2.5}$$

which asserts that there was a particle at the origin at t=o. From $G_s(\vec{r},t)$ we also have

$$\int G_s(\vec{r},t)\, d\vec{r} = 1 \qquad \text{for all } t \tag{2.6}$$

Equation (2.4) can easily be solved by introducing the Fourier transform

$$F_s(\vec{k},t) = \int d\vec{r}\, \exp(-i\vec{k}.\vec{r})\, G_s(\vec{r},t). \tag{2.7}$$

Noting that $F_s(k,t{=}o) = 1$, one immediately obtains from eq.(2.4) that

$$F_s(k,t) = \exp(-k^2 D|t|) \,. \tag{2.8}$$

The Fourier inversion of equation (2.7) gives

$$G_s(\vec{r},t) = (4\pi D\,|t|)^{-3/2} \exp\left(-\frac{r^2}{4D\,|t|}\right) \,. \tag{2.9}$$

Using equation (2.9) one easily finds that

$$\begin{aligned} \langle r^2(t)\rangle &= r^2\, G_s(\vec{r},t)\, d\vec{r} \\ &= 6\,D\,t \end{aligned} \tag{2.10}$$

Thus the diffusion constant is a measure of the variation with the time of the mean square displacement of the particle in the fluid.

We now proceed further to relate the diffusion coefficient in terms of $S_s(k,\omega)$ which is the Fourier-transform of $F_s(k,t)$ i.e.

$$S_s(k,\omega) = \int_{-\infty}^{\infty} dt \exp(i\omega t) F_s(k,t) \quad . \tag{2.11}$$

Using eq. (2.8) and (2.11) we obtain

$$S_s(k,\omega) = \frac{2 D k^2}{\omega^2 + (k^2 D)^2} \tag{2.12}$$

which is valid for low frequencies and long-wavelengths. Using equation (2.12) we obtain a general expression for the diffusion coefficient as

$$D = \lim_{\omega \to 0} \frac{\omega^2}{2} \lim_{k \to 0} \frac{S_s(k,\omega)}{k^2} \quad . \tag{2.13}$$

This equation is regarded as Kubo formula for the transport coefficient. It is clear that it relates macroscopic properties of the system to microscopic motion of the atoms in the fluid.

In order to make microscopic calculation of $S_s(k,\omega)$ we introduce the dynamical variable

$$n_o(\vec{k},t) = e^{ikx_1(t)} \tag{2.14}$$

which is the Fourier transform of the tagged particle density. In eq.(2.14) wavevector $\vec{k}$ is along the direction of x-axis and $x_1(t)$ is the x-component of the position of the tagged particle. In terms of tagged particle density, $F_s(k,t)$ is defined as

$$F_s(k,t) = \langle n_o(\vec{k},t) \; n_o(-\vec{k},o) \rangle \quad . \tag{2.15}$$

$S_s(k,\omega)$ is then obtained by equation (2.11). Differentiating eq.(2.15) twice w.r. to t we obtain

$$\frac{\partial^2 F_s(k,t)}{\partial t^2} = -k^2 \langle v_x(t) \; v_x(o) \; e^{ik(x_1(t)-x_1(o))} \rangle \tag{2.16}$$

Introducing the velocity autocorrelation function (VAF) as

$$Z(t) = \langle v_x(t) \; v_x(o) \rangle \tag{2.17}$$

we obtain from (2.16)

$$Z(t) = \lim_{k \to 0} \left[-\frac{1}{k^2} \frac{\partial^2 F_s(k,t)}{\partial t^2} \right] . \tag{2.18}$$

Fourier-transform of this equation gives

$$Z(\omega) = \lim_{k \to 0} \frac{\omega^2 S_s(k,\omega)}{k^2} \tag{2.19}$$

where $Z(\omega)$ is called the frequency spectrum of the VAF. Combining eq.(2.19) with (2.13) we obtain a relation for the diffusion coefficient in terms the VAF as

$$D = \int_o^{\infty} dt \langle v_x(t) \; v_x(o) \rangle = \int_o^{\infty} dt \, Z(t) \tag{2.20}$$

We now define the other dynamical variables of our interest such as number density, longitudinal and transverse current density fluctuations as

$$n(\vec{k},t) = n^{-1/2} \sum_i \exp(ikx_i(t)) \tag{2.21}$$

$$J_\ell(\vec{k},t) = n^{-1/2} \sum_i v_{ix}(t) \exp(ikx_i(t)) \tag{2.22}$$

$$J_t(\vec{k},t) = n^{-1/2} \sum v_{iy}(t) \exp(ikx_i(t)) \tag{2.23}$$

respectively. Here $x_i(t)$, $v_{ix}(t)$, $v_{iy}(t)$ are respectively the x-component of the position, x- and y-components of the velocity of the ith particle. As noted earlier wave-vector k is taken along x-axis. We define the corresponding correlation functions as $F(k,t)$, $C_\ell(k,t)$ and $C_t(k,t)$ similar to eq.(2.15). Their respective Fourier transforms viz. $S(k,\omega)$, $C_\ell(k,\omega)$ and $C_t(k,\omega)$, which give the fluctuation spectra are defined in the similar fashion to that of $S_s(k,\omega)$ from (2.11). It is important here to give the relation between the longitudinal current correlation function $C_\ell(k,\omega)$ with the dynamic structure function $S(k,\omega)$ as

$$C_\ell(k,\omega) = \frac{\omega^2}{k^2} S(k,\omega) \tag{2.24}$$

which is obtained by taking the Fourier transform of the second derivative of $F(k,t)$. In the free particle limit the expressions for $S(k,\omega)$ ($C_\ell(k,\omega)$) and $C_t(k,\omega)$ are quite simple and are given by

$$S^f(k,\omega) = \sqrt{\frac{2\pi m\beta}{k^2}} \exp\left(-\frac{m\beta\omega^2}{2k^2}\right) \tag{2.25}$$

$$C_t^f(k,\omega) = \sqrt{\frac{2\pi}{k^2 m\beta}} \exp\left(-\frac{m\beta\omega^2}{2k^2}\right) \tag{2.26}$$

Sum Rules

Sum rules, commonly known as frequency moment sum rules are exactly calculable properties of the system and are defined as the short-time expansion of the time-dependent correlation function. The knowledge of these rules is very important to check the internal consistency in a theory or experiment. We describe below some low-order sum rules for the longitudinal current correlation function $C_\ell(k,\omega)$ (hence of $S(k,\omega)$) and of the transverse current correlation function $C_t(k,\omega)$ for their use in these lectures.

We define the $2m^{th}$ moment of $C_\ell(k,\omega)$ as

$$\langle \omega_{\ell,t}^{2m} \rangle = \frac{1}{2\pi} \int_{-\infty}^{\infty} d\omega \; \omega^{2m} C_{\ell,t}(k,\omega) \Big/ C_{\ell,t}(k,t=0)$$

$$= (i)^{2m} \left. \frac{d^{2m} C_{\ell,t}(k,t)}{dt^{2m}} \right|_{t=0} \Big/ C_{\ell,t}(k,t=0) \qquad (2.27)$$

In terms of these moments, the short-time expansion of the correlation functions $C_{\ell,t}(k,t)$ are defined as

$$C_{\ell,t}(k,t) = 1 + \sum_{m=1}^{\infty} (i)^{2m} \left. \frac{d^{2m} C_{\ell,t}(k,t)}{dt^{2m}} \right|_{t=0} C^{-1}_{\ell,t}(k,t=0) \frac{t^{2m}}{2m!} \qquad (2.28)$$

It is easy to note that $C_{\ell,t}(k,t=0) = \frac{1}{m\beta}$. Starting from eqs. (2.22) and (2.23), making use of eq.(2.15) for definitions of $C_{\ell,t}(k,t)$ it is straightforward to derive the expressions for the low order moments (upto m=2) of the spectral functions of the current correlations. These results are known [8-10] but for the sake of completeness and their extensive use in these lectures, these results are quoted here

$$\langle \omega_\ell^0 \rangle = \langle \omega_t^0 \rangle = \frac{k^2}{m\beta} = \omega_o^{\,2} \qquad (2.29a)$$

$$\langle \omega_\ell^2 \rangle = 3\omega_o^2 + \frac{n}{2m} \int dr\, g(\vec{r})\, (1 - \cos \vec{k}.\vec{r})\, [(\hat{k}.\vec{\nabla})^2] \phi(r) \qquad (2.29b)$$

$$\langle \omega_t^2 \rangle = \omega_o^{\,2} + \frac{n}{m} \int d\vec{r}\, g(r)\, (1 - \cos \vec{k}.\vec{r}) [\nabla^2 - (\hat{k}.\nabla)^2] \phi(r) \qquad (2.29c)$$

$$\langle \omega_\ell^4 \rangle = 15\,\omega_o^{\,2} + \frac{n}{\beta m^2} \int d\vec{r}\, g(r)\, [\,15\, (\vec{k}.\nabla)^2\, \phi(r) +$$

$$6k \sin(\vec{k}.\vec{r})\, (\hat{k}.\nabla)^3 \phi(r) + 2\beta (1 - \cos \vec{k}.\vec{r})\, (\hat{k}.\nabla \nabla\, \phi(r))^2]$$

$$+ \frac{n^2}{m^2} \iint d\vec{r}\, d\vec{r}\,'\, g_3(\vec{r},\vec{r}\,')\, \{ 1 - 2\cos \vec{k}.\vec{r} + \cos[\vec{k}.(\vec{r}-\vec{r}\,')] \} \times$$

$$(\hat{k}.\nabla)\, (\hat{k}.\nabla')\, (\nabla.\nabla')\, \phi(r)\, \phi(r') \qquad (2.29d)$$

$$\langle \omega_t^4 \rangle = 3\omega_o^{\,2} + \frac{3n}{\beta m^2} \int d\vec{r}\, g(r)\, [\, k^2 \nabla^2 \phi(r) - \frac{2}{3} (\vec{k}.\nabla)^2\, \phi(r)\,]$$

$$+ \sin(\vec{k}.\vec{r}) [\nabla^2 - (\hat{k}.\nabla)^2]\, (\hat{k}.\nabla)\, \phi(r)$$

$$+ \frac{1}{3\beta} (1 - \cos \vec{k}.\vec{r}) [\, (\nabla \nabla \phi(r))^2 - (\hat{k}.\nabla \nabla\, \phi(r))^2\,] +$$

$$\frac{n^2}{2m^2} \iint d\vec{r}\, d\vec{r}\,'\, g_3(\vec{r},\vec{r}\,')\, \{ 1 + \cos[\vec{k}.(\vec{r}-\vec{r}\,')] - 2\cos \vec{k}.\vec{r} \} \times$$

$$\{ (\nabla.\nabla')^2 - (\hat{k}.\nabla)\, (\hat{k}.\nabla')\, (\nabla.\nabla') \}\, \phi(r)\, \phi(r') \qquad (2.29e)$$

In eqs.(2.29) $g(r)$, $g(\vec{r},\vec{r}')$ are respectively the pair correlation and static triplet correlation functions and $\phi(r)$ is the potential. Results for the moments for m = 3 have been also derived by Bansal and Pathak, for these we refer the original reference [11].

Equations (2.29b) and (2.29c) are extremely simple to be evaluated for a given potential g(r). These are plotted for liquid Ar and Rb along with their structure functions in Fig.1. The results shown in Fig.1 for liquid Ar and Rb are from molecular dynamics (MD) studies of Levesque et al and Rahman [12,13], respectively. The

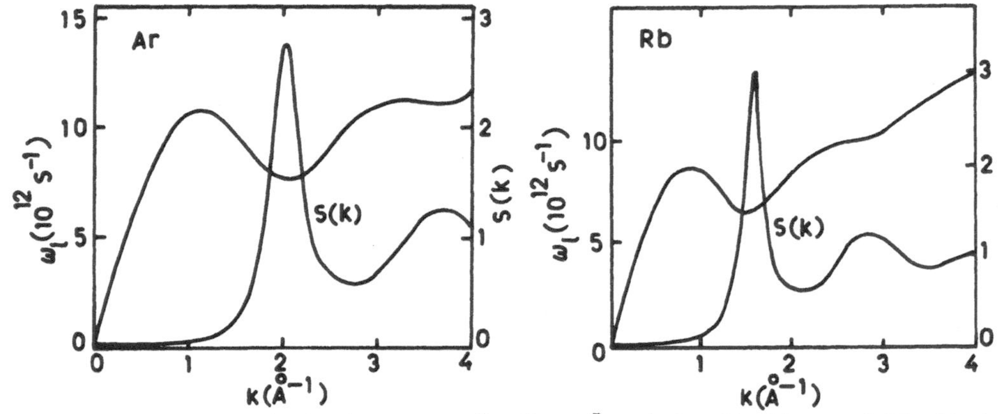

Fig. 1 The characteristic frequency ω_l[eq.(2.29b)] vs k for liquid Ar and liquid Rb obtained from the molecular dynamics (MD) values. The structure factor S(k) is also plotted for completeness.

evaluation of eqs. (2.29d) and (2.29e) was not possible till recently due to appearance of multi-dimensional integrals. These integrals were first simplified by Bansal and Pathak [14] and evaluated using the superposition approximation (SA) for $g_3(\vec{r},\vec{r}')$. Later their exact evaluation using MD method [15-16] justified the use of SA as far as the calculation of the frequency moments is concerned. These results for liquid Rb and Ar alongwith MD results are plotted in Figs. 2(a) and 2(b). Details of these are referred to original references [12,16].

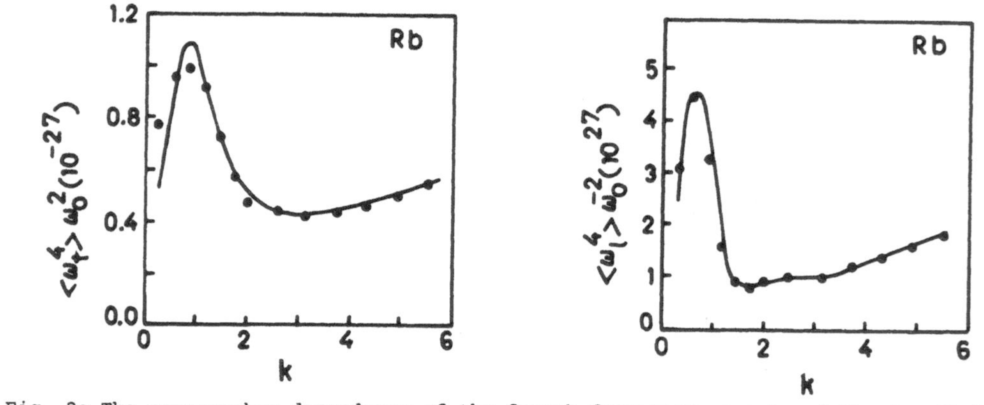

Fig. 2a The wave-number dependence of the fourth frequency moments of the longitudinal and transverse current correlation functions for liquid Rb, superposition approximation - solid curve, solid circles - MD [15].

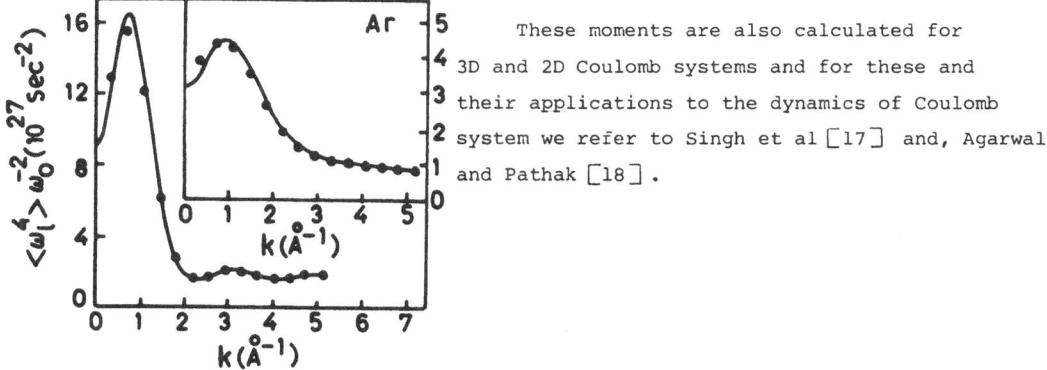

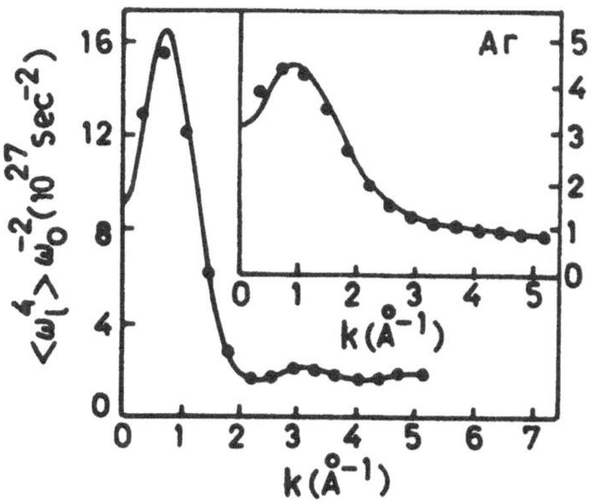

These moments are also calculated for 3D and 2D Coulomb systems and for these and their applications to the dynamics of Coulomb system we refer to Singh et al [17] and, Agarwal and Pathak [18].

Fig. 2b Same as in 2a for liquid Ar, MD results from [16].

3. Langevin Equation

The basic ideas for the discussion of the transport processes in liquids are often taken from the theory of Brownian motion, which describes the problem of self-diffusion. It starts with a simple Langevin equation of a Brownian particle

$$m\frac{d\vec{v}(t)}{dt} + \alpha\,\vec{v}(t) = \vec{f}(t) \tag{3.1}$$

where $\alpha\,\vec{v}(t)$ represents usual viscous drag on a particle of mass m moving with velocity $\vec{v}(t)$. $\vec{f}(t)$ is the randomly fluctuating force of average value zero, representing the effect of molecular impacts on the particle. This is also responsible for the viscous drag coefficient α. The random force is also uncorrelated with the initial velocity i.e.

$$\langle \vec{f}(t)\,.\,\vec{v}(o)\rangle = 0 \tag{3.2}$$

The solution of eq.(3.1) has been discussed in detail by Hansen and McDonald [5], here we simply note the Einstein's relation for the diffusion coefficient in terms of the friction coefficient :

$$D = \frac{1}{\beta\alpha} \quad . \tag{3.3}$$

A direct estimate of α can also be made from the Stoke's equation [5].

From eq.(3.1) we can easily obtain the expression for the VAF of the Brownian particle as

$$\begin{aligned} Z(t) &= \frac{1}{3}\langle \vec{v}(t)\,.\,\vec{v}(o)\rangle \\ &= \frac{1}{m\beta}\exp\left(-\frac{\alpha t}{m}\right) \end{aligned} \tag{3.4}$$

and the frequency spectrum of the VAF is given by

$$Z(\omega) = \frac{1}{D\,m^2\beta^2}\cdot\frac{2}{\omega^2 + (1/m\beta D)^2} \quad . \tag{3.5}$$

we shall see that this frequency spectrum is far from that one gets for a typical liquid. The exponential decay of the VAF is also neither applicable for short-times nor
for long-times. The former can be easily ascertained from the exact knowledge of the
short-time expansion of the correlation functions (2.28).

When the size of the Brownian particle becomes comparable with those of its neighbours, the Markovian approximation of the theory which sets the viscous drag force
on a particle at a given time proportional to the velocity of the particle only at
that time is no longer valid. Because it implies that the motion of the particle ad-
justs itself instantaneously to the changes in the surrounding medium. It is intuitively correct to assume that the viscous drag acting on a particle at a given time depends upon the previous history of the particle. In other words, we must associate
memory effects for the motion of the particle by introducing the time-dependent frictional force as α (t-t'). This leads to non-Markovian generalization of the Langevin
equation written as

$$m \; \frac{d\vec{v}(t)}{dt} + \int_0^t \alpha (t-t') \; \vec{v}(t') \; dt' = f(t) \tag{3.6}$$

A more systematic generalization of eq.(3.6) has been given by Zwanzig and Mori
using the projection operator technique. In fact their main contribution had been to
provide a statistical mechanical expression for the random force. This, we briefly
describe before we use them in the lectures.

In Zwanzig-Mori theory [6-7] we consider a linear vector space spanned by a set
of dynamical variable A($\vec{k}$,t) and define the scalar product in this space as

$$(A|B) = \beta < A^*B > \tag{3.7}$$

The time evolution of the dynamical variable is given by the Liouville equation which
has the formal solution

$$\exp (i \mathcal{L} t) \; A = A(t) \tag{3.8}$$

where $\mathcal{L}$ is the Liouville operator and A is the initial value. It is convenient to
define the Kubo relaxation function ϕ_A(k,t) as

$$\phi_A (k,t) = \beta < A(\vec{k},t) \; A^* \; (\vec{k},o) > . \tag{3.9}$$

It is also useful to define the response function associated with the dynamical variable A($\vec{k}$,t) as

$$\chi (k,t) = - \theta(t) \; \frac{d\phi_A (k,t)}{dt} \tag{3.10}$$

Defining the Fourier-Laplace transform (FLT) of ϕ_A(k,t) as

$$\tilde{\phi}_A (k,\omega) = i \int_0^\infty \exp (i\omega t) \; \phi_A (k,t) \tag{3.11}$$

where ω is in the upper-half plane, one obtains

$$\chi_A(k,\omega) = \Phi_A(k,t=o) + \omega\tilde{\Phi}_A(k,\omega) \tag{3.12}$$

where $\chi_A(k,\omega)$ is the Fourier transform of $\chi_A(k,t)$. In order to get the generalized Langevin equation it is convenient to define a projection operator [19] .

$$P = |A\rangle\langle A|A^*\rangle^{-1}\langle A| \tag{3.13}$$

in the linear vector space spanned by dynamical variable A. The projection operator satisfies $P^2 = P$ as it should. If Q denotes the projector orthogonal to P (i.e. $P+Q=1$, $PQ = QP = 0$, $Q^2=Q$) and the resolvent is defined as $\tilde{R}(\omega) = (\mathcal{L}-\omega)^{-1}$, we have an identity [20]

$$[P\mathcal{L}P - \omega - P\mathcal{L}Q(Q\mathcal{L}Q-\omega)^{-1}Q\mathcal{L}P] \quad PRP = P \tag{3.14}$$

The relaxation function defined through (3.9) can be seen to have FLT

$$\tilde{\Phi}_A(k,\omega) = \beta(A(k)\mid(\mathcal{L}-\omega)^{-1}\mid A(k)) \;. \tag{3.15}$$

Operating the identity (3.14) on $|A(k)\rangle$ and multiplying by $\langle A(k)|$ from left we obtain

$$\tilde{\Phi}_A(k,\omega) = \frac{\beta(A(k)\mid A(k))}{\Omega_A(k) - \omega - \tilde{M}_A(k,\omega)} \tag{3.16}$$

where

$$\Omega_A(k) = (A\mid\mathcal{L}\mid A) \quad (A(k)\mid A(k))^{-1}$$

and

$$\tilde{M}_A(k,\omega) = -(A(k)\mid A(k))^{-1}(Q\mathcal{L}A(k)\mid(Q\mathcal{L}Q-\omega)^{-1}\mid Q\mathcal{L}A(k)). \tag{3.17}$$

$\Omega_A(k)$ is zero in this case due to time reversal property of the variable A(k). Thus we have

$$\tilde{\Phi}_A(k,\omega) = -\frac{\Phi(k,t=o)}{\omega+\tilde{M}_A(k,\omega)} \tag{3.18}$$

It is noted that $\tilde{M}_A(k,\omega)$ is again a resolvent matrix element of a reduced Liouville operator $\mathcal{L}_1 = Q\mathcal{L}Q$. Therefore, an identity analogous to (3.15) can be used to generate the equation for $\tilde{M}_A(k,\omega)$. This procedure leads to Mori's continued fraction representation [7] for the relaxation kernel $\tilde{\Phi}(k,\omega)$.

4. Application to Liquid Dynamics

(a) Velocity Autocorrelation Function :

If we take the dynamical variable $A(\vec{k},t)$ to be tagged particle density $n_o(\vec{k},t)$ the relaxation kernel $\tilde{\Phi}_s(k,\omega)$, which is related to the self-correlation function $S_s(k,\omega)$ is obtained from eq.(3.18) as

$$\tilde{\phi}_s(k,\omega) = -\frac{\phi_{s(k,t=o)}}{\omega + \tilde{M}_s(k,\omega)} \tag{4.1}$$

here $\phi_s(k,t)$ is defined as (instead of eq. (3.9))

$$\phi_s(k,t) = F_s(k,t) \quad , \tag{4.2}$$

thereby $\phi_s(k,t=o) = 1$. We write eq.(4.1) for convenience as below

$$\tilde{\phi}_s(k,\omega) = \frac{-1}{\omega + k^2 \tilde{D}(k,\omega)} \tag{4.3}$$

where

$$k^2 \tilde{D}(k,\omega) = \tilde{M}_s(k,\omega)$$

$$= -(Q\mathcal{L}n_o(k) \mid (Q\mathcal{L}Q-\omega)^{-1} \mid Q\mathcal{L}n_o(k)) \tag{4.4}$$

which follows from eq.(3.17). From the definitions of $S_s(k,\omega)$ and $\tilde{\phi}(k,\omega)$ [eqs. (2.11) and (3.11)] it can be seen that

$$\phi''(k,\omega) = S_s(k,\omega)/2 \tag{4.5}$$

where $\phi''(k,\omega)$ is the imaginary part of $\tilde{\phi}(k,\omega+io)$. Using equation (4.5) in (2.19) it can be found that the Fourier transform of the VAF is given by

$$Z(\omega) = 2 \lim_{k\to o} \frac{\omega^2}{k^2} \phi''(k,\omega) \tag{4.6}$$

from which it immediately follows that the self-diffusion constant D is obtained as

$$D = \lim_{\omega\to o} \lim_{k\to o} \frac{\omega^2}{k^2} \phi''(k,\omega) \tag{4.7}$$

From eq.(4.3) one easily calculates

$$\phi''(k,\omega)_{k\to o} = \frac{k^2}{\omega^2} D''(o,\omega) \tag{4.8}$$

Combining eqs.(4.6) and (4.8) we can calculate the frequency spectrum of the normalized VAF $f(\omega)$ which is obtained as

$$f(\omega) = m\beta Z(\omega) = 2m\beta D''(o,\omega) \tag{4.9}$$

Now our job is to calculate first stage relaxation kernel given by eq.(4.4). Several phenomenological forms for the relaxation kernels have been used in the past for which we refer Copley and Ovesey [21] and latest book by Boon and Yip [22]. But we do this by calculating the relaxation kernel in microscopic mode-coupling approximation. In order to approximate the relaxation kernel given by (4.4) it is convenient to express it as

$$k^2 D(k,t) = (Q\mathcal{L}n_o(k) \mid e^{-iQ\mathcal{L}Qt} \mid Q\mathcal{L}n_o(k)). \tag{4.10}$$

It is easy to note that

$$D(k,t=o) = \frac{1}{m\beta} \tag{4.11}$$

Introducing two mode states as intermediate states in the evaluation of the matrix element of the reduced Liouville operator, using the factorization approximation and the fact that positions and velocities are classically independent, we get

$$k^2 D(k,t) = \sum_{\vec{p}\alpha\beta} k_\alpha \; k_\beta \, F_s(p,t) \; C_{\alpha\beta} \; (\vec{k} - \vec{p},t) \tag{4.12}$$

where $C_{\alpha\beta}(p,t)$ is the current correlation function and is given by

$$C_{\alpha\beta}(p,t) = \frac{p_\alpha \;\; p_\beta}{p^2} \; C_\ell(p,t) + [\, \delta_{\alpha\beta} - \frac{p_\alpha \; p_\beta}{p^2}] C_t(p,t) \tag{4.13}$$

In eq.(4.13) α and β are cartesian coordinates.

We now substitute $C_{\alpha\beta}(p,t)$ in eq.(4.12) and after a little bit of manipulation we get

$$D(k,t) = \frac{1}{3} \sum_p F_s(k-p,t) \quad [C_\ell(p,t) + 2C_t(p,t)] \tag{4.14}$$

which in limit $k \to o$ can be written as

$$D(k=o,t) = \frac{1}{3} \sum_p F_s(p,t) \; [C_\ell(p,t) + 2C_t(p,t)] \tag{4.15}$$

The dominant contributions to the large-time behaviour of the VAF can be obtained using the asymptotic behaviour

$$F_s(p,t) \sim \exp(-p^2 D \,|\, t \,|)$$

and

$$C_t(p,t) \sim \frac{1}{m\beta} \quad \exp \quad (-p^2 \; m \; n \; \eta \; | \; t \; |)$$

and replacing the summation over p by integration, we obtain the normalized VAF f(t) as

$$f(t) = \frac{1}{3\pi^2} \int_0^\infty p^2 \; dp \quad e^{-p^2 (mn\eta + D) \; |t|}$$

which reduces to

$$f(t) = 1/ \; 12 \; \pi^{3/2} \quad (mn\eta + D)^{3/2} \quad | \; t \; |^{3/2} \tag{4.16}$$

Therefore, it is clear that the mode coupling theory gives the correct long-time behaviour of the VAF.

It is noted that eq.(4.15) no more gives the correct value of $D(\vec{k},t)$ at t=o. This is because of mode coupling approximation. In order to get the correct initial value of $D(\vec{k},t)$ we multiply right hand side of eq.(4.15) by $p^3_{max} / 6\pi^2 n$, where p_{max} is some cutoff wave-number. Taking the imaginary part of FLT of $D(\vec{k},t)$ from eq.(4.14), we obtain the expression for $D''(k,\omega)$ in limit $k \to o$ as

$$D''(o,\omega) = \frac{1}{4\pi\, p_{max}^3} \int_o^{p_{max}} dp\, p^2 \int_{-\infty}^{\infty} d\omega'\, S_s(p,\omega') \left[C_\ell(p,\omega-\omega') + 2\, C_t(p,\omega-\omega') \right] \quad (4.17)$$

Using above expression we obtain the frequency spectrum of the normalized VAF from (4.9). For $C_\ell(k,\omega)$ and $C_t(k,\omega)$ mode coupling equations can also be developed and solved self-consistently [20]. However, in the present work we use models for $C_t(k,\omega)$ and $C_\ell(k,\omega)$ which give results closer to experimental values. For $F_s(k,t)$ we use the gaussian approximation. More details are given by Dubey et al [23]. In these calculations a quite good agreement has been achieved as shown in Fig. (3).

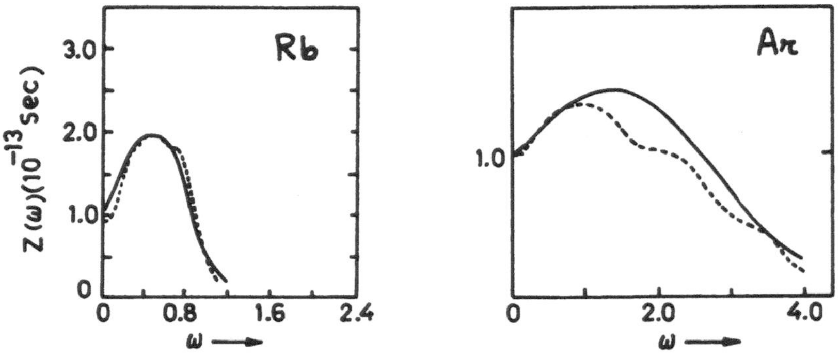

Fig. 3 Frequency spectrum $Z(\omega)$ vs ω for Rb and Ar, broken curves - MD results, continuous curves - mode coupling calculations [23,20].

We also present, in Fig.4, the molecular dynamics results for Z(t) for liquid Ar and Rb from [12] and [13] respectively. For details original references are suggested.

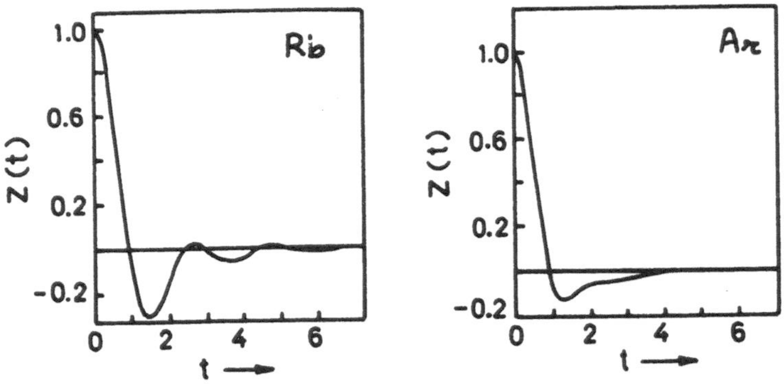

Fig. 4 MD results for Rb and Ar [13,12] for variation of the velocity autocorrelation function Z(t) with time.

In order to see more distinct features of the VAF we also present them for 3D and 2D classical Coulomb liquid in Fig. (5). The details of these calculations are given in original references [24,25] .

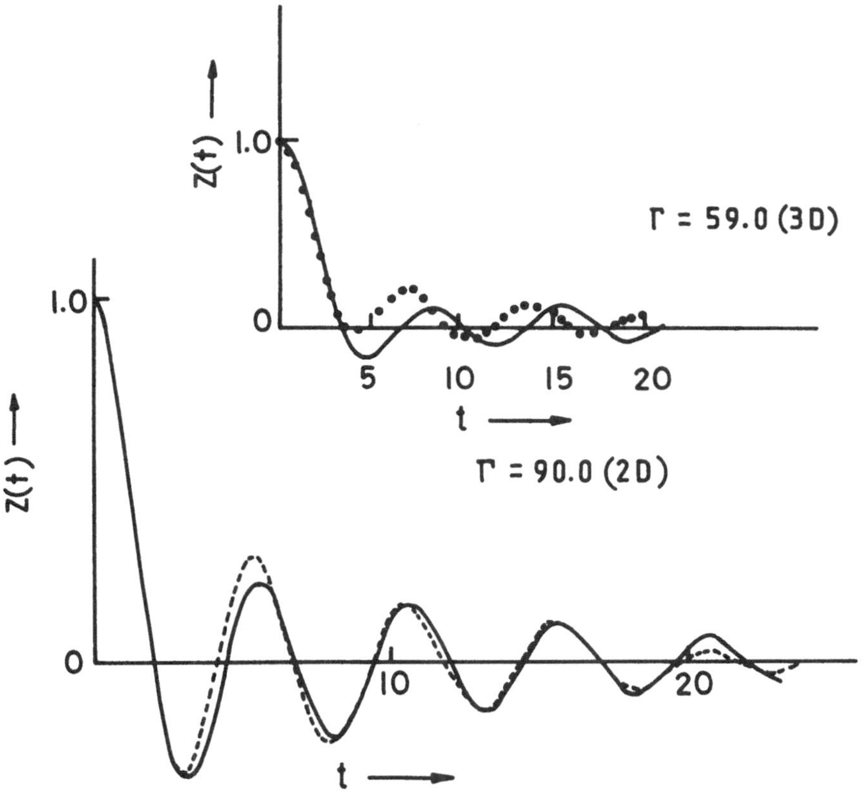

Fig. 5   Z(t) for three-dimensional and two-dimensional classical Coulomb liquids, full line curves - theoretical [24,25] , dots or broken curve - MD results.

(b) Collective Excitations :

If the dynamical variable is taken as density fluctuation operator (defined by eq. (2.21)), the relaxation kernel $\tilde{\phi}(k,\omega)$ is obtained as

$$\tilde{\phi}(k,\omega) \quad = \quad \frac{-\phi(k,t=o)}{\omega + \tilde{M}_1(k,\omega)} \quad . \tag{4.18}$$

In above equation $\phi(k,t=o) = \beta F(k,t=o) = \beta S(k)$, where S(k) is the static structure function and $\tilde{M}_1(k,\omega)$ is the first stage memory kernel. $\tilde{M}_1(k,\omega)$ is given as

$$\tilde{M}_1(k,\omega) = - (n(k) \mid n(k))^{-1}(Q \mathcal{L} n(k) \mid (\mathcal{L}_1 - \omega)^{-1} \mid Q \mathcal{L} n(k)) \tag{4.19}$$

Similar to   (4.18) one write the expression for $\tilde{M}_1(k,\omega)$ and obtains

$$\tilde{M}_1(k,\omega) = \frac{\delta_1}{\omega + \tilde{M}_2(k,\omega)} \tag{4.20}$$

and so on. In eq. (4.20)

$$\tilde{M}_2(k,\omega) = (Q\mathcal{L}n(k) \mid Q\mathcal{L}n(k))^{-1}(Q\mathcal{L}^2 n(k) \mid (Q_1\mathcal{L}Q_1-\omega)^{-1} \mid Q\mathcal{L}^2 n(k)) , \tag{4.21}$$

where Q_1 is orthogonal to $n(k)$ and $Q\mathcal{L}n(k)$ and we have used the relations $Q_1 Q\mathcal{L}Q\mathcal{L}n(k) = Q\mathcal{L}^2 n(k)$ and $Q_1 \mathcal{L}_1 Q_1 = Q_1\mathcal{L}Q_1$. The coefficients δ_k are exactly calculable quantities and are obtained in terms of sum rules [eq.(2.29)] . We record first two of them here

$$\delta_1 = \frac{\omega_0^2}{S(k)} = \Omega_0^2 \tag{4.22a}$$

$$\delta_2 = \langle \omega_\ell^2 \rangle - \Omega_0^2 \tag{4.22b}$$

Combining eqs. (3.12), (4.18) and (4.20) we obtain the density response function as

$$\chi(k,\omega) = \frac{-\beta S(k)\ \Omega_0^2}{\omega^2 - \Omega_0^2 + \omega\tilde{M}_2(k,\omega)} \tag{4.23}$$

The dynamical structure function $S(k,\omega)$ is obtained in terms of the imaginary part of $\chi(k,\omega)$ using eqs. (3.11) and (3.12) and the definition of $S(k,\omega)$. It is given by

$$S(k,\omega) = \frac{2}{\beta\omega}\ \chi''(k,\omega) \tag{4.24}$$

$C_\ell(k,\omega)$ can also be calculated knowing $S(k,\omega)$ and using the relation (2.24).

In order to proceed further one has to make approximation for the relaxation kernel $\tilde{M}_2(k,\omega)$. The various phenomenological approximations for $M_2(k,t)$ i.e. exponential, gaussian and others have been made [26,27,21,22] . The parameters in these are fixed through sum rules. For the dynamics of liquid Ar essentially a single relaxation time kernel has been applied quite successfully, which is due to Pathak and Singwi [28]. In this approximation one replaces $M(k,\omega)$ by its renormalized form as [18,29]

$$\tilde{M}_r(k,\omega) = -\omega + \frac{1}{\omega}\left[\Omega_{or}^2 - \frac{\beta S(k)\ \Omega_0^2}{\tilde{\chi}_r(k,\omega)}\right] \tag{4.25}$$

where $\tilde{\chi}_r(k,\omega)$ is the renormalized free-particle response function [18]. Numerical results for the longitudinal current excitation spectra $C_\ell(k,\omega)$ are shown in Fig. (6a). The excitation current frequencies are shown in Fig.(6b). Similarly the current excitation frequencies in liquid sodium [30] are shown in Fig.(6c).

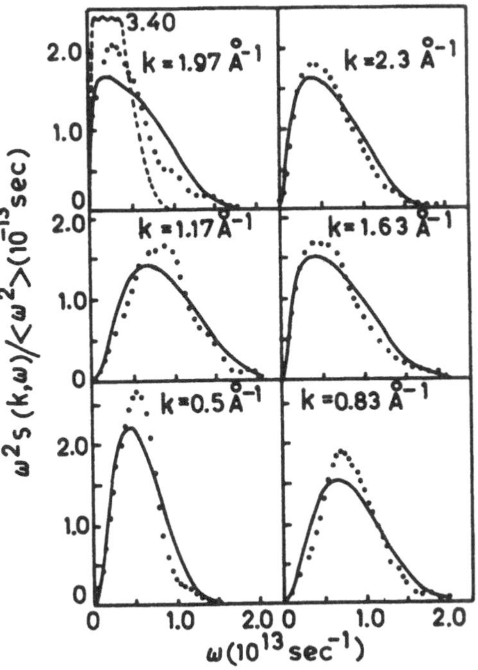

Fig.6(a) Spectral function of the longitudinal current correlations $\omega^2 S(k,\omega)/\omega^2$ plotted as a function of the frequency. Area of each curve is normalized to $\pi/2$ solid circles - MD calculation of Rahman.

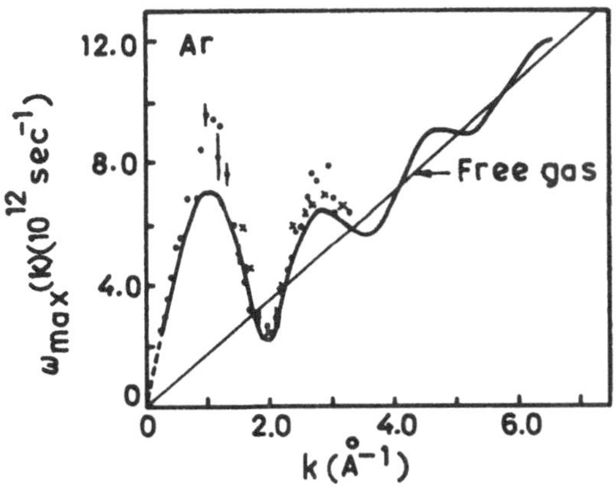

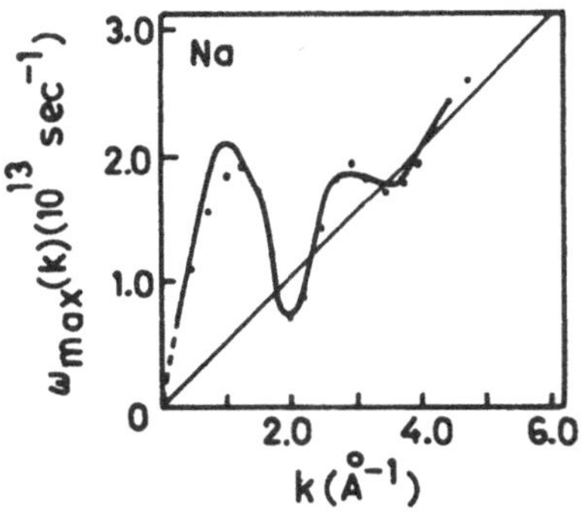

Fig. 6(b,c) Dispersion curve $\omega_{max}(k)$ versus k for the longitudinal current fluctuations, solid circles - MD calculation of Rahman, crosses-experimental results [37].

The experimental observations for liquid metals are more successfully explained in terms of relaxation function having two characteristic relaxation times. For example we choose for M(k,t) [36].

$$M(k,t) = A(k)\ e^{-t^2 / \tau_1^2(k)} + [1 - A(k)]\ e^{-t^2/\tau_2^2(k)} \tag{4.26}$$

where

$$A(k) = [<\omega_\ell^2> - \gamma(k)\ \Omega_o^2]/[<\omega_\ell^2> - \Omega_o^2] \tag{4.27}$$

The parameters are chosen as discussed by Kahol et al [36] and the results of the calculations are shown in Figs. (7) and (8). It is evident from the Fig.7 that a good agreement with the experimental data is achieved. The results of density fluctuation spectra obtained by Jacucci and Mc Donald [17] for liquid Na and K, and their alloys are shown in Fig.9. All these curves show the existence of the collective excitations in liquid metals. The dispersion curves for the density waves are shown in Fig. 10(a) and 10(b).

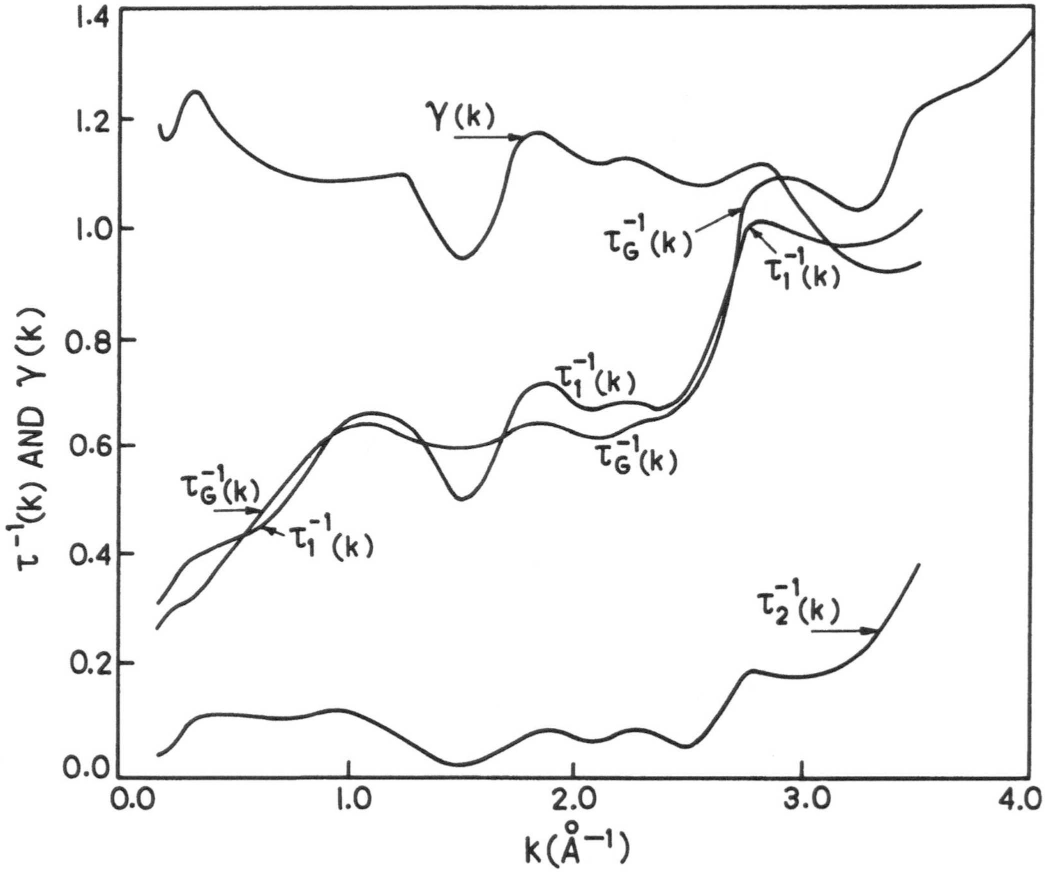

Fig. 7 Inverse relaxation times $\gamma(k)$ in units of 10^{-13} sec^{-1} versus wavenumber k.

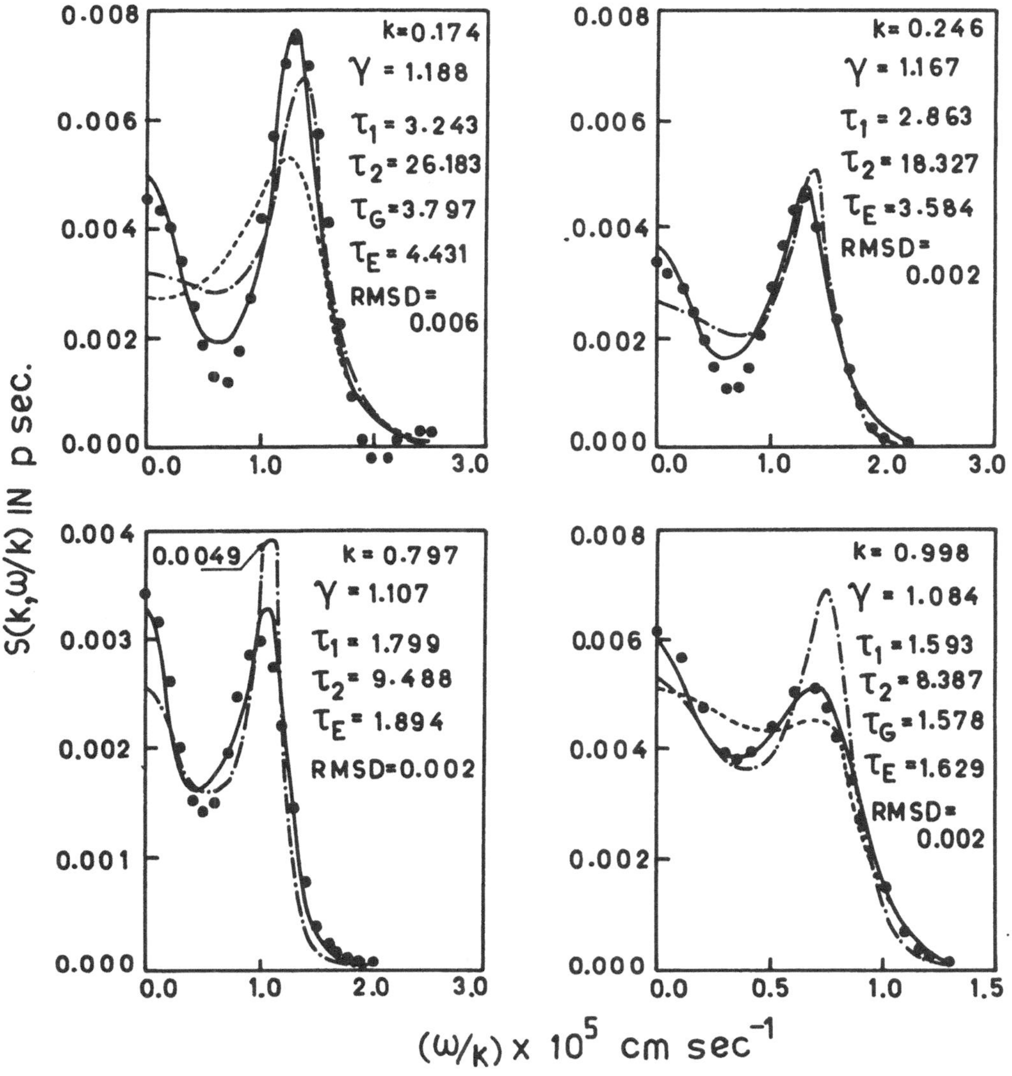

Fig. 8(a) $S(k,\omega/k)$ as a function of velocity ω/k, solid circles - MD calculation [13] solid and broken curves - results using single and double Gaussian memory functions respectively, dash-dot curves - results from exponential form of memory function.

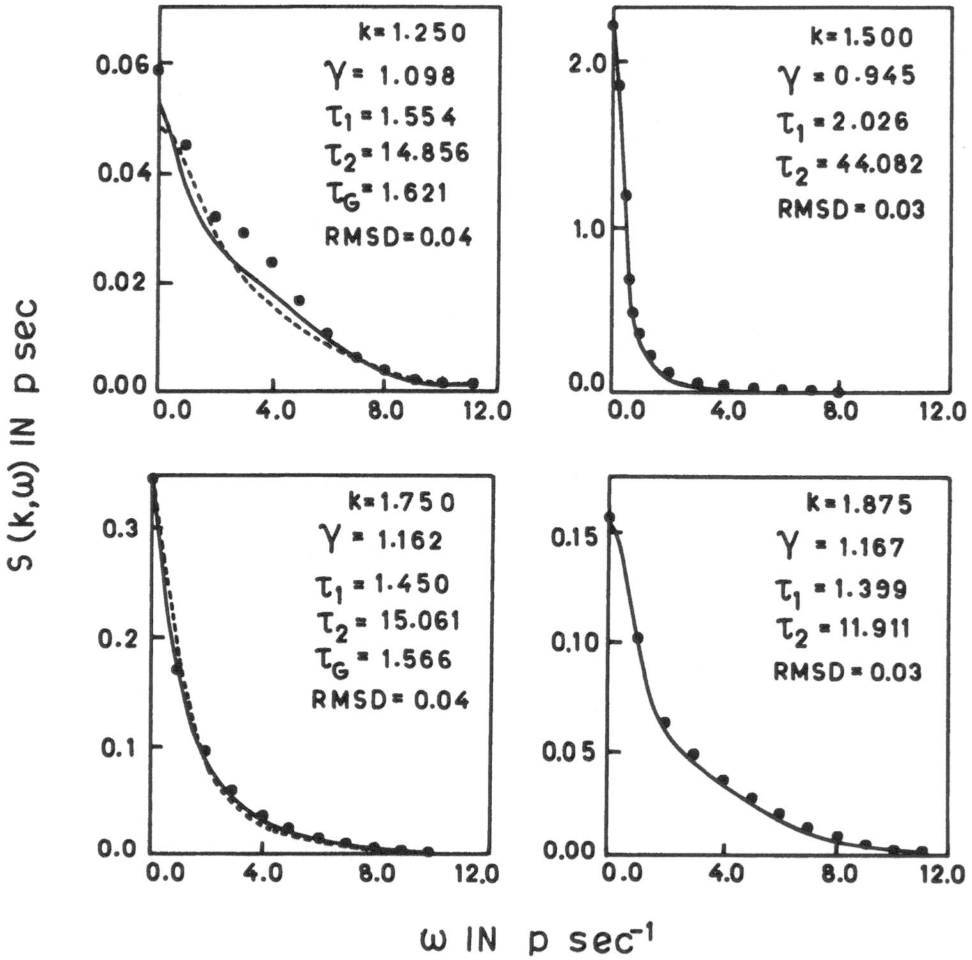

Fig. 8(b) S(k,ω) versus ω - solid circles - the neutron scattering experiments [38].

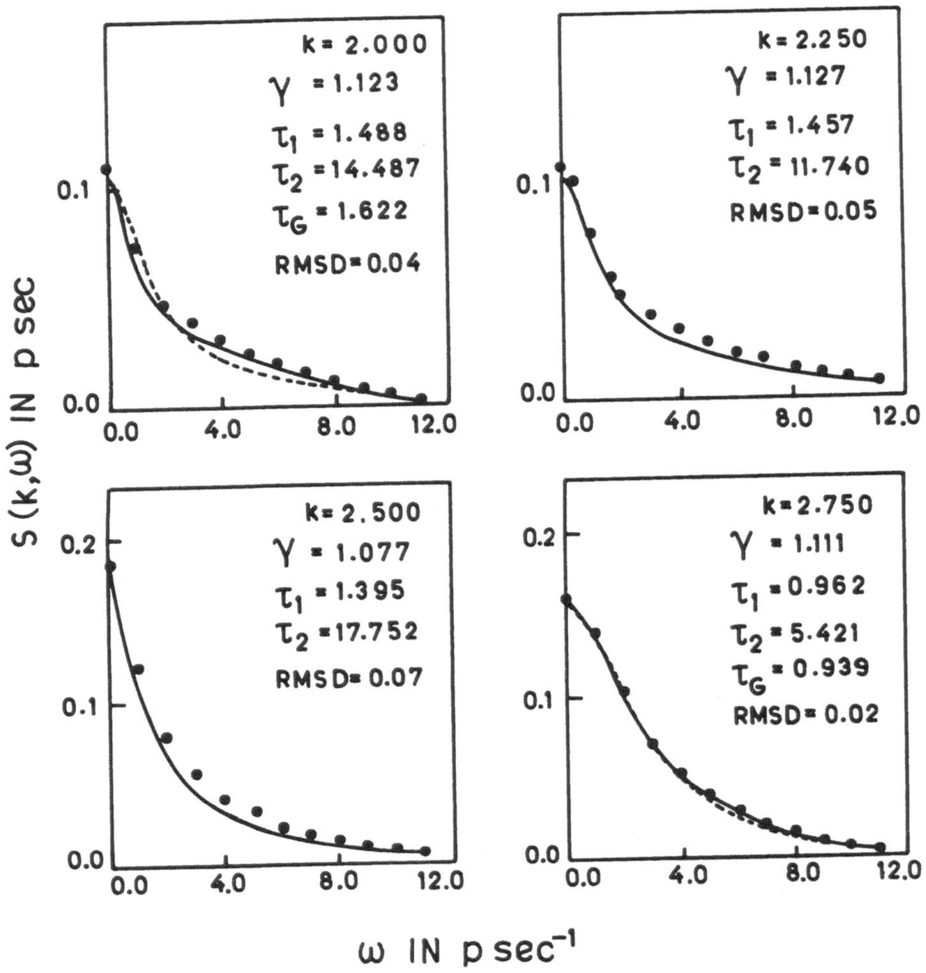

Fig. 8(c) S(k,ω) versus ω-solid circles - the neutron scattering experiments [38] .

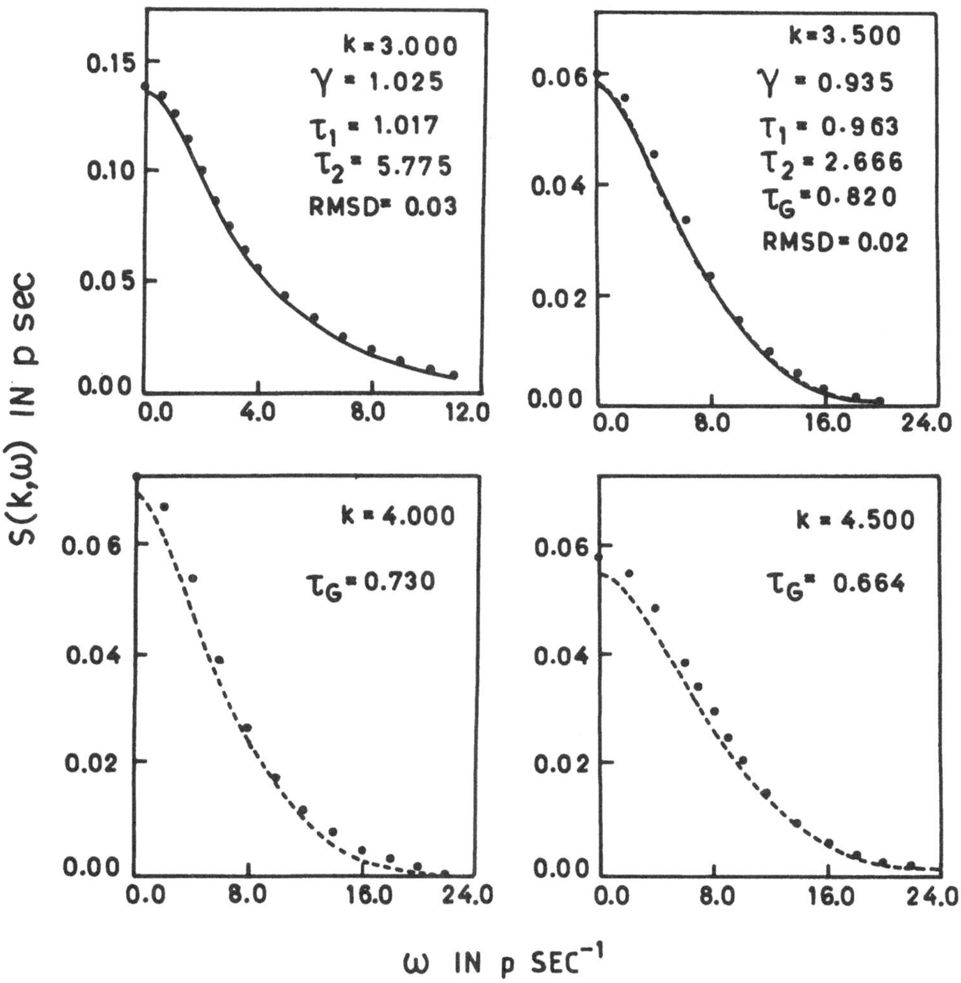

Fig. 8(d) $S(k,\omega)$ versus ω-solid circles - the neutron scattering experiments [38] .

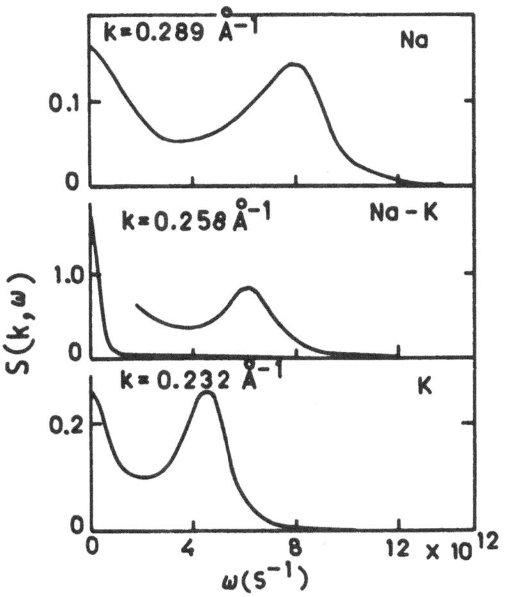

Fig. 9 The density fluctuation spectra $S(k,\omega)$ vs ω. High frequency region for Na-K alloys is shown on a scale amplified twenty times. The area of each curve is normalized to unity.

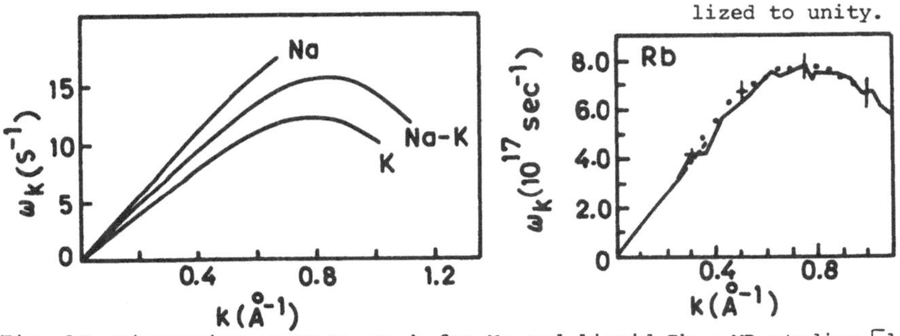

Fig. 10 Dispersion curve ω_k vs k for Na and liquid Rb - MD studies [17,13].

It is also found that the transverse current excitations exist in liquids for wavenumber k greater than certain minimum values. These are discussed in the paper by Kahol et al [29] but more exhaustive reference for transverse current correlation function can be found in Boon and Yip [22].

In recent years attempts have been made to develop microscopic theory for the relaxation kernel, the major effort has been made by Götze and collaborators [31-32], and Sjölander and collaborators [33]. Götze and his collaborators have used microscopic mode coupling theory to calculate the fluctuation spectra of the various correlation functions and have obtained a reasonably good agreement with the experimental data. The approximation due to Sjölander and collaborators is based on the kinetic theory and the results obtained are also in reasonably good agreement with the experimental results. More formal framework known as fully renormalized kinetic theory has been developed by Mazenko [34,35] , but it still remains to make microscopic calculations of the collision kernel.

5. Summary and Conclusion

In these lectures I have introduced various relevant correlation functions which are essential in the study of atomic motion of liquids. The generalized Langevin equation has been discussed and procedure to make calculations from it is outlined. A mode coupling theory for the VAF has been discussed which quite successfully explains the experimental data, as well as known long time tail. Collective excitations are discussed within the framework of phenomenological theory. It is emphasized that the liquid Argon type system can be understood in terms of essentially one relaxation time, whereas for liquid metal type systems, two distinct relaxation times are essential. Recent microscopic calculations are also pointed out. In our opinion it would be of interest to relate the two relaxation times to microscopic processes and interatomic potential. In conclusion it can be fair to say that microscopic dynamics of liquid state is still an open problem.

I am grateful to Mr. G.K. Agarwal without whose help it would have been impossible to prepare this draft. He has done an excellent job of organising and editing the manuscript.

References

1. K.N. Pathak, Phys. Rev. 139A, 1569 (1965).
2. R.S. Tripathi and K.N. Pathak, Nuovo Cimento 21B, 289 (1974).
3. P.F. Choquard, 'The Anharmonic Crystal' (Benjamin; 1967 N.Y.).
4. P.A. Egelstaff, 'An Introduction to the Liquid State' (Academic Press:1967).
5. J.P. Hansen and I.R. Mc Donald, 'Theory of Simple Liquids'(Academic Press:1976).
6. R. Zwanzig, Lectures in Theoretical Physics, Boulder Vol.III, ed. W.E. Britton (Interscience : 1961 N.Y.).
7. H. Mori, Phys. Rev. 112, 1892 (1958); Prog. Theo. Phys. 28, 763 (1962); 33, 423 (1965); 34, 399 (1965).
8. G. Placzek, Phys. Rev. 86, 377 (1952).
9. P.G. de Gennes, Physica 25, 825 (1959).
10. D. Forster, P.C. Martin and S. Yip, Phys. Rev. 170, 155 (1968).
11. R. Bansal and K.N. Pathak, Phys. Rev. A9, 2773 (1974).
12. D. Levesque, L. Verlet and J. Kurkijarvi, Phys. Rev. A7, 1690 (1973).
13. A. Rahman, Phys. Rev. Lett. 32, 52 (1974); Phys. Rev. A9, 1667 (1974).
14. R. Bansal and K.N. Pathak, Phys. Rev. A9, 2773 (1974).
15. R. Bansal, G.S. Dubey and K.N. Pathak, Phys. Rev. A24, 3239 (1981).
16. Ravinder Bansal and W. Bruns, Phys. Rev. A18, 1637 (1978).
17. H.B. Singh, Aruna Sharma and K.N. Pathak, Phys. Rev. A19, 899 (1979).

18. G.K. Agarwal and K.N. Pathak, J. Phys. C15, 5063 (1982).
19. D. Forster, 'Hydrodynamic Fluctuations, Broken Symmetry and correlation Functions'
20. T. Munarkata and A. Igarashi, Prog. Theo. Phys. 58, 1345 (1977); 60,45 (1978).
21. J.R.D.Copley and S.W. Lovesey, Rep. Prog. Phys. 38, 461 (1975).
22. J.P. Boon and S. Yip, 'Molecular Hydrodynamics' (Mc Graw Hill, 1980).
23. G.S. Dubey, V.K. Jindal and K.N. Pathak, Prog. Theo. Phys. 64, 1893 (1980).
24. G.K. Agarwal and K.N. Pathak, J. Phys. C (In press).
25. Rajini Varma, M. Phil. Thesis (Panjab University, Chandigarh) 1982 (unpublished).
26. N.K. Ailawadi, A. Rahman and R. Zwanzig, Phys. Rev. A4, 1616 (1971).
27. Ravinder Bansal and K.N.Pathak, Phys. Rev. A15, 2519 (1977) and A15, 2531 (1977).
28. K.N. Pathak and K.S. Singwi, Phys. Rev. A2, 2427 (1970).
29. P.K. Kahol, D.K. Chaturvedi and K.N. Pathak, Physica 87A, 192 (1977).
30. K.N. Pathak, G. Cubiotti, K.S. Singwi and M.P. Tosi, Nuo Cimento 13B, 185 (1973).
31. W. Götze and M. Lucke, Phys. Rev. A11, 2173 (1975).
32. J. Bosse, W. Götze and M. Lucke, Phys. Rev. A17, 434 (1978).
33. L. Sjögren and A. Sjölander, Ann. Phys. 110, 122 (1978).
34. G.F. Marenko, Phys. Rev. A3, 2121 (1971).
35. G.F. Marenko and S. Yip "Modern Theoretical Chemistry", ed. B.J. Berne Vol. 6 Plenum Press, p.181 (1977).
36. P.K. Kahol, Ravinder Bansal and K.N. Pathak, Phys. Rev. A14, 408 (1976).

STOCHASTIC MODELING OF MOLECULAR DYNAMICS*

Jack H. Freed
Baker Laboratory of Chemistry
Cornell University
Ithaca, New York 14853/USA

We outline here a method, due to Stillman and Freed,[1] for the stochastic modeling of the non-Markovian many-body features of diffusing molecules. We believe this approach is a particularly useful one in that it introduces, in a transparent manner, the basic physics of the relevant degrees of freedom and their couplings, and it is not restricted to linear transport laws. Furthermore, the stochastic features of the bath variables are introduced in a simple and physically transparent fashion. Also, this approach permits consideration of either equilibrium or non-equilibrium dynamics in that the modeling allows for physically relevant choice of equilibrium or stationary-state, and it subjects the expressions to the constraints of detailed balance with respect to this state.

In this method the set of relevant dynamical variables is first augmented with stochastic bath variables which are assumed to obey simple Markovian laws. The augmented set then represents a multidimensional Markov process which obeys a stochastic Liouville equation (SLE) that is, in general, incomplete because it ignores the back reaction of the molecule (i.e. the relevant dynamical variables) on the bath variables. The back reaction effects are incorporated into the model by adding terms to the SLE which are obtained by the constraints required for detailed balance. The resulting augmented Fokker-Planck equation (AFPE) properly describes relaxation to thermal equilibrium, and, for the appropriate limiting conditions, reduces to a classical Fokker-Planck equation. Augmented Langevin equations (ALE) may readily be obtained from the AFPE, and because of the constraint of detailed balance, they automatically obey the fluctuation-dissipation theorem.

The AFPE, which is in general complex, can be solved by means of modern computational algorithms. In particular, it has been shown in recent work,[2,3] that the Lanczos algorithm,[4] which is closely related to the method of moments,[5] is a powerful approach. It is intriguing to note that the projection operator of the method of moments, which can be applied as a mathematical technique in any Hilbert space, is closely related to the projection operator that Mori[6] introduced to obtain a continued fraction representation of the generalized Langevin equation (GLE) for linear physical laws.[3]

We consider a set of independent dynamical variables $\underset{\sim}{\Delta}$ whose equation of motion may be written as

$$\frac{d}{dt}\underset{\sim}{\Delta} = \underset{\sim}{F}(\underset{\sim}{\Delta};\underset{\sim}{\Xi}(t),\underset{\sim}{\lambda}) \tag{1}$$

where $\underset{\sim}{\Xi}(t)$ denotes a set of independent stochastic bath variables and $\underset{\sim}{\lambda}$ is a set of externally determined parameters such as temperature. In general $F(\underset{\sim}{\Delta};\underset{\sim}{\Xi},\underset{\sim}{\lambda})$ may be a non-linear function of the variables. The stochastic process for $\underset{\sim}{\Xi}$ is assumed to be stationary and Markovian with an associated master equation

$$\frac{\partial}{\partial t}P(\underset{\sim}{\Xi},t\underset{\sim}{\lambda}) = -\underset{\sim}{\Gamma}_{\Xi}P(\underset{\sim}{\Xi},t;\underset{\sim}{\lambda}) \tag{2}$$

the SLE may be written as[7]

$$\frac{\partial}{\partial t}P(\underset{\sim}{\Delta},\underset{\sim}{\Xi},t;\lambda) = -[\underset{\sim}{\nabla}_{\Delta}\cdot F(\underset{\sim}{\Delta};\underset{\sim}{\Xi},\underset{\sim}{\lambda}) + \underset{\sim}{\Gamma}_{\Xi}]P(\underset{\sim}{\Delta},\underset{\sim}{\Xi},t;\underset{\sim}{\lambda}) \tag{3}$$

Here ∇_Δ represents the divergence over the space spanned by $\underset{\sim}{\Delta}$ and the first term on the rhs of Eq. (3) represents the Liouville equation form of Eq. (1). Also $P(\underset{\sim}{\Delta},\underset{\sim}{\Xi},t;\lambda)$ is the joint probability distribution in the combined set of variables: $\underset{\sim}{\Delta}$ and $\underset{\sim}{\Xi}$. It should be emphasized that Eq. (3) is incomplete in that the back-reaction effects of $\underset{\sim}{\Delta}$ on the diffusion of $\underset{\sim}{\Xi}$ do not appear in Eq. (2). Thus the stationary solution of Eq. (3) will, in general, yield the correct Boltzmann distribution only in the limit of infinite temperature. Equivalently we may say that the joint probability density of $\underset{\sim}{\Delta}$ and $\underset{\sim}{\Xi}$ as defined by Eq. (3) does not relax to thermal equilibrium. In order to obtain the physically correct stationary solution, additional terms which have been neglected in Eq. (3) need to be found. We note that the SLE will relax to thermal equilibrium if we require that it obey the principle of detailed balance. We therefore seek additional terms to Eq. (3) subject to the constraint that detailed balance be obeyed.

It is convenient to incorporate $\underset{\sim}{\Delta}$ and $\underset{\sim}{\Xi}$ into a new set of augmented dynamical variables $\underset{\sim}{q}$. The SLE may then be written as

$$\frac{\partial}{\partial t} P(\underset{\sim}{q},t;\underset{\sim}{\lambda}) = -\underset{\sim}{\Gamma}(\underset{\sim}{q})P(\underset{\sim}{q},t;\underset{\sim}{\lambda}) \tag{4}$$

If $\underset{\sim}{\Gamma}_{\Xi}$ contains only first and second derivative terms, then $\Gamma(q)$ is of the form of an AFPE:

$$-\underset{\sim}{\Gamma}(q) = -\sum_i\left(\frac{\partial}{\partial q_i}\right)\cdot K_i(\underset{\sim}{q};\underset{\sim}{\lambda}) + \frac{1}{2}\sum_{i,k}\left(\frac{\partial^2}{\partial q_i\partial q_k}\right)K_{ik}(\underset{\sim}{q};\underset{\sim}{\lambda}) \tag{5}$$

where the K_i and K_{ik} are drift and diffusion coefficients, respectively, and except for their time independence are otherwise quite general functions of $\underset{\sim}{q}$ and $\underset{\sim}{\lambda}$. $\underset{\sim}{\Gamma}_{\Xi}$ may in general contain higher order derivative terms or be an integral operator. This will complicate the analysis below without adding any fundamentally new features. Irreversible and reversible drift coefficients may be respectively defined by

$$D_i(\underset{\sim}{q};\lambda) \equiv \frac{1}{2}\left[K_i(\underset{\sim}{q};\underset{\sim}{\lambda}) + \varepsilon_i K_i(\underset{\sim}{\tilde{q}};\underset{\sim}{\tilde{\lambda}})\right] \tag{6}$$

and
$$J_i(\underset{\sim}{q};\underset{\sim}{\lambda}) \equiv \frac{1}{2}\left[K_i(\underset{\sim}{q};\underset{\sim}{\lambda}) - \varepsilon_i K_i(\underset{\sim}{\tilde{q}};\underset{\sim}{\tilde{\lambda}})\right] \tag{7}$$

where
$$\tilde{q} = \{\varepsilon,q_1\ \varepsilon_2,\ldots,\varepsilon_n q_n\} \tag{8}$$

and $\varepsilon_i = \pm 1$ depending on whether q_i changes sign upon time reversal. The necessary and sufficient conditions for detailed balance D_i are given by Haken as[8]

$$K_{ik}(\underset{\sim}{q};\underset{\sim}{\lambda}) = \varepsilon_i\varepsilon_k K_{ik}(\underset{\sim}{\tilde{q}};\underset{\sim}{\tilde{\lambda}}) \tag{9a}$$

$$D_i - \frac{1}{2}\sum_k \frac{\partial K_{ik}}{\partial q_k} = -\frac{1}{2}\sum_k K_{ik}\frac{\partial \Phi}{\partial q_k} \tag{9b}$$

and
$$\sum_i\left(\frac{\partial J_i}{\partial q_i} - J_i\frac{\partial \Phi}{\partial q_i}\right) = 0, \tag{9c}$$

where $\Phi(\underset{\sim}{q};\underset{\sim}{\lambda})$ is the generalized thermodynamic potential defined by the stationary solution of Eq. (4):

$$P_0(q;\lambda) = N\exp(-\Phi) \tag{9d}$$

with N a normalization constant. The AFPE is obtained by adding (or modifying) J_i and/or D_i terms to Eq. (5) so that Eqs. (9) are all fulfilled subject to a particular form for $\Phi(q;\lambda)$ which must be determined by physical considerations. Since by Eq. (9c) the choice of additional J_i is not unique, physical considerations are also required here. In this manner, the back-reaction effects of Δ or Ξ are implicitly included in $\Gamma(q)$, and relaxation to thermal equilibrium is ensured.

It is now possible to generate the ALE from the complete $\Gamma(q)$.[8] The ALE may be written for each dynamical variable q_ℓ as

$$\dot{q}_\ell = k_\ell(q,\lambda) + \sum_{j=1}^{m} g_{\ell j}(q,\lambda)\,\xi_j(t) \tag{10}$$

where the $\xi_j(t)$ are independent Gaussian δ-correlated random functions with

$$\langle \xi_j(t) \rangle = 0 \tag{11a}$$

and $$\langle \xi_\ell(t+\tau)\xi_j(t) \rangle = \delta_{\ell j}\delta(\tau) \tag{11b}$$

k_ℓ and $g_{\ell j}$ are related to the drift and diffusion coefficients by

$$K_\ell(q;\lambda) = k_\ell(q;\lambda) + \frac{1}{2}\sum_{k,j} (\partial g_{\ell j}/\partial q_k)g_{kj} \tag{12a}$$

and $$K_{\ell k}(q;\lambda) = \sum_j g_{\ell j}(q;\lambda)g_{kj}(q,\lambda) \tag{12b}$$

In general the matrix $K = (K_{\ell m})$ is symmetric. Usually K is also nonnegative definite. If K is nonnegative definite, then there exists a real symmetric nonnegative definite matrix $G = (g_{\ell k})$ such that $G^2 = K$. Then

$$G = U^{-1}[(K_i^0)^{1/2}\delta_{ij}]U \tag{13}$$

where $K_i^0\delta_{ij}$ is the eigenvalue matrix of K and U is the corresponding eigenvector matrix. Then the $K_\ell(q;\lambda)$ are obtained from Eq. (12a).

The method may be illustrated by a planar model of torque fluctuations.[1] Let γ, $\dot{\gamma}$, and I be the angular orientation, angular momentum and moment of inertia of the rotator respectively, while $T(\gamma,\dot{\gamma},t)$ is the fluctuating part of the torque and $N(\gamma)$ is the mean-field torque. The Liouville equation for this stochastic process is just:

$$\frac{\partial}{\partial t} P(\gamma,\dot{\gamma},t) = -[\dot{\gamma}\frac{\partial}{\partial \gamma} + I^{-1}\frac{\partial}{\partial \dot{\gamma}}(N(\gamma)+T(\gamma,\dot{\gamma},t)]P(\gamma,\dot{\gamma},t) \tag{14}$$

corresponding to the first term on the rhs of Eq. (3). A simple but useful form for modeling the stochastic properties of $T(\gamma,\dot{\gamma},t)$ is to set it equal to $IV\sqrt{\frac{kT}{I}}f(\gamma-\phi)$ where $f(\gamma-\phi)$ is periodic in $(\gamma-\phi)$, and regard the angle $\phi(t)$ corresponding to the direction that minimizes the torque acting on the rotator as stochastic. If we choose a simple diffusional model in ϕ we may write:

$$\frac{\partial}{\partial t} P(\phi,t) = -\tau_\phi^{-1}\frac{\partial^2}{\partial \phi^2} P(\phi,t) \tag{15}$$

corresponding to Eq. (2). Eqs. (14) and (15) may then be combined as in Eq. (3), to give the SLE for $P(\gamma,\dot{\gamma},\phi,t)$. This SLE is incomplete because it is found to violate

Eq. (9c) when we let the equilibrium Eq. (9d) be a Boltzmann distribution in $\dot{\gamma}$ and a Boltzmann distribution in the mean potential $U_N(\gamma)$, where $N(\gamma) = -\frac{\partial}{\partial \gamma} U_N(\gamma)$. That is

$$\Phi(\gamma,\dot{\gamma}) = I\dot{\gamma}^2/2kT + U_N(\gamma)/kT \tag{16}$$

One readily finds that we may correct this by introducing a reversible drift term related to the fluctuation in ϕ of form $J_\phi = \sqrt{\frac{I}{kT}}\, Vg(\gamma-\phi)\dot{\gamma}$ where $f(\gamma-\phi) = \frac{\partial}{\partial \gamma} g(\gamma-\phi)$. This is the back-reaction of the rotator on the bath variable ϕ. Then the complete AFPE operator corresponding to Eqs. (4) and (5) becomes:

$$\Gamma(\gamma,\dot{\gamma},\phi) = \dot{\gamma}\frac{\partial}{\partial \gamma} + I^{-1}N(\gamma)\frac{\partial}{\partial \dot{\gamma}} + \sqrt{\frac{kT}{I}}\, Vf(\gamma-\phi)\frac{\partial}{\partial \dot{\gamma}} + \dot{\gamma}\sqrt{\frac{I}{kT}}\, V\frac{\partial}{\partial \phi} g(\gamma-\phi) - \tau_\phi^{-1}\frac{\partial^2}{\partial \phi^2} \tag{17}$$

For the simpler form of Eq. (18) with $N(\gamma) = 0$, the spectral densities:

$$j(\omega) = \frac{1}{2\pi} \operatorname{Re}\langle e^{i\gamma}|[i\omega+\Gamma]^{-1}|P_o e^{i\gamma}\rangle \tag{18}$$

(real part of the Fourier transform of the auto-correlation function $\langle e^{-i\gamma(t)}e^{+i\gamma(o)}\rangle$) were found to reduce to the Brownian form in the limit of large τ_ϕ^{-1}. For slower torque diffusion rates more complicated behavior was observed. In particular, for small ω the spectral density approaches the form $j(\omega) = \tau_R/(1+\varepsilon\omega^2\tau_R^2)$ with $\varepsilon > 1$, where τ_R is the Brownian rotational diffusion correlation time. Such a form has been found useful in the interpretation of molecular dynamics studies by ESR[9-12] and NMR[13] methods. Care must be taken, however, in interpreting 3D results in terms of planar models because of periodicity effects.

When, however, the equilibrium of Eq. 9d is taken to the instantaneous value of the potential, i.e.

$$\Phi(\gamma,\dot{\gamma},\phi) = I\dot{\gamma}^2/2kT + U_T(\gamma-\phi)/kT \tag{19}$$

where $U_T(\gamma-\phi) = \sqrt{kTI}\, Vg(\gamma-\phi)$ and we have let $N(\gamma)=0$ for simplicity, then one finds that Eq. (9b) is not obeyed. One readily finds that we must introduce the irreversible drift coefficient: $D_\phi = \tau_\phi^{-1}T(\gamma-\phi)/kT$, so that the AFPE operator is now:

$$\Gamma(\gamma,\dot{\gamma},\phi) = \dot{\gamma}\frac{\partial}{\partial \gamma} + I^{-1}T(\gamma-\phi)\frac{\partial}{\partial \dot{\gamma}} - \tau_\phi^{-1}\left(\frac{\partial^2}{\partial \phi^2} + \frac{\partial}{\partial \phi}\frac{T(\gamma-\phi)}{kT}\right) \tag{20}$$

This form will <u>not</u> reduce to the Brownian form in any limit, since the fluctuating torques do <u>not</u> lead to friction. The equivalent ALE corresponding to Eqs. 17 and 20 may be found in the paper of Stillman and Freed.[1] It was suggested that Eqs. (16) and (17) apply to more rapidly fluctuating torques, while Eqs. (19) and (20) apply to torque components fluctuating at a rate slower than the motion of the planar rotator.[1,9,10,14] This latter case has been called the slowly-relaxing local structure (SLRS) model,[10,14] and it has been useful in analyzing magnetic resonance relaxation studies on molecules in liquid crystals and model membranes. Appropriate 3D forms are given elsewhere.[1,14]

We now address a method of solution of the AFPE, Eqs. (4) and (5) appropriate for the general non-linear case (Eqs. 17 and 20 are non-linear). It is frequently convenient to "symmetrize" Γ by the similarity transformation:

$$\tilde{\Gamma} \equiv P_o(\underset{\sim}{q};\lambda)^{-1/2}\,\Gamma P_o(\underset{\sim}{q},\lambda)^{+1/2}. \tag{21}$$

Then any correlation function can be written as:[2]

$$g(t) \equiv \overline{f^*(t)f(o)} = \langle f|e^{-\Gamma t}P_o|f\rangle = \langle fP_o^{\frac{1}{2}}|e^{-\tilde{\Gamma}t}|fP_o^{\frac{1}{2}}\rangle \tag{22}$$

where $f(t) = f(\underset{\sim}{\Delta}(t))$. Therefore the correlation functions can be obtained once the eigenvalue problem for $\tilde{\Gamma}$ is solved. One may first expand the Hilbert space of the variables $\underset{\sim}{q}$ in a complete ortho-normal basis set so chosen that $\tilde{\Gamma}$ is a complex-symmetric matrix. One chooses $|fP_o^{\frac{1}{2}}\rangle \equiv |z\rangle$ as the starting vector and forms the $(\tilde{\Gamma})^n|z) \equiv |z_{k+1}\rangle$ for k=0 to n-1. Then by Schmidt orthonormalization the linearly independent vectors $|z_{k+1}\rangle$ are transformed into an n-dimensional basis set represented by $|k\rangle$ for k=1 to n. One obtains the recursion relation characteristic of the Lanczos algorithm:[2-4]

$$\beta_k|k\rangle = (\tilde{\Gamma} - \alpha_{k-1}\underset{\sim}{1})\,|k-1\rangle - \beta^*_{k-1})k-2\rangle$$

where $\alpha_k = \langle k|\tilde{\Gamma}|k\rangle = \langle k|\tilde{\Gamma}_n|k\rangle$ and $\beta_k = \langle k|\tilde{\Gamma}|k-1\rangle = \langle k|\tilde{\Gamma}_n|k-1\rangle$. Here $\tilde{\Gamma}_n$ is the n dimensional approximation of $\tilde{\Gamma}$ in the space spanned by the n $|k\rangle$'s. (That is, we may regard $\tilde{\Gamma}_n = P_n\tilde{\Gamma}P_n$ where P_n is the operator that projects any vector in the full Hilbert space onto the n-dimensional sub-space.) In the $|k\rangle$ representation $\tilde{\Gamma}_n$ is tri-diagonal (i.e. $\underset{\sim}{T}_n$) with k^{th} diagonal element given by α_k and off-diagonal elements given by β_k. Then the nth approximation to the spectral density function of Eq. 22 is

$$j_n(\omega) \equiv \mathrm{Re}\int_o^\infty e^{-i\omega t}g_n(t)dt = \mathrm{Re}\{\langle 1|[i\omega\,1 + \underset{\sim}{T}_n]^{-1}|1\rangle\} \tag{23}$$

where $|1\rangle$ is the starting vector in the $|k\rangle$ representation. In this form $j_n(\omega)$ can be calculated as a continued fraction, or alternatively $\underset{\sim}{T}_n$ can be diagonalized by standard methods. The Lanczos tri-diagonalization is extremely efficient for computations.[2] Furthermore, by means of Lanczos methods for (complex) non-symmetric matrices one can work directly with the unsymmetrized AFPE operator, Γ, in Eq. 22.

The above examples illustrate some of the applications of our AFPE approach. We believe that this method has application for Stochastic molecular dynamics calculations.[15] In fact, our approach appears to us to be even more convenient than the present use of generalized Langivin equations (GLE), because the ALE have time-independent coefficients rather than the memory kernels of the GLE, and also non-linear couplings are easily included in either the AFPE or the ALE. The absence of memory kernels is due to the fact that the set of relevent variables has been extended to the required level to make this possible; i.e. the remaining "bath" variables are truly "irrelevant." The Mori[6] concept of a heirarchy of memory functions in which one couples to increasingly faster variables, would have its analogue in the present method. Thus, one would construct a Markov process for the fastest variables, coupling in via the SLE the set of next fastest variables. This, in turn, would serve as the more general Markov process to which the next set of slower variables couple, etc. In this manner, one would go "up the chain" in model building. Lastly, we believe, our method offers the appeal of a direct approach to stochastic modeling such that the physical assumptions being made are quite transparent.

REFERENCES

*Supported by NSF Grants: CHE 8024124 and DMR 8102047

1. A.E. Stillman and J.H. Freed, J. Chem. Phys. 72, 550 (1980).
2. G. Moro and J.H. Freed, J. Chem. Phys. 74, 3757 (1981).
3. G. Moro and J.H. Freed, J. Chem. Phys. 75, 3157 (1981).
4. C. Lanczos, J. Res. Natn. Bur. Stand. 45, 255 (1950); 49, 33 (1952),B.N. Parlett, The Symmetric Eigenvalue Problem (Prentice Hall, N.J. 1980).
5. Yu V. Vorobyev, Method of Moments in Applied Mathematics (Gordon and Breach, N.Y. 1965).
6. H. Mori, Prog. Theor. Phys 34, 399 (1965).
7. N.G. van Kampen, Phys. Rept. 24c, 171 (1976); U. Frisch, in Probabilistic Methods in Applied Mathematics, edited by A.T. Bharucha-Reid (Academic, New York, 1968); A. Brissaud and U. Frisch, J. Math. Phys. 15, 524 (1974).
8. H. Haken, Rev. Mod. Phys. 47, 67 (1975).
9. J.S. Hwang, R.P. Mason, L.-P. Hwang, and J.H. Freed, J. Phys. Chem. 79, 489 (1975).
10. C.F. Polnaszek and J.H. Freed, J. Phys. Chem. 79, 2283 (1975).
11. J.S. Hwang, K.V.S. Rao, and J.H. Freed, J. Phys. Chem. 80, 1490 (1976); W.-J. Lin and J.H. Freed, J. Phys. Chem. 83, 379 (1979).
12. S.A. Zager and J.H. Freed, J. Chem. Phys. 77, 3344, 3360 (1982).
13. H.A. Lopes Cardozo, J. Bulthius, J.H. Freed and W.M.M.J. Bovee, Chem. Phys. Lett. 60, 335 (1979).
14. J.H. Freed, J. Chem. Phys. 66, 4183 (1977).
15. S.A. Adelman and J.D. Doll, Accts. Chem. Res. 10, 378 (1977) and references cited therein.

NON EQUILIBRIUM PHASE TRANSITIONS

NONEQUILIBRIUM PHASE TRANSITIONS - A REVIEW

G. Venkataraman and K. Neelakantan
Reactor Research Center, Kalpakkam 603 102
Madras, India

1. Introduction

This is an elementary and tutorial review on nonequilibrium phase transitions. The tremendous progress in understanding equilibrium phase transitions is well known, and even as these advances were being made, concepts like order-parameter, order parameter fluctuations, etc began to find a place in the study of nonequilibrium phase transitions. It is these developments which we shall discuss.

2. Bifurcation

Let us suppose we start with a system initially not coupled to outside world so that it is in a (thermodynamic) equilibrium state. We now switch on the coupling and apply some stationary external 'force' whose strength is at our disposal. The system will naturally respond, as schematically illustrated in Fig.1. For a small applied force, the system is driven from a state of thermodynamic equilibrium to a steady state. The sequence of nonequilibrium states is usually called a branch, (often, as the thermodynamic branch). When the system is driven farther and farther away from equilibrium, nonlinearities begin to creep in, eventually dominating to the point the thermodynamic branch becomes unstable. At this stage one or more totally new options become available to the system, some of which may be stable while others may be unstable. This availability of two or more new states following an instability is usually referred to as bifurcation [1] . An example of bifurcation is given in Fig. 1. We shall mostly be concerned with this type of bifurcation as it has close analogies to a second-order phase transition.

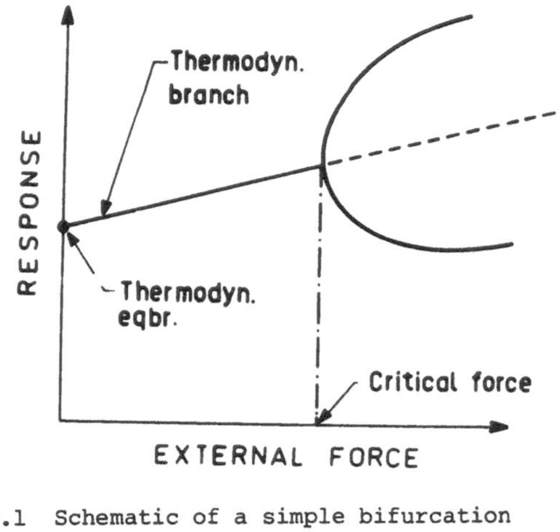

Fig.1 Schematic of a simple bifurcation

The new state that emerges following bifurcation usually has either temporal order, spatial order or even both. The resulting ordered structures are often referred to as dissipative structures [2]. The states on the thermodynamic branch will have the symmetry of the thermodynamic equilibrium state and the applied force. At the point of bifurcation there is a breaking of symmetry [3,4] similar to what happens in phase transitions.

Although we shall be primarily concerned with the type of bifurcation illustrated in Fig.1, we also note that as the external force is increased, the process of bifurcation usually repeats itself many times, resulting in more and more structured and highly complicated nonequilibrium steady states. Eventually, a statistical description of this complicated structure may be necessary for practical reasons.

III. Instabilities

We have noted that when a system is driven far from equilibrium, instabilities set in. We now consider how these instabilities may be analyzed. Initially the description of the process will be deterministic.

To start with, we suppose that our system is described by the macrovariables

$$\bar{X}(r,t) = \left\{ X_1(r,t),\ X_2(r,t), \dots \right\} \tag{1}$$

and a set of 'control' parameters

$$\lambda = \left\{ \lambda_1,\ \lambda_2 \dots\dots \right\}. \tag{2}$$

In the problems we shall be interested in, the evolution equations for $\bar{X}(r,t)$ typically have the form

$$\dot{\bar{X}}(r,t) = F\,(\bar{X}(r,t), \lambda). \tag{3}$$

F is in general a nonlinear operator, and describes the various processes involved energy exchange, matter exchange, transport phenomena, feedback etc. Observe that $\dot{\bar{X}}(r,t)$ depends only on the present state $\bar{X}(r,t)$.

Given below is an example of (1) and (3)

$$\bar{X}(r,t) = \left\{ \rho\,(r,t),\ V(r,t),\ T(r,t) \right\}$$

where ρ , V and T are respectively the density field, velocity field and temperature field. Equation (3) in this case comprises of the well-known trio, namely, the continuity equation, the Navier-Stokes equation and the heat-conduction equation.

Going back to (3), on the thermodynamic branch, the steady states X_s are deduced from

$$F(\bar{X}_s, \lambda) = 0. \tag{4}$$

The emergence of a dissipative structure or self organization as it is sometimes called, is essentially a transition from X_s to a new type of solution. Presently we wish to explore the instability that precedes such a transition. To investigate instabilities, we visualize small disturbances $\bar{x}$ added to the steady state $\bar{X}_s$ and then study the behaviour of the resultant state

$$\bar{X} = \bar{X}_s + \bar{x}\ . \tag{5}$$

When (5) is introduced into (3) one gets a nonlinear dynamical problem. Linearizing, one obtains

$$\dot{\bar{x}} = \bar{\bar{L}}(\lambda)\ \bar{x}, \tag{6}$$

where $\bar{\bar{L}}$ is an appropriate linear operator.

To recapitulate, when the control parameter is gradually increased, the system emerges from thermodynamic equilibrium and 'rides' the thermodynamic branch. At each stage, one has a steady state appropriate to the value of λ involved. What we are now trying to do is to perturb each one of these states and see how the system behaves. The behaviour of the perturbed state $\bar{X}(r,t)$ for $t \longrightarrow \infty$ will determine stability of the reference state $\bar{X}_s(r,\lambda)$.

Under the conditions defined above, $\bar{\bar{L}}$ is time independent and (6) admits solutions of the form

$$\bar{x} = x_o \exp(\omega t), \tag{7}$$

with

$$\bar{\bar{L}}(\lambda)\,\bar{x}_o = \omega\,\bar{x}_o. \tag{8}$$

Everything now depends on ω. The normal-mode frequency ω will in general be complex i.e.

$$\omega(\lambda) = \mathrm{Re}\ \omega(\lambda) + i\ \mathrm{Im}\ \omega(\lambda) \tag{9}$$

from which we see that the stability of the system depends on the sign of Re $\omega(\lambda)$. If it is negative, the oscillations induced by the perturbation will decay and the reference state will be stable. Occurence of instability is therefore connected with a situation as in Fig.2 where Re $\omega(\lambda)$ changes sign at a critical value λ_c of the control parameter. For $\lambda > \lambda_c$, the reference state $\bar{x}_s$ is unstable.

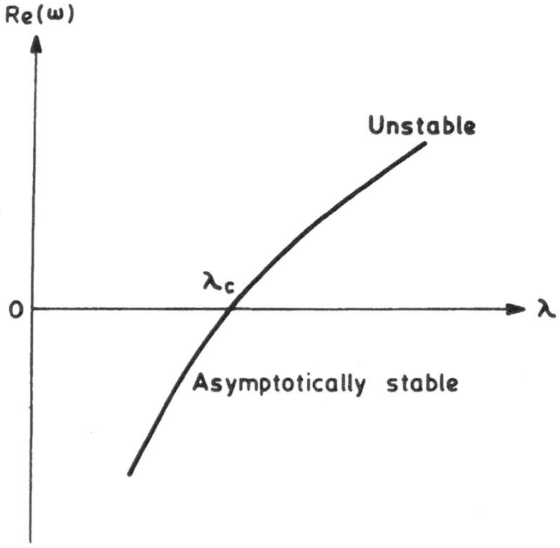

Fig.2 Transition to instability at a critical value of the control parameter.

Directing attention now to Im $\omega(\lambda)$, two possibilities exist:

$$\text{(i)} \quad \text{Im } \omega(\lambda) = 0 \quad \text{for all } \lambda . \tag{10a}$$

$$\text{(ii)} \quad \text{Im } \omega(\lambda_c) \neq 0. \tag{10b}$$

Instability of type (10a) is referred to as soft-mode instability while that corresponding to (10b) is referred to as hard-mode instability.

In some problems like those involving chemical reactions, the operator $\bar{\bar{L}}$ has a term D ∇^2 to deal with diffusive effects [5]. In such cases, we have (considering a 1-D system for simplicity)

$$\bar{x}(r,t) = \bar{x}_o e^{\omega(\lambda, k)t} e^{ikr}; \ k = \frac{2\pi}{L} n; n = 0,1,2, \ldots. \tag{11}$$

where L $\sim$ linear dimension of the system. It is conceivable that the instability is associated with a mode of wave vector $k \neq 0$. If the instability is of the soft-mode type (i.e. $\omega(\lambda, k) = \text{Re } \omega(\lambda, k)$; $\omega(\lambda_c, k) = 0$), then a static, spatially inhomogeneous pattern with wave vector k will result. With hard-mode instability, both spatial and temporal oscillations arise.

4. Role of Nonlinearities

The linear stability analysis does not obviously tell the full story. We must now extend consideration to nonlinearities and also to another equally important entity namely, fluctuations or noise.

Let us ignore noise for the time being and just consider (3). Displaying explicitly the linear and nonlinear parts,

$$\frac{\partial}{\partial t} \bar{x} = \bar{\bar{K}} \bar{x} + \bar{g}(\bar{x}). \tag{12}$$

The nonlinear part will typically have the form

$$g_i(x) = \sum_{\mu} C^{(2)}_{i\mu\upsilon} x_\mu x_\upsilon + \sum_{\mu\upsilon\lambda} C^{(3)}_{i\mu\gamma\lambda} x_\mu x_\upsilon x_\lambda + \ldots. \tag{13}$$

When ths system rides the thermodynamic branch, all the $x_i(t)$'s will decay rapidly. Near the point of bifurcation, one mode i = u say, becomes unstable. Since Re $\omega_u(\lambda \sim \lambda_c) \approx 0$, the decay constant of this mode will be very long i.e. this will be a slow mode as compared to all the other modes. Under these circumstances, one may set

$$\frac{\partial x_i}{\partial t} \approx 0 \quad i \neq u. \tag{14}$$

This is quite reasonable on the time scale on which x_u exhibits variations. Using (14) we can write the r.h.s of

$$\frac{\partial x_u}{\partial t} = \sum_i K_{ui} x_i + g_u(\bar{x}) \tag{15}$$

entirely in terms of x_u. For instance, in the laser problem one has [6]

$$E \approx (G - K)E - C|E|^2 E \tag{16}$$

where G is proportional to the pump power, K is the loss constant and C is a constant depending on the laser material.

The essential physical content of the foregoing steps is that near instability one exploits the fact that there are two vastly different time scales associated with the stable and unstable modes respectively. The stable modes being fast, continuously adjust to the much slower unstable modes. This enables us to write (14) and eventually eliminate the fast modes in the equation for the unstable mode, a process known as adiabatic elimination. It is clear that the mode x_u is connected with the dissipative structure that emerges beyond λ_c and therefore is the analogue of the order parameter of equilibrium phase transition. The slowing down of x_u near bifurcation is thus the analogue of critical slowing down. The dominating role played by x_u is referred to by Haken as the 'slaving principle'. Equation (15) will guarantee the $x_u(t)$ does not increase without bounds.

5. Role of Noise - A First Look

Having identified the analogue of the order parameter, we now ask whether there is the analogue of critical fluctuation also. To discuss this question, we generalize (12) to include fluctuations, i.e., write

$$\frac{\partial}{\partial t}\,\bar{x} = \bar{\bar{K}}\,\bar{x} + g(\bar{x}) + \bar{F}(t) \tag{17}$$

where F(t) is the noise term. One must then go through the adiabatic elimination procedure as before, leading eventually to

$$\frac{\partial x_u}{\partial t} = (\text{terms involving } x_u \text{ only}) + \mathcal{F}_u(t) \tag{18}$$

where $\mathcal{F}_u(t)$ is a fluctuating term. In other words, instead of a deterministic equation as in (16), we now have a Langevin equation. If a Langevin equation occurs, can the corresponding Fokker-Planck equation be far behind! We shall say something more about fluctuations later.

6. Phase Transitions of an OPAMP

We now consider an explicit example of a system exhibiting a non equilibrium phase transition, i.e., an operational amplifier with positive feedback. It is instructive to examine first the role of feedback, especially since feedback is important also in equilibrium phase transitions [7,8]. From Fig.3

$$A_{eff} = \frac{v_o(t)}{v_i(t)} = \frac{A}{1-\beta A} \tag{19}$$

(shades of RPA!). Also sketched in Fig.3 are the transfer characteristics (input vs output) for various values of β .

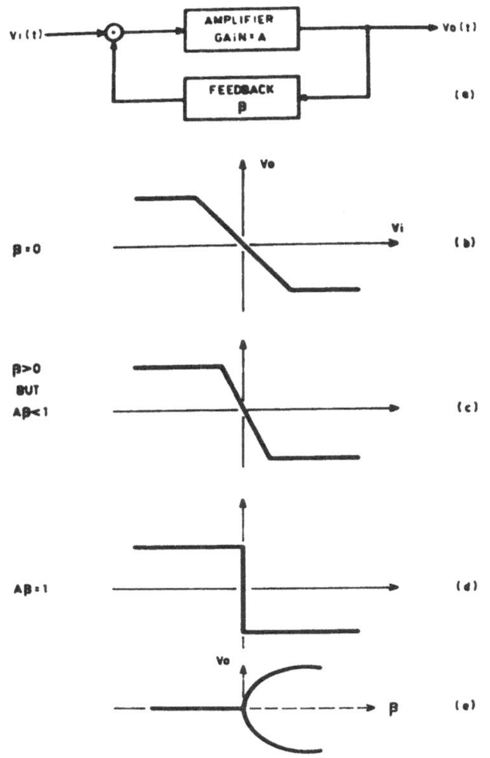

Fig.3 a) Amplifier feedback loop
b) - d) Transfer characteristics for various values of
e) Bifurcation pattern

Suppose v_i is set $\equiv$ 0 and we atart the system from the condition $\beta = 0$. From Fig.3 we see that the output voltage will also be zero. A linear stability analysis (which we skip), will show that this in fact is a stable state. Indeed this state is stable even if $\beta > 0$, as long as $A\beta < 1$. When $A\beta = 1$, the system has two new options. When $A\beta$ exceeds unity, the system settles down into one of the available options, leading in effect to the bifurcation pattern of Fig.3e.

The stability of the output voltage v_o depends essentially on what happens to the energy of the OPAMP capacitor, when the voltage changes by a small amount from the previously occupied state. A little reflection shows that the energy stored W is given by.

$$S = \frac{Cv_o^2}{2} - C \int (v_i - \beta v_o) \, A dv_o . \tag{20}$$

The gain A depends on the differential input voltage which, for the OPAMP used, has the form

$$A(v_i) = A_o (1 - \alpha v_i^2) \tag{21}$$

where A_o and α are constants. Hence, for $v_i = 0$,

$$W = \frac{Cv_o^2}{2} (1 - A_o\beta) + \frac{CA_o\alpha\beta^3}{4} v_o^4 \tag{22}$$

which has the familiar Landau form. The manifestation of a second-order phase transition by the OPAMP is immediately understandable.

In Fig.4a we give some observed results. The OPAMP also exhibits a first-order transition when $v_i \neq 0$; see Figs. 4b and 4c. (In the case of the laser too one has

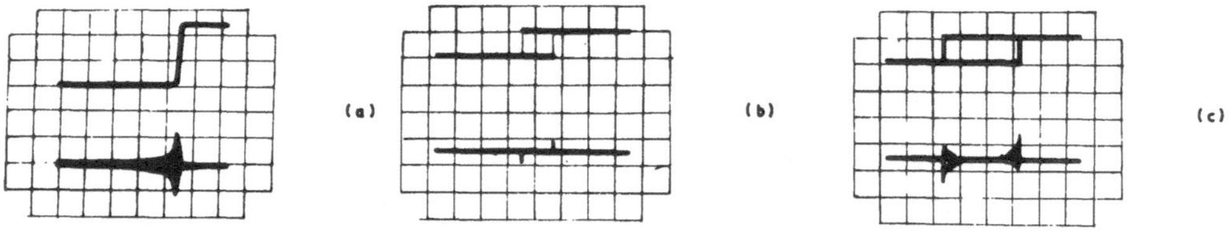

Fig.4 Phase transitions of OPAMP. a) shows the second-order transition. The enhancement of critical fluctuations near the transition is clearly visible. b) and c) show the first-order transition.

a similar behaviour for, with an injected signal, the laser exhibits a first-order transition instead of the usual second-order one). An interesting sidelight relates to the fluctuations. No fluctuations were found in the experiment corresponding to Fig.4b but when the system time constant τ_s exceeds the correlation time τ_c of the input noise, one sees fluctuations.

7. Role of External Noise

We come back again to the question of noise. Firstly we note that near an instability, noise plays a delicate role in switching the system to the 'other phase' even as it does in equilibrium phase transitions. Once the system acquires a dissipative structure, noise loses its dominant role. (For example, the laser emits only monochromatic radiation).

The role of noise in triggering a switching action deserves a closer look when noise enters in a multiplicative way. The rate equations now have the form

$$\dot{x} = \alpha(x) + \beta(x)\,\eta(t) \tag{23}$$

where $\eta(t)$ is the random force. Particularly interesting examples of (23) are [9]

$$\dot{x} = \gamma x - gx^2 + x\,\eta(t) \tag{24a}$$

$$\dot{x} = \gamma x - gx^m + x^m \eta(t) \tag{24b}$$

An experiment to study the role of external noise has been done by Kabashima and Kawakubo [10]. These authors used a degenerate parametric amplifier connected to an external noise source having a flat spectrum between 0.01 and 10^5 Hz. The pumping frequency was 50 KHz and the subharmonic 25 KHz. Figure 5 sketches how the transition was affected by the noise frequency. Kabashima and Kawakubo have also performed several other experiments clarifying various aspects of the role played by external noise.

To paraphrase, multiplicative noise produces nontrivial changes in the behavioural pattern of the system concerned. Such effects are of particular interest

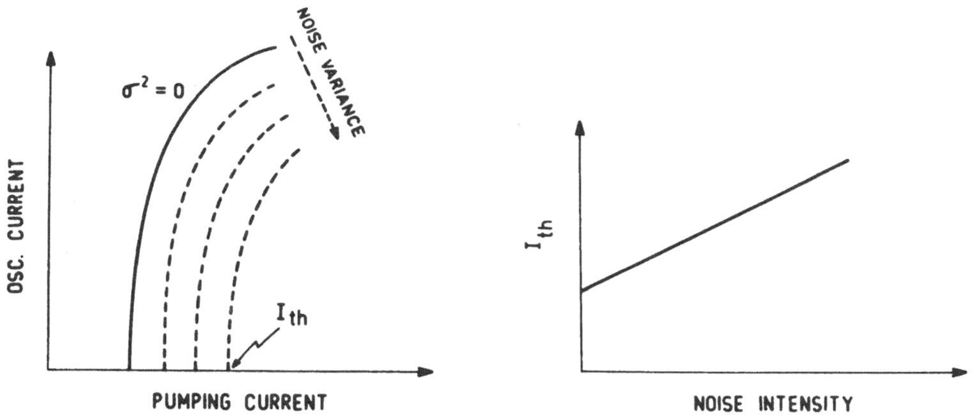

Fig.5 Effect of multiplicative noise on the behaviour of a parametric amplifier. The sketches are schematic. (After ref.[10]

in many biological and ecological problems. We ourselves are interested in external noise with reference to a metallurgical phenomenon known as serrated yielding.

8. Chaos

Although in this review we have been primarily concerned with the very first bifurcation, it is perhaps worth making a few remarks about the random behaviour usually exhibited under high excitation. Such random phenomena often called chaos, are not only seen in hydrodynamical systems (which are the ones usually cited) but also in several other systems like the laser. We shall introduce the concept of chaos via the driven anharmonic oscillator, especially as it is amenable to study in the laboratory.

The oscillator we consider consists of a LCR circuit driven by a sinusoidal voltage, the capacitance C being that of a varactor diode. The special characteristic of C is that it is nonlinear, the law of variation being

$$C(V) \quad \frac{C_o}{(1+V/\phi)^{r}}$$

where V is the voltage across the diode, and ϕ and r are suitable parameters.

The experiment consists in tuning the signal generator to the fundamental frequency f_1 of the LCR circuit and gradually increasing the amplitude of the applied sine wave [11]. For low applied voltages, the normal mode frequency f_1 alone is excited (labelled 0 in Fig. 6). On increasing the driven voltage, the first subharmonic ($f_1/2$) appears (marked 1 on the figure). With further increase in the excitation, there are further manifestations of period doubling. The asymptotic situation is shown in Fig. 6d. The spectrum corresponds to chaotic behaviour, and is definitely not a white noise spectrum.

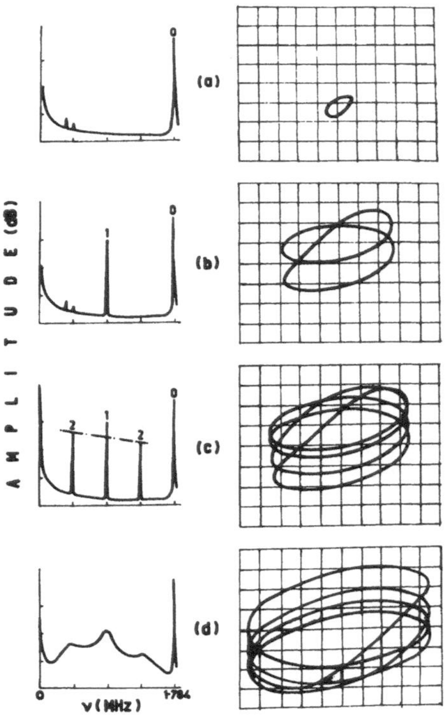

Fig.6 Response of the anharmonic oscillator. On the left are shown the power spectra obtained by Linsay (ref.11). On the right are the corresponsing phase-space plots obtained by us.

We have studied the approach to chaos by observing the phase-space plots electronically. Initially, the phase-space plot has just one loop. As successive period doublings occur, more and more loops get added; after n doublings, there are 2^n loops. When n becomes infinite (which happens when λ exceeds λ_c), we no longer see individual loops but just a blur. In this situation, the solutions $x_i(t)$ of (3) are nonperiodic. Despite this feature, we notice that the representative point always stays around a given region of phase-space. For this reason this region is called an attractor.

We now turn to universalities in nonlinear systems exhibiting period doubling. The benchmark work in this area is due to Feigenbaum [12-14]. In the problems he considers, the value assumed by the variable x at some step (j+1) in its evolution depends on the value it had at the previous step j, i.e.

$$x_{j+1} = f(x_j, \lambda) \tag{25}$$

Making rather general assumptions about the nature of f, Feigenbaum deduced that the control parameter λ should asymptotically satisfy the recurrence relation

$$\frac{\lambda_{n+1} - \lambda_n}{\lambda_{n+2} - \lambda_{n+1}} = \delta$$

where index n refers to the nth period doubling and δ is universal constant that depends only on the value of f near an extremum. For a quadratic extremum, Feigenbaum deduced $\delta = 4.699 \ldots$. In the case of the anharmonic oscillator, Linsay [11] measured δ by observing the voltage thresholds for the successive bifurcations. One value he obtained for δ was 4.5 ± 0.6 which is in reasonable agreement with theory.

Another universal feature deduced by Feigenbaum related to the amplitude of the Fourier components of the oscillations. Let $x^n_{(2k+1)}$ be the (complex) amplitude of

(2k+1)th Fourier component of x(t) when it is periodic with period $T_n = 2^n T_o$. Feigenbaum showed that there is relationship connecting the amplitudes of the odd components x^n_{2k+1} featuring a universal constant $\alpha = 2.5029 \ldots$. Translated to the anharmonic oscillator experiment, what this means is that if a line is drawn joining the peaks labelled 2 (see Fig.6), then the line is drawn joining the n = 3 and n = 4 peaks occur respectively 8.2 and 16.4 dB below it. Linsay found that the third and fourth generation peaks are in remarkable agreement with the theoretical prediction. Gollub [15]has noted a similar feature in his hydrodynamical experiments.

To sum up :

(i) Chaos is a situation where a seemingly random behaviour is exhibited by a deterministic system.

(ii) The chaotic state arises due to the existence of nonlinearities.

(iii) The power spectrum of the chaotic state is quite different from that of thermal noise.

(iv) Experiments have confirmed certain universalities when the route to chaos is through bifurcations involving period doubling.

References

1. D.H. Sattinger, in Lecture Notes in Mathematics, Vol.309: Topics in Stability and Bifurcation Theory, ed. A. Dold and B. Eckmann (Springer-Verlag, Berlin), 1973
2. P. Glansdorff and I. Prigogine, Thermodynamic Theory of Structure, Stability and Fluctuations (Wiley, New York), 1971
3. G. Nicolis, in Systems Far from Equilibrium, ed. by L. Garrido (Springer-Verlag, Berlin), 1980
4. V. Srinivasan, this volume
5. G. Nicolis and I. Prigogine, Self Organization in Nonequilibrium Systems, (Wiley-Interscience, New York), 1977
6. H. Haken, in Synergetics : Far from Equilibrium, ed. by A. Pacault and C. Vidal (Springer-Verlag, Berlin), 1979
7. H. Thomas, IEEE Trans. on Magnetics MAG-5, 874 (1969)
8. H. Thomas, in Noise in Physical Systems, ed. by D. Wolf (Springer-Verlag, Berlin), 1978
9. A. Schenzle and H. Brand, Phys. Rev. A20, 1628 (1979)
10. S. Kabashima and T. Kawakubo, in Systems Far from Equilibrium, ed. by L. Garrido (Springer-Verlag, Berlin), 1980
11. P.S. Linsay, Phys. Rev. Lett. 47, 1349 (1981)
12. M.J. Feigenbaum, J. Stat. Phys. 19, 25 (1978); 21, 665 (1979)
13. M.J. Feigenbaum, Comm. Math. Phys. 77, 65 (1980)
14. M.J. Feigenbaum, Phys. Lett. 74A, 375 (1979)
15. J.P. Gollub, in Systems Far from Equilibrium, ed. by L. Garrido (Springer-Verlag, Berlin), 1980.

ANALOGUE OF OPTICAL BISTABILITY IN DRIVEN JOSEPHSON JUNCTIONS

S.R. Shenoy
School of Physics, University of Hyderabad
Hyderabad - 500134, India

1. Introduction

When open systems are pushed away from equilibrium by an external drive, they settle down to steady states, with the external energy flux balanced by dissipations. Two competing types of dissipations can combine to produce nonlinearities and multiple stable roots in the steady state equations. Transitions between these states are called nonequilibrium or dissipative phase transitions [1,2]. The stationary solution of the Fokker-Planck equation serves to define a free energy-like potential. One can then talk about first-order and second-order transitions, just as in the equilibrium case.

The first example of a second-order dissipative phase transition was the single-mode laser [3], with the incoherent pump intensity being the drive, the coherent photon number being the order parameter, and the competing dissipations being the atomic decay rate and the cavity photon leakage rate. A much-studied recent example of a first-order dissipative phase transition is optical bistability [4]. Here the drive and order parameter are the incident and transmitted coherent radiation intensity, respectively. The competing atomic decay processes $\gamma_{\perp}$, $\gamma_{\parallel}$ and photon leakage κ, produce an effective nonlinearity of the refractive index that yields two possible transparencies, for a given value of the incident intensity. The mathematical formalism involves two-level atoms or spins linearly coupled to photons.

In superconductivity, it is well known that certain combinations of fermion operators, acting on occupied/unoccupied pair states, can be regarded as pseudo-angular momentum operators acting on spin up/spin down states [5]. This pseudo-spin formalism has been applied to Josephson junctions, where the (total) raising or lowering operator $S^{\pm}$ transfers a pair from one superconductor to the other. The total z component operator S^{z} counts the number of excess pairs on one side.

Given the formal description of optical bistability in terms of two-level spins, it is natural to ask if an analogous radiation-induced dissipative transition can occur in a Josephson junction, with external photons coupling to the tunneling pairs.

It turns out that, within a simple model, the answer is 'yes', but with important differences that simplify the problem. These differences arise because the underlying algebra is that of fermi operators, and not actual angular momenta—certain commutators are manifestly negligible if physically appropriate normalizations are chosen. The problem, thus simplified, could be of pedagogic usefulness as an introduction to bistability and hysteresis [6,7].

The system is a Josephson tunnel junction, of capacitance C and with an external resistance R across it. The oxide layer forms an electromagnetic resonance cavity of lowest resonant frequency ω_c. The order parameter is a self-consistently developed d.c. voltage across the junction. (By the Josephson relation, this is also the ac current frequency, $\bar{\omega} = 2e\bar{V}/\hbar$). An external microwave source $\Omega \simeq \bar{\omega}$ is a drive, but it turns out that there could be other drives, that act additively.

The results have been published [6,7] and we will only outline them below, emphasizing some aspects not treated elsewhere.

In Section 2 we present the coupled pseudo spin and photon operator equations, and reduce the problem to one dimension by adiabatic elimination of fast modes. Various possible drives are considered. In Section 3 we examine the conditions for experimental observability of bistability and consistency of the assumptions made. Section 4 considers noise and dynamic correlation effects at the limit of metastability. Finally in Section 5, we place the effect in perspective with other d.c. jump and dissipation effects in Josephson junctions.

2. The Langevin Equations

The hamiltonian for the system is [6,8]

$$H = H_C + H_J + H_{cav} + H_{ext} , \tag{2.1}$$

where the capacitance or charging energy of the junction is

$$H_C = \frac{1}{2}\frac{(2eS^z)^2}{C} ; \tag{2.2}$$

the pair coupling hamiltonian that gives rise to the Josephson tunneling current is

$$H_J = -\frac{\hbar I_J}{4e}(S^+ + S^-) ; \tag{2.3}$$

the hamiltonian describing cavity photons and their interaction with the tunneling current is

$$H_{cav} = i\hbar T(\omega_c)(S^- - S^+)(a+a^+) + \hbar\omega_c a^+ a ; \tag{2.4}$$

and the coupling of the current to external radiation is

$$H_{ext} = i\hbar\int d\Omega' T(\Omega')(S^- - S^+)\, W(\Omega\Omega')\, N_{ext}^{\frac{1}{2}} \cos\Omega' t. \tag{2.5}$$

In the above equation, $T(\omega_c)$ is a coupling constant between the current and the electromagnetic field coming from the familiar $\int d^3x\, \vec{J}.\vec{A}$ interaction, In terms of junction parameters

$$T(\omega_c) = \frac{1}{4}\left(\frac{\pi}{\hbar\omega_c C}\right)^{\frac{1}{2}} I_J \tag{2.6}$$

where I_J is the maximal Josephson current in zero dc magnetic field. The operators a, a^+ are cavity photon descruction and creation operators and S^-, S^+ are fermion operator combinations that obey [6,8]

$$[S^z, S^{\pm}] = \pm S^{\pm} \tag{2.7a}$$

$$[S^+, S^-] \approx \frac{2}{m^2} S^z \tag{2.7b}$$

In this physically preferred normalization

$$\langle S^{\pm} \rangle = e^{i\,\Theta(t)}, \tag{2.8}$$

where the angular brackets are steady state averages and $|\langle S^{\pm} \rangle| = 1$, with the current from (2.3) being $I_J \sin\Theta$. In (2.7b) $m \gg 1$ is of order the system size. N_{ext} is the number of externally supplied photons in the cavity and is not treated as an operator. $W(\Omega, \Omega')$ is the external photon spectral distribution, centred on Ω, with width covering the hysteresis region. A dc magnetic field of wavenumber $\propto \Omega$ is applied, to provide current-photon coupling.

The nonequilibrium average $\langle S^z \rangle$ is related to the quantum phase difference $\Theta(t)$ and voltage $V(t)$ across the junction, by the Josephson relation

$$\frac{2e}{\hbar} \frac{2e\langle S^z \rangle(t)}{C} = \frac{2e}{\hbar} V(t) = \dot{\Theta}(t) \tag{2.9}$$

The other Langevin equations are

$$\dot{a} = -(i\omega_c + K)a - T(\omega_c)(S^+ - S^-) + f_a(t) \tag{2.10}$$

$$\dot{S}^z = -\frac{S^z}{RC} - \frac{I_J}{4ei}(S^+ - S^-) - T(\omega_c)(a + a^+)(S^+ + S^-)$$

$$- 2N_{ext}^{1/2} \int d\Omega'\, T(\Omega')\, W(\Omega, \Omega')(S^+ + S^-)\cos\Omega' t + f_z(t) \tag{2.11}$$

Dissipative terms representing the photon leakage $-Ka$, and capacitor discharge through the external resistance, $-S^z/RC$ have been added to the Heisenberg equations of motion. (In optical bistability language, $\gamma_\perp = 0$, $\gamma_\parallel = (RC)^{-1}$). The quality factor Q of the cavity is $Q = \omega_c/2K$. Noise operator terms f_a, f_z represent photon shot noise and generalized Johnson noise in the resistor that includes zero point effects [9]. They obey

$$\langle f_a(t)\, f_a^+(o) \rangle = 2K\,\delta(t) \tag{2.12}$$

with others $\langle f_a f_a \rangle = 0 = \langle f_a^+ f_a^+ \rangle$ and

$$\langle f_z(t)\, f_z(o) \rangle = \frac{\hbar\omega_c}{4e^2 R} \operatorname{Coth} \frac{\hbar\omega_c}{2k_B T}\, \delta(t) \tag{2.13}$$

Now the following approximations are made: (i) replace operators by nonequilibrium averages; (ii) put $a(t) = \tilde{a}(t)\, e^{-i\Theta(t)}$, where $\tilde{a}$ is slowly varying on a scale Ω^{-1}, and time average on this scale; (iii) assume

$$K \gg RC \tag{2.14}$$

so the photons are fast modes, and $\dot{\tilde{a}} \approx 0$.

Then, in terms of dimensionless scaled variables,

$$f \quad = \quad \omega/\omega_c \quad = \quad 2\ eV/\hbar\omega_c \quad = \quad (2e/\hbar\omega_c)(2e\langle S^z\rangle/C) \tag{2.15}$$

and, on a slow time scale $\gg \Omega^{-1} \approx \omega^{-1}$,

$$\dot{f}(t) \quad = \quad -\frac{1}{RC}\left\{ f + \frac{2Q\alpha}{\left[1+4Q^2(f-1)^2\right]} - \mu \right\} + F_f(t) \tag{2.16}$$

Here the I_J Sin θ term has averaged to zero. The random force obeys

$$\langle F_f(t)\ F_f(o)\rangle \quad = \quad \mathcal{T}^{-1}\ \delta(t) \tag{2.17}$$

where, approximating f-dependence of by its f = 1 value,

$$\mathcal{T}^{-1} \quad = \quad \frac{8e^2 Q\alpha}{\hbar\omega_c\ RC^2}\ (1+\frac{1}{2Q\alpha}) \tag{2.18}$$

The parameters that set the scale are $(\Omega \approx \omega_c)$

$$\alpha \ = \ \frac{8e^2}{\hbar}\left(\frac{T(\omega_c)}{\omega_c}\right)^2 R\ ; \qquad N_c^{-\frac{1}{2}} \simeq \frac{8e^2}{\hbar}\ \frac{T(\Omega)}{\omega_c}\ W(\omega_c,\Omega) \tag{2.19}$$

The first term in (2.16) is the Joule loss term, the second, resonant term, comes from fast mode elimination. It says that photon escape loss is largest when the ac Josephson current is on resonance with the cavity frequency, $f = \omega/\omega_c = 1$. The last term, with opposite sign, is the total drive term

$$\mu \ = \ (N_{ext}/N_c)^{\frac{1}{2}} + 2eV_b/\ \hbar\,\omega_c \tag{2.20}$$

Motivated by the analogy with optical bistability we have considered [6] an external photon drive. Another possibility is as follows. From (2.9), the first term $\sim$ -V/RC says that, in the absence of all other terms, the voltage across the capacitance decays to zero. With a battery V_b in series with the resistance and (junction) capacitance, the charging up of the capacitance, in the absence of other terms would lead to $V \to V_b$ at long times. The decay term for the R.C. circuit is $-(V-V_b)/R\,C$. Hence (2.20) is obtained: the two drives are additive. From (2.5), an oscillatory voltage of maximum amplitude $\propto N_{ext}^{\frac{1}{2}}$, across the junction with a d.c. blocking capacitor, is another possible drive.

The main point to be kept in mind is that for bistable effects discussed below, the d.c. voltage across the junction is the order parameter, and must not be pinned down.

3. The conditions for Bistability

From (2.16) a stochastically equivalent Fokker-Planck equation can be written down. The stationary solution $P_0(f,\mu) = e^{-(2\mathcal{T}/RC)U(f,\mu)}$ defines a potential $U(f,\mu)$ with extrema at $f = \bar{f}$ satisfying the deterministic version of (2.16),

$$\mu - \bar{f} = \frac{2Q\alpha}{1+4Q^2(\bar{f}-1)^2} \ . \qquad (3.1)$$

Within a range of parameters defined by

$$\mu_{c2} < \mu < \mu_{c1} \qquad (3.2a)$$

$$Q^2 \alpha \geqslant 2/3\sqrt{3} \qquad (3.2b)$$

there are three solutions of (3.1), $\bar{f}_1 < \bar{f}_3 < \bar{f}_2$ with $\bar{f}_{1,2}$ corresponding to local minima of $U(f,\mu)$ and $\bar{f}_3$ being a maximum. This simple bistability condition should be compared with that of optical bistability [4] where the full angular momentum commutation relations enter. $\bar{f}(\mu)$ is an S-shaped curve with size and shape determined by the external resistance, for fixed junction parameters.

From (3.2b) it is at first sight easy to satisfy the bistability conditions: for given R, C, I_J, the oxide layer should be well-formed making a good cavity, with large Q. However, for $\alpha = \alpha_0 + \delta$, ($Q^2\alpha_0 = 2/3\sqrt{3}$) and $\delta \ll 1$ the spinodal points ($\mu_{c1,2}$ $\bar{f}(\mu_{c1,2})$) can be evaluated to leading order in δ. The result is $\mu_{c1} - \mu_{c2} \sim \delta^{3/2}/Q$ and $\bar{f}(\mu_{c2}) - \bar{f}(\mu_{c1}) \sim \delta^{\frac{1}{2}}/Q$. Thus 'good' cavities, with $Q \gg 1$ have a small bistability region. On the other hand, very poor cavities, $Q \sim 1$ have diffuse, overlapping cavity modes, washing out the hysteresis. The junction parameters have to be carefully chosen, for the phenomena to be seen.

For (2.16) to be valid, the fast-mode assumption of (2.14) should hold and that is most easily satisfied for small R. For the hysteresis curve to be well described by the extrema $\bar{f}(\mu)$ alone, and not be washed out by fluctuations the stationary probability must be sharply peaked,

$$2\tau/RC = \frac{\hbar\omega_c C}{2e^2}(1+2Q\alpha) \gg 1 \qquad (3.3)$$

This is most easily satisfied for large R. (Note that this condition arises from a Fokker-Planck description and would not arise from a study of steady states alone).

The compatibility of the sharp-peaking and fast-moding conditions depend on the physics of the problem. Using (2.19), (3.3) can be written as

$$1 > 2\ (e^2/C\ \hbar\omega_c)\ [(2Q(T(\omega_c)/\omega_c)^2(e^2R/\hbar) + 1] \qquad (3.4)$$

each bracket of which is less than or around unity, for a range of accessible physical parameters. The first bracket is the ratio of the charging energy to the photon energy; the second is the ratio of the $\vec{j}.\vec{A}$ field-current coupling to the photon energy, and the third is the ratio of the resistance to $\hbar/e^2$ that sets the scale for localization or insulator behaviour. Thus although (3.3), (2.14) seem to impose opposing constraints, there can still be enough freedom to satisfy both, with reasonable parameters [6].

An estimate of the relaxation and metastable decay rates for the system has also been made [6]. This is useful in gauging whether hysteresis will be easily achieved or not [10].

4. Dynamic Correlations

The deterministic relaxation rate within a given well, from linearizing (2.16) is

$$\frac{1}{T_1} = 1 - \frac{16Q^3\alpha(\bar{f}-1)}{[1+4Q^2(\bar{f}-1)^2]} \qquad (4.1)$$

$\frac{1}{T_1}$ vanishes like $\sim |\mu - \mu_{cl,2}|^{\frac{1}{2}}$, analogous to 'critical slowing down' but here at spinodal values $\mu_{cl,2}$ where the metastable well flattens and disappears. The relaxation rate T_1^{-1} appears in noise and dynamic correlations that can be experimentally measured.

Voltage fluctuations across the junction, from (2.16), (2.17) are given by

$$\langle(f(t)-\bar{f})\ (f(o)-\bar{f})\rangle = (T_1/2\tau)\ e^{-t/T_1} \qquad (4.2)$$

in a gaussian approximation. This yields a Lorentzian noise spectrum, $[(\omega T_1)^2+1]^{-1}$ that narrows and rises at the transitions at $\mu_{cl,2}$: the time scale of the fluctuations in voltage increases.

The Josephson radiation line shape is related to the current-current correlation and hence to

$$\langle S^+(t)S^-(o)\rangle = \langle e^{i(\theta(t)-\theta(o))}\rangle \approx e^{i\omega_c\bar{f}t}\ e^{-\frac{1}{2}\langle(\theta(t)-\theta(o))^2\rangle} \qquad (4.3)$$

This yields a gaussian spectrum $\sim \exp[-(\omega-\omega_c\bar{f})^2/(\omega_c^2 T_1/\tau)]$ that broadens and falls at transition : the extent of the fluctuations in voltage, increases. From (4.3) the dispersion in ω is $\langle(\Delta\omega)^2\rangle \sim (T_1/\tau)$. As $\langle(\Delta\omega)^2\rangle \sim \langle(f-\bar{f})^2\rangle$ through the Josephson relation, this is consistent with the static limit of (4.2).

A numerical study of the dynamic correlations through eigenfunction expansions and continued fraction methods [11] has been done [12]. It confirms the validity of the approximations made.

Besides the (mean-field) exponent occurring in the relaxation time, other critical and spinodal exponents can be defined [7].

5. Comparison with other dc effects

The Josephson junction exhibits a family of dc effects [13] when suitably driven and it is useful to place our results, with a radiation drive, in perspective.

Three 'good cavity' effects rely on frequency matching and not on photon leakage, i.e. they exist in the $Q \to \infty$ limit. The Shapiro steps in the dc characteristic [14] occur when $\bar{\omega} = n\Omega$, n = 1,2,3. Fiske steps occur when $\bar{\omega} = n\,\omega_c$ a cavity resonance [15]. In both cases, a d.c. current source is applied to the junction. A d.c. voltage effect observed by Chen et al [16] involves irradiating an unbiassed junction but this is also a good cavity effect, best seen for $Q \to \infty$. Moreover, it is small in our parameter regime [6] .

In another effect [17] a peak occurs in the tunneling current as a function of magnetic field for very poor cavities $Q \sim 1$ with no standing waves. The dc voltage across the junction is pinned down.

In contrast the effect we have considered [6] involves both cavity modes and dissipation, in an intermediate Q regime, with both $Q \to 0$ and $Q \to \infty$ killing it. The effect is radiation intensity driven, assuming a broad $\Omega \sim \omega$ spectrum and is not just a frequency matching effect. It crucially depends on a resonant dissipation, with the dc voltage free to change. It differs from other effects, has connections with optical bistability, and seems worth investigating, experimentally.

I would like to thank M.R. Beasley for a useful comment.

References

1. H. Haken, Synergetics, An Introduction, (Springer, Berlin, 1977)
2. G. Nicolis and I. Prigogine, Self-Organization in Nonequilibrium Systems, (Wiley, New York, 1977)
3. V. de Giorgio and M.O. Scully, Phys. Rev. A2, 1170, (1970)
4. R. Bonifacio and L.A. Lugiato, Phys. Rev. A18, 1129, (1978); H.M. Gibbs, S.L. McCall and T.N.C. Venkatesan, Phys. Rev. Lett. 36, 1135 (1976); see also Optical Bistability, ed. by C.M. Bowden, M. Ciftan, and H.R. Robl, Plenum (1981)
5. P.W. Anderson, Phys. Rev. 112, 1900 (1958); P.R. Wallace and M.J. Stavn, Can. J. Phys. 43, 411, (1965)
6. S.R. Shenoy and G.S. Agarwal, Phys. Rev. Lett. 44, 1525 (1980); 45, 401(E) 1980; Phys. Rev. B23, 1977 (1981)
7. G.S. Agarwal and S.R. Shenoy, Phys. Rev. B25, 1879 (1982)
8. P.A. Lee and M.O. Scully, Phys. Rev. B3, 769 (1971)
9. R.H. Koch, J.D. Van Harlingen and J. Clarke, Phys. Rev. 26, 74 (1982); Phys. Rev. Lett. 45, 2132 (1980)
10. G.S. Agarwal and S.R. Shenoy, Phys. Rev. A23, 2719 (1981)
11. H. Risken, and H.D. Vollmer, Z. Physik B33, 297 (1979)
12. K. Voigtlander, Diplom. Thesis, Universitat Ulm, 1982
13. L. Solymar, Superconductive Tunneling and Applications, (Chapman and Hall, London 1972) and references therein
14. S. Shapiro, Phys. Rev. Lett. 11, 80 (1963)
15. M.D.Fiske, Rev. Mod. Phys. 36, 221 (1964); R.E. Eck, D.J. Scalapino and B.N. Taylor, Phys. Rev. Lett. 13, 15 (1964)
16. J. Chen, R.S. Todd and Y.W. Kim, Phys. Rev. B5, 1843 (1972)
17. D.N. Langenberg, D.J. Scalapino and B.N. Taylor, Proc. IEEE 54, 560 (1966)

NONLINEAR PHENOMENA IN CHEMICAL KINETICS

S. Chaturvedi
School of Physics, University of Hyderabad
Hyderabad - 500 134, India

1. Introduction

Recent years have seen an upsurge of activity in the area of non-linear phenomena in open systems. Numerous phenomena in a wide variety of fields such as hydrodynamics, quantum optics, lasers, electrical networks, ecology, chemical reactions have been investigated by several authors so as to determine features that might be common to all these seemingly diverse systems[1,2] . Indeed this has led to the emergence of a new discipline in physics called Synergetics[1]. Chemical reactions have been of particular interest because one hopes that they might offer a clue to the formation of ordered structures, spatial and temporal oscillations that occur in many biological systems. Main interest in this area has been in setting up simple models which reproduce the essential features of the phenomena observed in nature. Our modest aim in these lectures is to make the uninitiated somewhat familiar with the models and the theoretical techniques used in this area.

2. Chemical Kinetics : Rate Equations :

Consider a chemical reaction involving n chemicals $X_1 \ldots\ldots X_n$ taking place in a vessel of volume V

$$s_1X_1 + s_2X_2 + \ldots\ldots\ldots s_nX_n \xrightarrow{k} r_1X_1 + r_2X_2 \ldots\ldots\ldots + r_nX_n \tag{1}$$

The macroscopic variables of interest are obviously the number of molecules X_i of the chemical X_i. Each time a reaction occurs the number of molecules X_i of type X_i change from X_i to $X_i + r_i - s_i$. The question we ask now is : what are the equations that describe the change in number of molecules of the reactants during the course of a chemical reaction? The answer to this question has been known since the early days of chemical kinetics : The rate at which a chemical reaction occurs is proportional to the product of concentrations of the reactants

$$k \prod_{i=1}^{n} x_i^{s_i} \quad ; \quad x_i = \frac{X_i}{V} \tag{2}$$

Here k is a constant called the rate constant and involves the cross section for a reactive collision between the reactant molecules. More precisely (2) is the number of collisions per unit time per unit volume in which the number of molecules X_i of type X_i change from X_i before the collision to $X_i + r_i - s_i$. The rate equations are therefore:

$$\frac{dX_i}{dt} = k(r_i - s_i)\ V \prod_i (x_i)^{s_i} \tag{3}$$

or

$$\frac{dx_i}{dt} = k\,(r_i - s_i) \prod_i (x_i)^{s_i} \tag{4}$$

These considerations allow us to write down the rate equations for any arbitrary chemical reaction. Thus, for instance, for the reaction

$$2X \underset{k_2}{\overset{k_1}{\rightleftharpoons}} Y \tag{5}$$

The rate equations are

$$\frac{dx}{dt} = -2k_1x^2 + 2k_2y$$

$$\frac{dy}{dt} = k_1x^2 - k_2y \tag{6}$$

The above description is valid provided a number of physical requirements are satisfied[3]. It is a good description for an isothermal homogeneous chemical reaction in which Maxwell distribution is maintained through sufficiently frequent non-reactive collisions.

3. Some Examples of Open Chemical Systems Exhibiting Interesting Non-Linear Phenomena:

The reaction (5) considered above is an example of a closed chemical system as it does not involve any flow of matter from outside. The stationary states of such systems do not exhibit anything particularly striking. However a wide variety of interesting phenomena become possible in open chemical systems i.e. chemical systems in which certain chemicals are externally maintained at fixed controllable concentrations. Some examples of such chemical reactions which have been extensively studied in the literature are given below

(1) $$A+X \underset{k_4}{\overset{k_2}{\rightleftharpoons}} 2X \quad ; \quad B+X \underset{k_3}{\overset{k_1}{\rightleftharpoons}} C \tag{7}$$

Here the chemicals A,B and C are held at a fixed concentration denoted by a,b and c respectively. This model due to Schlögl[4]is an example of an open chemical system which exhibits a second order phase transition behaviour in the steady state. The rate equation for this model is

$$\frac{dx}{dt} = (k_2a - k_1b)x - k_4x^2 + k_3 \tag{8}$$

from which it follows that in the limit $k_3 \to o$, $x=o$ is a stable steady state for $k_2a < k_1b$ and for $k_2a > k_1b$, $x = (k_2a - k_1b)/k_4$ becomes the stable steady state.

(2) A second model also due to Schlögl which exhibits a first order phase transition behaviour is as follows

$$A + 2X \underset{k_2}{\overset{k_1}{\rightleftharpoons}} 3X \quad ; \quad B+X \underset{k_4}{\overset{k_3}{\rightleftharpoons}} C \tag{9}$$

The rate equation for this reaction is

$$\frac{dx}{dt} = k_1ax^2 - k_2x^3 - k_3bx + k_4c \quad . \tag{10}$$

In the steady state one finds that depending on the parameter values one has either one stable steady state or two stable and one unstable steady state.

(3) A two variable example known as the Brusselator[2] which models the oscillations observed in Belusov-Zhabotinsky reaction is as follows :

$$A \xrightarrow{k_1} X \ ; \ B + X \xrightarrow{k_2} Y + D \ ; \ 2X + Y \xrightarrow{k_3} 3X \ ; \ X \xrightarrow{k_4} E \tag{11}$$

The corresponding rate equations are

$$\frac{dx}{dt} = k_1a - k_2bx + k_3x^2y - k_4x$$

$$\frac{dy}{dt} = k_2bx - k_3x^2y \tag{12}$$

A detailed analysis of these equations[2] shows that as one varies the parameter this model exhibits a transition from a homogeneous steady state to a limit cycle behaviour i.e. to sustained temporal oscillations.

(4) Recently Escher[5] has constructed a chemical reaction scheme

$$A+2X \longrightarrow 3X, \ B+2Y \longrightarrow X+3Y, \ X+Y \longrightarrow C, \ D+X \longrightarrow X+Y+E \ ; \ F \longrightarrow X \ . \tag{13}$$

the rate equations for which involve, in contrast to the Brusselator, only quadratic nonlinearities. This model exhibits sustained oscillations on an ellipse in the x-y plane.

We note here that the last three models involve trimolecular reactions which hardly occur in nature. However these reaction schemes are to be understood as effective reaction mechanisms corresponding to a more elaborate reaction system. Thus for instance the trimolecular step in (9) may be thought of as arising from two bimolecular reactions involving a short lived intermediate Y :

$$A+Y \rightleftharpoons X+Y \ , \qquad Y \rightleftharpoons 2X \quad . \tag{14}$$

Similarly a model called the Oregonator[6] with the same features as the Brusselator but without the trimolecular step is

$$A+Y \longrightarrow X \ ; \ X+Y \longrightarrow P \ ; \ B+X \longrightarrow 2X+Z \ ; \ 2X \longrightarrow Q \ ; \ Z \longrightarrow fY \ . \tag{15}$$

The rate equations above apply to homogeneous systems. If we drop this requirement and phenomenologically add diffusion terms in accordance with Fick's law then the resulting equations open up possibilities for new kinds of instabilities. For instance, the reaction diffusion equations for the Brusselator

$$\frac{\partial x}{\partial t} = D_x \nabla^2 x + k_1a - k_2bx + k_3x^2y - k_4x$$

$$\frac{\partial y}{\partial t} = D_y \nabla^2 y - k_3x^2y + k_2bx \ . \tag{16}$$

exhibit, depending on the parameters, a hard mode instability leading to spatial oscillations [1,2].

Chemical Kinetics : Stochastic Desctiption [7] :

In a stochastic description of a chemical reaction system the question before us is as follows. Given a chemical reaction such as (1) what is the evolution equation for the probability $P(X_1, X_2 \ldots\ldots t)$ that at time t we have X_1 molecules of type X_1, X_2 molecules of type X_2 etc present in the reaction system? Two obvious requirements on such a description are, that one should be able to write down an equation for $P(X_1, X_2 \ldots\ldots t)$ from simple kinematic considerations and that this equation should reproduce the macroscopic rate equations in the thermodynamic limit. Such a description is known as a mesoscopic description[3]. Let us now outline how one writes an evolution equation - the master equation for $P(\underline{X},t)$. As an illustrative example, we consider the reaction

$$2X \longrightarrow A \tag{17}$$

The master equation for (17) must clearly have the following structure

$$\frac{dP(X,t)}{dt} = W(X \mid X+2)\ P(X+2,t) - W(X-2 \mid X)\ P(X,t) \tag{18}$$

where $W(X \mid X+2)$ is the transition probability per unit time that (X+2) molecules of X change by 2 to X. We now fix the form of $W(X\ \ X+2)$ from simple combinatorial considerations. Let us divide V into cells of volume ΔV with sides of the order of the interaction range between the molecules. Now clearly

$W(X \mid X+2) \propto$ (probability that exactly two molecules chosen from (X+2) molecules would find themselves within a cell of volume ΔV somewhere in V) . (no. of ways in which such a pair may be chosen from a total X+2 molecules)

$$\propto V \left(\frac{\Delta V}{V}\right)^2 \left(1 - \frac{\Delta V}{V}\right)^X \frac{(X+2)\ (X+1)}{2!}$$

$$= k\ V \left(\frac{1}{V}\right)^2 (X+2)\ (X+1) \qquad \text{for } \frac{\Delta V}{V} \ll 1\ . \tag{19}$$

Hence the master equation for (17) becomes

$$\frac{dP(X,t)}{dt} = kV^{-1} \left[(X+2)\ (X+1)\ P(X+2,t) - X(X-1)\ P(X,t) \right] . \tag{20}$$

From (20) we have for $\langle x \rangle$, as V ,

$$\frac{d\langle x \rangle}{dt} = -2k_1 \langle x^2 \rangle \tag{21}$$

If we now heuristically neglect fluctuations by replacing $\langle x^2 \rangle$ by $\langle x \rangle^2$ then we obtain the macroscopic rate equation for (17). A careful analysis based on system size expansion[3] shows that the macroscopic rate equation do indeed emerge from the master equation such as (20) in the thermodynamic limit.

The considerations given above can be easily extended to an arbitrary chemical reaction system. Thus for the reaction system (7) the master equation reads

$$\begin{aligned}\frac{dP(X,t)}{dt} &= k_2 A V^{-1}\left[(X-1)\ P(X-1,t) - X\ P(X-1,t)\right] \\ &+ k_1 B V^{-1}\left[(X+1)\ P(X+1,t) - X\ P(X,t)\right] \\ &+ k_4\ V^{-1}\left[(X+1)\ X\ P(X+1,t) - X(X-1)\ P(X,t)\right] \\ &+ k_3\ C\left[P(X-1,t) - P(X,t)\right] \end{aligned} \tag{22}$$

Master equations of the type (20) or (22) go by the name of birth-death master equations.

The above description applies to systems where homogeneity is maintained either externally through stirring or internally through rapid diffusion. When this is not the case we have to incorporate effects of spatial diffusion in our description [8]. To do this we divide the system into cells of volume ΔV with side length l. The cells are labelled by an index i and the number of molecules of a chemical X inside cell i are denoted by X_i. Inside each cell we assume homogeneity so that the reaction is described by a master equation such as (22). We model diffusion as a birth death process in which a molecule is transferred from cell i to cell j with probability per unit time $d_{ij}\ X_i$ i.e. the probability of transfer from cell i to cell j is proportional to the number of molecules in the cell i. The master equation for the reaction diffusion systems then becomes

$$\begin{aligned}\frac{dP(\underline{X})}{dt} &= \sum_{ij} d_{ij}\left[(X_{i+1})\ P(X_1 \ldots . X_{i+1}, \ldots . X_{j-1}, \ldots X_{n,t}) \right. \\ &\left. - X_i\ P(X_1 \ldots . X_{n,t})\right] + \sum_i \left(\frac{\partial P}{\partial t}\right)_{i,\text{chemical}}\end{aligned} \tag{23}$$

Let us first look at only the diffusion part of (23). It follows from (23) that for $\langle x_i(t)\rangle$ we have

$$\frac{d}{dt}\langle x_i\rangle = \sum_j D_{ij}\langle x_j(t)\rangle \tag{24}$$

where $$D_{ij} = d_{ij} - \delta_{ij}\sum_k d_{jk} \tag{25}$$

We now look at the continuum limit of (25). Suppose that the centre of the cell i is located at r_i

$$x(\underline{r}_{i,t}) = \frac{X_i(t)}{l^3} \tag{26}$$

We further assume that

$$d_{ij} = \begin{cases} 0 & \text{i,j not nearest neighbours} \\ d & \text{i,j nearest neighbours} \end{cases} \tag{27}$$

Then in the limit $l \longrightarrow 0$, (25) becomes

$$\frac{\partial}{\partial t} \langle x(\underline{r},t)\rangle = D\nabla^2 \langle x(\underline{r},t)\rangle ; \quad D = dl^2 \tag{28}$$

and we recover the usual diffusion equation.

Systematic Expansion Methods:

For any reaction diffusion system master equations such as (23) are easily written down. However solving them to get some information about the fluctuation, particularly when nonlinearities are present is an altogether different proposition. A large number of methods have been developed to this end [9]. A systematic expansion procedure of wide applicability known as system size expansion has been developed by van Kampen [3] to handle such situations. As this method has been discussed at length by Ananthakrishna in these proceedings we shall not follow this method directly but by first transforming the master equation into a more convenient form using the Poisson representation [8,10,11].

The Poisson representation method is based on expanding the probability distribution in terms of Poisson distributions. Thus in a single variable case one writes

$$P(X,t) = \int d\alpha \; e^{-\alpha} \frac{\alpha}{x!} \; f(\alpha,t). \tag{29}$$

$f(\alpha,t)$ is called the quasi probability distribution. It is easily seen that the moments of $f(\alpha,t)$ are equal to the factorial moments of $P(X,t)$.

$$\langle x^n\rangle_f = \langle\alpha^n\rangle \tag{30}$$

Substituting such an expansion in the master equation for any realistic chemical reaction (i.e. a reaction involving only bimolecular steps) one obtains, on integrating by parts, an exact Fokker-Planck type equation for $f(\alpha,t)$. Thus, for instance, for the master equation (22) we get

$$\frac{\partial f(\alpha,t)}{\partial t} = -\frac{\partial}{\partial\alpha}\left[k_3CV+(k_2a-k_1b)\alpha + k_4V^{-1}\alpha^2\right] f(\alpha,t) + \frac{1}{2}\frac{\partial^2}{\partial\alpha^2}\left[2(k_2a\alpha - k_4V^{-1}\alpha^2)\right] f(\alpha,t) \tag{31}$$

This in turn may be written more conveniently as a stochastic differential equation (Itô type) as

$$\frac{d\alpha}{dt} = k_xCV+(k_2a-k_1b)\alpha + k_4V^{-1}\alpha^2 + \sqrt{2(k_2a\alpha - k_4V^{-1}\alpha^2)}\;\xi(t) \tag{32}$$

where $\xi(t)$ denotes a Gaussian white noise. Defining concentration variables in the Poisson representation as $\eta = \alpha/V$ we get

$$\frac{d\eta}{dt} = k_3C+(k_2a-k_1b)\eta + k_4\eta^2 + \frac{1}{\sqrt{V}}\sqrt{2(k_2a\eta - k_4\eta\;)}\;\xi(t) \tag{33}$$

Note that in this representationthe fluctuating part automatically turns out to have the correct scaling i.e. it is down by a factor of $1/\sqrt{V}$ compared to the non fluctuating part. The equation (33) is in a form ideally suited for an expansion in powers of $1/\sqrt{V}$

$$\eta = \eta_o + \left(\frac{1}{\sqrt{V}}\right)\eta_1 + \left(\frac{1}{\sqrt{V}}\right)\eta_2 + \dots \tag{34}$$

Substituting (34) in (33) and solving for η_i iteratively we can solve for η and hence calculate the moments of $P(X,t)$ to an arbitrary power in $1/V$ in a straight forward systematic way.

For reaction diffusion master equations one generalises (29) to

$$P(\underline{X},t) = \int d\underline{\alpha} \left(\prod_i \frac{\alpha_i^{X_i}}{X_i!} e^{-\alpha_i}\right) f(\underline{\alpha},t) \tag{35}$$

Thus for the reaction diffusion system corresponding to (7) one obtains

$$\frac{d\eta_i}{dt} = \sum_j D_{ij}\eta_j + k_3C + (k_2a-k_1b)\eta_i + k_4\eta_i^2 + \frac{1}{\sqrt{\Delta V}}\sqrt{2(k_2a\eta_i-k_4\eta_i^2)} \quad \xi_i(t) \tag{36}$$

where $\eta_i(t) = \frac{\alpha_i}{\Delta V}$ and $\xi_i(t)$ are Gaussian white noise sources

$$\langle \xi_i(t)\rangle = 0 \; ; \quad \langle \xi_i(t)\, \xi_j(t')\rangle = \delta_{ij}\,\delta(t-t') \tag{37}$$

In a continuum notation

$$\eta_i(t) \longrightarrow \eta(\underline{r}_i,t), \quad \frac{\xi_i(t)}{(\Delta V)^{\frac{1}{2}}} \longrightarrow \xi(\underline{r}_i,t) \tag{38}$$

(37) goes over to

$$\frac{d\eta}{dt} = D\nabla^2\eta + k_3C + (k_2a-k_1b)\eta + k_4\eta^2 + \sqrt{(k_2a\eta - k_4\eta^2)}\; \xi(r,t) \tag{39}$$

which may again be solved iteratively by introducing a formal expansion parameter λ

$$\eta = \eta_o + \lambda\eta_1 + \lambda^2\eta_2$$

Detailed calculation of single time and two time correlation functions based on these methods may be found in [10-12].

We must mention that all techniques that have been developed so far turn out not to be capable of handling fluctuations near a critical point. They merely tell us that the fluctuations become large near a critical point. Much work needs to be done in developing techniques for dealing with critical fluctuations.

Effect of External Perturbations

Effects of external perturbations sotchastic or nonstochastic on chemical reaction systems have attracted much attention in recent years. For instance Tomita and Kai [12] have shown that a periodically driven Brusselator

$$\frac{dx}{dt} = k_1 a - k_2 bx + k_3 x^2 y - k_4 x + \alpha \cos \omega t$$
$$\frac{dy}{dt} = -k_3 x^2 + k_2 bx \tag{40}$$

exhibits chaos. A number of authors have investigated chemical reaction systems on noisy environments which have the effect of making the parameters appearing in the rate equations stochastic. This sometimes leads to what are known as 'noise induced transitions'. For details we refer the reader to ref. [13].

References

1. H. Haken, "Synergetics, An Introduction" Berlin-Heidelberg - New York, Springer (1977).
2. G. Nicolis and I. Prigogine, "Self Organisation in Non equilibrium Systems" John Wiley and Sons (1971).
3. N.G. van Kampen "Stochastic Processes in Physics and Chemistry", Amsterdam, New York, Oxford, North Holland (1981).
4. F. Schlogl, Z. Physik 253, 147 (1972).
5. C. Escher, Z. Physik B35, 351 (1979).
6. R.J. Field and R.M. Noyes, J. Chem. Phys. 60,1877 (1974).
7. D.A. Mcquarrie, J. Appl. Prob. 4, 413 (1967).
8. For a detailed discussion on diffusion master equations see C.W. Gardiner, "A Handbook of Stochastic Methods for Physics, Chemistry and Natural Sciences."
9. C. Blomberg, J. Stat. Phys. 25, 73 (1981).
10. C.W. Gardiner and S. Chaturvedi, J. Stat. Phys. 17, 429 (1977); 18, 501 (1978).
11. S. Chaturvedi, C.W. Gardiner, I.S. Matheson and D.F. Walls, J. Stat. Phys. 17, 469 (1977).
12. K. Tomita and T. Kai, J. Stat. Phys. 21, 1 (1979).
13. See articles by W.Horsthemke and R. Lefever and by M. San Miguel and J.M. Sancho in "Stochastic Non linear systems in Physics, Chemistry and Biology". L Arnold and R. Lefever eds. Synergetic Series Vol.8, Berlin, Heidelberg, New York Springer (1981).

GOLDSTONE MODES IN NON-EQUILIBRIUM PHASE TRANSITIONS

V. Srinivasan
School of Physics, University of Hyderabad
Hyderabad - 500134, India

Spontaneous symmetry breaking is a very important concept that pervades almost every branch of physics. Of course success of this concept in one branch stimulates one to apply it to another. When one does discover that this indeed works in a new area it reinforces our faith in the concept itself. The spectacular success of this concept in high-energy physics, condensed matter physics and even cybernetics makes spontaneous symmetry breaking a universal fundamental concept. The success of this concept also reinforces our belief that there is an underlying unity among the various branches of physics.

The history of the subject itself is quite interesting. T. Holstein and H. Primakoff [1] in 1940 introduced a method to diagonalize the Hamiltonian in the exchange interaction model of a ferromagnet by introducing second quantized creation and annihilation operators a_ℓ^+ and a_ℓ for magnons. They wrote the spin operators in terms of the magnon creation annihilation operators and the spins satisfy the usual SU(2) algebra. They then approximated the Hamiltonian by neglecting all terms that were not bilinear in a_ℓ and a_ℓ^+. This meant that in this approximation spin operators no longer satisfied the SU(2) commutation relation but E(2). They found a magnon mode whose frequency went to zero as $k \to o$. Holstein and Primakoff did not worry too much about the fact that rotational invariance present in the Hamiltonian is absent for the ground state. Nor did they link the zero frequency magnon mode to the asymmetric ground state. Nor did they bother about checking the conservation of the current that follows from rotational invariance.

The next person to choose an asymmetric ground state is N.N. Bogoliubov in his theory for liquid helium [2]. He chose a ground state that did not conserve the particle number and also obtained a phonon mode whose frequency went to zero as $k \to o$. But he again did not link this phonon mode to the asymmetric ground state.

The theory of superconductivity of Bardeen, Cooper and Schrieffer [3] is probably the forerunner to this theorem. In their epoch making paper, probably one of the greatest contributions to physics in the twentieth century, Bardeen, Cooper and Schrieffer chose a ground-state wave function which was not gauge invariant as their trial wave function. They diagonalized their Hamiltonian with this wave function, and showed that almost all properties of superconductors follow from their theory. The lack of gauge invariance of the ground state worried physicists for a while for it was thought that this would mean violation of charge conservation locally. P.W. Anderson and Y. Nambu [4] investigated the B.C.S. theory and showed that all is right with the B.C.S. theory. If one went beyond the B.C.S. approximation the current so calculated is conserved. Nevertheless there appears a phonon mode $\omega(k) \longrightarrow o$ as

$k \rightarrow o$. Also they showed that when the Coloumb interactions were taken into account this mode disappeared and became the plasma mode.

Later Nambu extended this idea to elementary particle physics in what is now known as the Jona Lasino-Nambu model. In fact it is Nambu who realized significance of the mechanism of spontaneous symmetry breaking. Later J. Goldstone [6] using a simple model illustrated this concept. It was finally proved as a theorem by S. Weinberg, A. Salam and J. Goldstone [7].

The Goldstone theorem states that if a Lagrangian is invariant under a continuous transformation while the ground-state is not, as a consequence of the conservation of the current that follows from the symmetry there exists a massless boson in the theory. [In the context of condensed matter physics there exists a mode $\omega(k) \rightarrow o$ as $k \rightarrow o$.] [8]

It is now known that in a large class of driven systemsin optics, chemical systems etc. "nonequilibrium phase transitions (N.P.T.) are observed between competing steady states. The similarities between N.P.T. and its equilibrium cousin has been well studied and a new area called synergetics has now emerged [9]. In what follows we shall show [10] that the Goldstone theorem in equilibrium phase-transition in condensed matter physics can be now extended to N.P.T. The analog of this theorem in N.P.T. is as follows:

"Whenever the order parameter is invariant under a continuous transformation while the steady state is not, there appears, in the limit the drive $J \rightarrow o$, a mode whose complex frequency goes to zero as $k \rightarrow o$." We shall now demonstrate this result by considering a few examples.

Consider the following order parameter equation:

$$\left[\frac{\partial}{\partial t} + \omega(\nabla^2)\right] \phi_\alpha = -\frac{\partial V}{\partial \phi_\alpha} + J_\alpha \tag{1}$$

Here J_α is a coherent external driving field and V is the potential which is a functional of the ϕ's. V transforms like a scalar under SU(N). $\phi = \begin{bmatrix} \phi_1 \\ \vdots \\ \phi_N \end{bmatrix}$ is an n component field which transforms like the fundamental representation of SU(N).

Let $-\frac{\partial V}{\partial \phi_\alpha} = F(\phi^+\phi)\ \phi_\alpha$

Then equation (1) becomes

$$\left[\frac{\partial}{\partial t} + \omega(\nabla^2)\right] \phi_\alpha = F(\phi^+\phi)\ \phi_\alpha + J_\alpha \tag{2}$$

We now look for a steady state solution of equation (2) of the form

$$\begin{aligned} \langle\phi_\alpha\rangle &= \lambda \qquad \text{for} \qquad \alpha = 1, \ldots\ldots m \\ \langle\phi_\alpha\rangle &= 0 \qquad \text{for} \qquad \alpha = m+1, \ldots\ldots n \end{aligned} \tag{3}$$

Since the steady state is space-time independent it is obvious that as $J \rightarrow o$

$$F(m\,|\lambda|^2)\,\lambda = 0 \qquad \Longrightarrow \qquad F(m\,|\lambda|^2) = 0 \tag{4}$$

since $\lambda \neq 0$ for the asymmetric steady state. Let us now look for the eigenmodes of the problem. To this effect we write

$$\phi_\alpha = \lambda + \psi_\alpha \qquad \alpha = 1 \ldots\ldots m$$

$$\phi_\alpha = \chi_\alpha \qquad \alpha = m+1 \ldots\ldots n \tag{5}$$

The linearisation of the steady state equations lead to

$$\left[\frac{\partial}{\partial t} + \omega(\nabla^2)\right]\psi_\alpha = F(m|\lambda|^2)\psi_\alpha + J_\alpha + \sum_\beta \left(\frac{\partial F}{\partial \phi_\beta}\right)_{\phi_\beta = \lambda} \psi_\beta \tag{6}$$

$$\left[\frac{\partial}{\partial t} + \omega(\nabla^2)\right]\chi_\alpha = F(m|\lambda|^2)\chi_\alpha + \ldots\ldots \tag{7}$$

Equation (7) with (4) shows that in the limit $J \to 0$ the frequency of the oscillation associated with χ field goes to zero as $k \to 0$. In this limit the frequency associated with ϕ is

$$\omega_\phi \xrightarrow[k \to 0 \; J \to 0]{} 2\lambda^2 mF'$$

where F' denotes the derivative of F w.r.t. its argument. Thus we have demonstrated the theorem, using a linearized approximation.

One might, instead of a local interaction, consider some non-local interaction which is short range. So let us consider the following order parameter equation

$$\left[\frac{\partial}{\partial t} + \omega(\nabla^2)\right]\psi = F(|\psi|^2)\psi - i\,\psi(x)\int d^3x'\,|\psi(x')|^2\,\Omega(x-x') \tag{8}$$

$F(|\psi|^2)$ is a real function of ψ.

We now look for a symmetry breaking steady state

$$\psi = \psi_0 e^{-iE_0 t} \qquad E_0 = \int d^3x\,\Omega(x)\,|\psi_0|^2$$

$$F(|\psi_0|^2) = 0 \tag{9}$$

To obtain the excitation frequencies we can linearize ψ around the steady state solution

$$\psi = (\psi_0 + \delta\psi)\, e^{-iE_0 t} \tag{10}$$

From (8) and (9) we obtain

$$\left[\frac{\partial}{\partial \tau} + \omega(\nabla^2)\right]\delta\psi = |\psi_0|^2\left[F'\delta\psi - i\int d^3x'\,\delta\psi(x')\,\Omega(x-x')\right] + |\psi_0|^2\left[F'\,\delta\psi^* - i\int d^3x\,\Omega(x-x')\,\delta\psi^*(x')\right] \tag{11}$$

A simple Fourier analysis of (11) and its conjugate shows that the system admits Goldstone modes provided the interaction is short range. i.e.

$$\underset{k \to o}{\mathrm{Lt}} \quad [\mathrm{Im}\,\omega(-k^2)]\,\Omega_k \quad = \quad 0$$

Note that generally $\mathrm{Im}\,\omega(-k^2) \longrightarrow k^2$ if it is non-zero.

We have thus shown that the generalized Goldstone theorem is valid in N.P.T. An exact proof of this has been now obtained by us and will be published elsewhere. We also have found that the Higgs mechanism goes through [11]. Finally it should be borne in mind that the general belief is that in two dimensions the Goldstone theorem is not true. However Nakanishi [12] has shown that one can construct a massless field in two dimensions if one uses indefinite metric. Since in our case we have complex energies probably Nakanishi's results can be extended here.

References

1. T. Holstein and H. Primakoff, Phy. Rev. 58, 1098 (1940)
2. N.N. Bogoliubov, J. Phys. U.S.S.R. 1123 (1947)
3. J. Bardeen, L.N. Cooper and J.R. Schrieffer, Phys. Rev. 106, 162 (1957); 108, 1175 (1957)
4. P.W. Anderson, Phys. Rev. 110, 827 and 112, 1900 (1958), Y. Nambu, Phys. Rev. 117, 648 (1960)
5. Y. Nambu and G. Jona Lasino, Phy. Rev. 122, 345 (1961), Y. Nambu and G. Jona Lasino, Phy. Rev. 124, 246 (1961)
6. J. Goldstone, Nuovo Cimento 19, 154 (1961)
7. J. Goldstone, A. Salam and S. Weinberg, Phy. Rev. 127, 965 (1962)
8. G.S. Guralnik, C.R. Hagen and T.W.B. Kibble eds. R.L. Cool and R.E. Marshak (Interscience, New York 1968) Vol.2, p.567
9. H. Haken, Synergetics (Springer-Verlag, Berlin 1977)
10. G.S. Agarwal and V. Srinivasan, to be published
11. G.S. Agarwal, S. Chaturvedi and V. Srinivasan, to be published
12. Nakanishi, Prog. Theor. Phys. 57, 269 (1977); 38, 158 (1977); 58, 1927 (1977) and 59, 607 (1978)

PHASE TRANSITIONS IN A SYSTEM OF ATOMS INTERACTING WITH A COHERENT FIELD

S.V. Lawande
Theoretical Reactor Physics Section
Bhabha Atomic Research Centre
Bombay-400 085, India

1. INTRODUCTION

Co-operative behaviour in systems driven far from thermal equilibrium is of great current interest. For two recent reviews of this rapidly growing subject see the books by Haken[1] and by Prigogine and Nicolis[2]. As an example of such a system we consider a collection of N identical two-level atoms interacting with a coherent field and located in a volume of dimensions smaller than the wavelength. Without the external field the model was first explored by Dicke[3] as an illustration of co-operation in quantum optics. The cooperative effect manifests itself in the form of a spontaneously emitted pulse with intensity proportional to N^2 (superradiance) for suitably prepared initial atomic states. The Dicke model is an example of exactly solvable quantum mechanical model[4]. It is also of interest because of its mathematical equivalence to spatially averaged extended systems[5].

The driven Dicke model has also been the object of much recent discussions[5-7] mainly because it involves both resonance fluorescence and cooperative emission. The model admits exact steady state solutions both in the case of resonant[8-12] and off-resonant[13] coherent driving field. These exact solutions show that at resonance the system undergoes as $N \to \infty$ a non-equilibrium phase transition similar to a second order phase transition (predicted also by numerical solutions for finite N[14]). This transition exhibits itself in the form that upto a certain critical amplitude of the driving field the system behaves in pure classical manner beyond which the quantum effects become dominant. At off-resonance there is no such critical behaviour. This is discussed further in Section 2.

Next, we consider the Dicke system placed in a cavity tuned to the frequency of the atomic transition. The cavity mode is excited by an external field in resonance with the atoms and is coupled to an output mode. In this case the quantum fluctuations and the cavity feedback together lead in the limit $N \to \infty$ to a bistable behaviour[15], manifested in the relation between the output and the input fields. This is briefly outlined in Section 2.

Finally, we include interaction between atoms in the Dicke system (in a special manner)[16]. The interaction has the effect of increasing the coherence region with respect to the amplitude of the driving field. But now there is a competition between the dephasing effect of the frequency detuning and the synchronizing effect of the interaction of the atoms which leads to a first order phase transition for certain values of the parameters involved. This is presented in Section 3.

2. THE DRIVEN DICKE MODEL

Master Equation

We consider a collection of N identical two-level atoms, each with a transition frequency ω_0, confined to a volume V of dimensions small compared to the wavelength and driven by a single mode coherent field of frequency ω:

$$\vec{E}(\vec{r},t) = \vec{\varepsilon}_0 \exp[-i(\omega t-\vec{K}_0.\vec{r})] + c.c. \; . \tag{1}$$

The atoms decay by virtue of their coupling to the vacuum radiation field. In the electric dipole approximation the Hamiltonian reads as

$$H = \sum_{j=1}^{N} \hbar\omega_0 S_{zj} + \sum_{j=1}^{N} \hbar(\Omega e^{-i\omega t} S_{+j} + \Omega^* e^{i\omega t} S_{-j}) + \sum_k \sum_{j=1}^{N} \hbar[g_k b_k (S_{+j}+S_{-j}) + h.c.] + \sum_k \hbar\omega_k b_k^+ b_k \tag{2}$$

where

$$g_k = -i\left(\frac{2\pi c|\vec{k}|}{\hbar V}\right)^{1/2} (\vec{d}.\vec{e}_k) e^{i\vec{k}.\vec{r}_a} ; \; \Omega = -\frac{(\vec{d}.\vec{\varepsilon}_0)}{\hbar} e^{i\vec{k}_0.\vec{r}_a} \; . \tag{3}$$

For simplicity, we have suppressed the polarization index. The two levels atoms are represented by the pseudo-spin operators S_{zj} and $S_{\pm j} = S_{xj} \pm iS_{yj}$ obeying the usual commutation relations. Each atom has a dipole moment $\vec{d}$ and $\vec{r}_a$ represents the centre of the atomic collection. The vacuum field modes are characterized by annihilation and creation operators b_k and b_k^+ corresponding to the wavelength $\vec{k}$, frequency ω_k and polarization $\hat{e}_k$. The summation over j in the Hamiltonian may be absorbed by defining the collective spin operators $S_\pm = \sum_j S_{\pm j}$, $S_z = \sum_j S_{zj}$ obeying the angular momentum commutation relations

$$[S_\pm, S_z] = \mp S_\pm, \; [S_+, S_-] = 2S_z. \tag{4}$$

It is clear that the $\hat{S}^2$ commutes with the Hamiltonian, a consequence of the fact that the phase factors $e^{i\vec{k}.\vec{r}_j}$ at different atomic sites are ignored.

From Hamiltonian (1), the master equation for the reduced atomic density operator $\rho_a = \text{Tr}\, \rho$ (where the trace is taken over the vacuum radiation field variables) is obtained by standard techniques[4]. In the rotating wave, Born and Markov approximations and in the frame rotating at the frequency ω this master equation reads as[4]

$$\frac{d\rho_a}{dt} = -i\Omega[S_+ + S_-, \rho_a] + i\delta_0[S_z, \rho_a] + \gamma_0([S_-, \rho_a S_+] + [S_- \rho_a, S_+]) \tag{5}$$

where $\delta_0 = \omega - \omega_0$ is the frequency detuning between the coherent field and the atom 2Ω is the Rabi frequency, $2\gamma_0$ is the Einstein A coefficient ($\gamma_0 = 2d^2\omega_0^3/3hc^3$).

Steady State Solution

It is readily seen that corresponding to $\hat{S}^2$ conservation in the Hamiltonian the master equation conserves $\langle S^2 \rangle$. Hence, the Dicke states $|S,m\rangle$ form a good set of basis states. These states are the simultaneous eigenstates of $\hat{S}^2$ and S_z obeying

$$\hat{S}^2|S,m\rangle = S(S+1)|S,m\rangle, \quad S_z|S,m\rangle = m|S,m\rangle \,. \tag{6}$$

$$S \pm |S,m\rangle = [(S \mp m)(S \pm m+1)]^{1/2} \, |S,m \pm 1\rangle$$

The quantum number S is the Dicke cooperation number $(0 < S < N/2)$ and m is half the difference between the number of excited and unexcited atoms $(m < S)$. For convenience, we assume the atoms to be in the ground state initially (just before the field is switched on). Thus $m = N/2$ and $S = N/2$ at $t = 0$ and by conservation of $\hat{S}^2$, ρ_a evolves over an (N+1)-dimensional manifold of collective atomic states $|N/2,m\rangle$ where $m = -N/2,\ldots,N/2$. For subsequent work, it is convenient to write $p = S-m$ and label the states $|S, S-p\rangle$ by $|p\rangle$.

Having fixed the space in which ρ_a evolves we proceed to derive the steady state solution. Note that the master equation does not obey the principle of detailed balance and the standard methods are ruled out. Nevertheless, at least, in the case of exact resonance $(\delta_0 = 0)$ it is easy to guess this solution, if we first convert the master equation into a C-number form. For this purpose, we introduce the atomic coherent states[17] $|\mu\rangle$ which are related to the Dicke states $|p\rangle$ by

$$|\mu\rangle = (1+|\mu|^2)^{-S} e^{\mu S_+}|2S\rangle = (1+|\mu|^2)^{-S} \sum_{p=0}^{2S} \left(\frac{2S!}{(2S-p)!p!}\right)^{1/2} \mu^{2S-p}|p\rangle \tag{7}$$

These states have the following properties

$$\langle\lambda|\mu\rangle = (1+\lambda^*\mu)^{2S}(1+|\lambda|^2)^{-S}(1+|\mu|^2)^{-S}$$

$$\frac{2S+1}{\pi}\int \frac{d^2\mu}{(1+|\mu|^2)^2} |\mu\rangle\langle\mu| = 1$$

$$\langle\lambda|S_+|\mu\rangle = (2S)(1+|\lambda|^2)^{-S}(1+|\mu|^2)^{-S}\lambda^*(1+\lambda^*\mu)^{2S-1} = (\langle\mu|S_-|\lambda\rangle)^* \tag{8}$$

$$\frac{2S+1}{\pi}\int \frac{d^2\lambda}{(1+|\lambda|^2)^{2S+2}} \lambda^{*m}\lambda^n = \left(\frac{(2S-m)!m!}{(2S)!}\right)\delta_{mn} \quad (0 < m,\, n < 2S).$$

These states are used to construct a diagonal representation for the matrix ρ_a. Writing

$$\rho_a(\mu^*,\mu,t) = \langle\mu|\rho_a|\mu\rangle(1+|\mu|^2)^{2S} \tag{9}$$

and using the above properties, it is easy to cast the master equation $(\delta_0=0)$ in the form

$$\frac{\partial\rho_a}{\partial\tau}(\mu^*,\mu,t) = \left(\frac{\partial}{\partial\mu} - N\mu^* + \mu^{*2}\frac{\partial}{\partial\mu^*}\right)\left(\frac{\partial}{\partial\mu^*} + g\right)\rho_a(\mu^*,\mu,\tau) + \text{C.C.} \tag{10}$$

where $g = i\Omega/\gamma_0$ and $\tau = \gamma_0 t$. It is now easy to verify that the steady state solution

$(\frac{\partial\rho_a}{\partial\tau} = 0)$ of Eq.(10) is given by

$$\rho_a^{ss}(\mu^*,\mu,t) = \frac{1}{D_o}\sum_{m,n=0}^{N}(g^*)^{-m}(g)^{-n}(\frac{\partial}{\partial\mu^*})^m(\frac{\partial}{\partial\mu})^n(1+|\mu|^2)^N$$

which may also be cast back in the operator form

$$\rho_a^{ss} = \frac{1}{D_o}\sum_{m,n=0}^{N}(g^*)^{-m}(g)^{-n}S_-^mS_+^n . \tag{11}$$

The normalization constant D_o determined by setting Tr $\rho^{ss} = 1$ is given by

$$D_o = \sum_{m=0}^{N}|g|^{-2m}H_{Nm}$$

$$H_{Nm} = \sum_{q=0}^{N-m}\frac{(N+q)!(q+m)!}{(N-q-m)!q!} \tag{12}$$

This solution is valid for all N and g. Taking the clue from this, the steady state solution for the case of finite detuning ($\delta_o \neq 0$) may be easily derived and reads as

$$\rho_a^{ss} = \frac{1}{D}\sum_{m,n=0}^{N}a_{mn}(g^*)^{-m}(g)^{-n}S_-^mS_+^n \tag{13}$$

where

$$D = \sum_{m=0}^{N}a_{mm}H_{Nm}|g|^{-2m}$$

$$a_{mn} = \frac{\Gamma(m+\Delta^*+1)\Gamma(n+\Delta+1)}{m!n!\Gamma(1+\Delta^*)\Gamma(1+\Delta)} \tag{14}$$

Here $\Delta = i\delta_o/\gamma_o$ and $\Gamma(z)$ is the usual Γ function of the complex argument z. Note that when $\Delta = 0$ (resonant case), $a_{mn} = 1$ for all m, n and one recovers from Eqs.(13) and (14), the steady state solution displayed in Eqs.(11) and (12).

Atomic Observables and Correlation Functions

The exact form (13) for the density operator ρ_a may now be used to obtain the expressions for the atomic operator expectation values, the atomic correlation functions and the fluctuations. These can all be derived from the operator average:

$$\begin{aligned}\langle S_+^pS_z^nS_-^q\rangle_N &= \mathrm{Tr}[S_+^pS_z^nS_-^q\rho_a^{ss}] \\ &= D^{-1}\sum_{n=\max(p,q)}^{N}(g^*)^{q-n}(g)^{p-n}a_{n-p,n-q}\sum_{m=0}^{N-n}\frac{(m+n)!(N-m)!}{(N-m-n)!m!}(\frac{1}{2}N-n-m)^r.\end{aligned} \tag{15}$$

From this general formula, it is easy to obtain the following expressions for various atomic and radiation field observables as a function of the driving field parameter g for a given N:

$$\langle S_z\rangle = -\frac{1}{2D}\sum_{r=0}^{N}r|g|^{-2r}a_{rr}H_{Nr} = \frac{|g|}{4N}\frac{\partial}{\partial|g|}\ln D \tag{16}$$

$$\langle S_+ \rangle = \frac{1}{D} \sum_{r=1}^{N} (g^*)^{-r} (g)^{-r+1} a_{r\ r-1} H_{Nr} = \langle S_- \rangle^* \tag{17}$$

$$G^{(m,n)}(0) = \langle S_+^m S_-^n \rangle = \frac{1}{D} \sum_{r=\max(m,n)}^{N} (g^*)^{n-r} (g)^{m-r} a_{r-n,r-m} H_{Nr} \tag{18}$$

We remark that $\langle S_z \rangle$, the expectation value of the atomic population inversion is also a measure of the power absorbed by the atomic system. The mean atomic polarization $\langle S_y \rangle$ and dispersion $\langle S_x \rangle$ are provided by formula (17). The functions $G^{m,n}_{(0)}$ of Eq.(18) are proportional to the correlation functions $\langle \hat{E}_-^m \hat{E}_+^n \rangle$ of the scattered radiation. Of these, the function $G^{(1,0)} = \langle S_+ \rangle$ is proportional to the average amplitude of the scattered radiation, $G^{(1,1)}$ is proportional to the intensity and $G^{(2,2)}$ is proportional to the intensity correlation function. The correlation properties of the scattered radiation are usually determined by the normalized functions of first and second order defined as

$$g^{(1)} = G^{(1,1)}/|G^{1,0}|^2, \quad g^{(2)} = G^{(2,2)}/|G^{(1,1)}|^2 \,. \tag{19}$$

Note that at $g^{(1)} = 1$, the scattered radiation is fully coherent and its deviation from unity implies partial coherence. The function $g^{(2)}$ determines the probability that two photons will be simultaneously emitted by the system. The value $g^{(2)} > 1$ indicates the presence of bunching of the scattered photons, and $g^{(2)} < 1$ shows anti-bunching while $g^{(2)} = 2$ implies that the properties of the scattered light are equivalent to the properties of the equilibrium thermal radiation.

We might add that to obtain the properties of the atomic system at resonance, we have to set $a_{mn} = 1$ in the above formulae. An immediate consequence of this is that $\langle S_x \rangle \equiv 0$ at resonance.

Second Order Phase Transition

We now examine the behaviour of the atomic observables and their fluctuations when the driving field is in exact resonance ($\delta_o = 0$). We are, in particular, interested in the thermodynamic limit $N \to \infty$, $|g| \to \infty$ with $\theta = \frac{2|g|}{N}$ remaining finite. To this end we shall require the behaviour of the quantity $D_o \equiv D_o(N,\theta)$ for large N. The asymptotic expression for $D_o(N,\theta)$ for large N turns out to be

$$D_o(N,\theta) = \begin{cases} \dfrac{N\pi}{2\theta^{2N}} \dfrac{(1+\sqrt{1-\theta^2})^{2N+2} \exp(-2N\sqrt{1-\theta^2})}{\sqrt{1-\theta^2}} & (\theta < 1) \\[2ex] \dfrac{N\theta^2}{\sqrt{\theta^2-1}} \mathrm{Sin}^{-1}(1/\theta) & (\theta > 1) \end{cases} \tag{20}$$

Equations (15)-(17) may now be combined with (20) to arrive at the limiting forms of the quantities of interest. In particular, writing

$$m_\mu = \lim_{N\to\infty} \langle S_\mu\rangle/N, \quad \sigma_{\mu\nu} = [\langle S_\mu S_\nu\rangle - \langle S_\mu\rangle\langle S_\nu\rangle]/N^2 \quad (\mu,\nu \equiv x,y,z) \tag{21}$$

we obtain

$$m_x = 0 \qquad 0 < \theta < \infty$$

$$m_y = \begin{cases} \theta/2 & 0 < \theta < 1 \\ \theta(1-f(\theta))/2 & 1 < \theta < \infty \end{cases} \tag{22}$$

$$m_z = \begin{cases} -\frac{1}{2}\sqrt{1-\theta^2} & 0 < \theta < 1 \\ 0 & 1 < \theta < \infty \end{cases}$$

$$\sigma_{zz} = \begin{cases} 0 & 0 < \theta < 1 \\ [1-\theta^2(1-f(\theta))]/4 & 1 < \theta < \infty \end{cases} \tag{23}$$

$$f(\theta) = \sqrt{\theta^2-1}/[\theta^2 \sin^{-1}(1/\theta)] \tag{24}$$

It is thus seen that in the "thermodynamic" limit the atomic observables and their fluctuations are continuous at $\theta=1$, but show a discontinuity in their derivatives (with respect to θ) at $\theta=1$. This behaviour is typical of a second order phase transition at the critical bifurcation point $\theta=1$. We show in Figs.1 and 2 the variation of $\langle S_z\rangle/N$ and the fluctuations σ_{zz} with θ for several values of N and for $N \to \infty$. Note the smooth behaviour of these quantities with θ for finite N and their approach to the asymptotic limit as N increases.

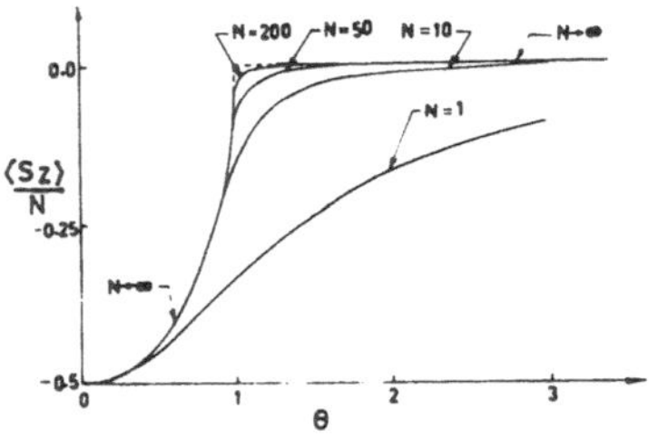

Fig.1. Steady state atomic population difference (per atom) as a function of θ for several values of N.

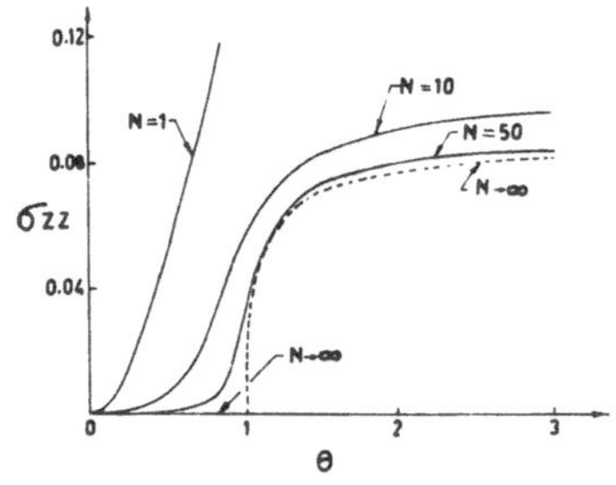

Fig.2. Fluctuations in the steady state atomic population difference as a function of θ for various N.

Further the fluorescent intensity $G^{(1,1)}(0)/(N/2)^2$ and the normalized intensity-intensity correlation function $g^{(2)}(0)$ also show the discontinuity in slope at $\theta=1$ in the limit $(N \to \infty)$. The analytical expressions are[12]

$$G^{(1,1)}(0)/(\tfrac{N}{2})^2 = \begin{cases} \theta^2 & \theta < 1 \\ \theta^2(1-f(\theta)) & \theta > 1 \end{cases} \tag{25}$$

$$g^{(2)}(0) = \begin{cases} 1 & \theta < 1 \\ \theta^4(1 - (1 + \frac{2}{3\theta^2})f(\theta)] & \theta > 1 \end{cases} \tag{26}$$

The behaviour of steady state intensity and the correlation function $g^{(2)}(0)$ is shown in Figs.3 and 4. It is clear from Eq.(26) that the scattered radiation is completely coherent for $\theta < 1$ whereas for $\theta > 1$ the radiation is partially coherent. Note also that $g^{(2)}(0) = 1.2$ in the limit $N \to \infty$. Curves of $g^{(2)}(0)$ also indicate photon antibunching around $\theta=1$ for finite N.

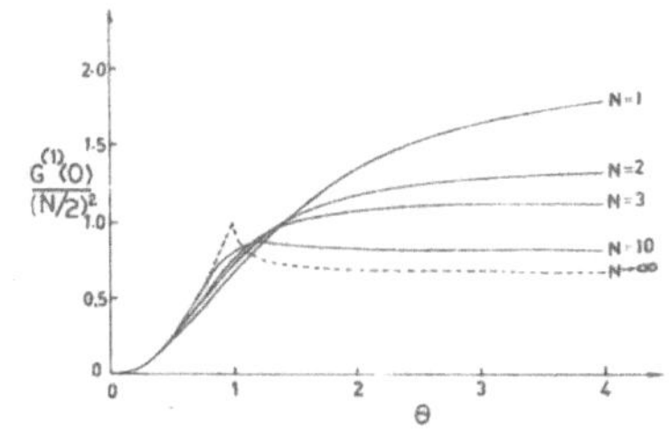

Fig.3. Behaviour of steady state fluorescent intensity with θ for several values of N.

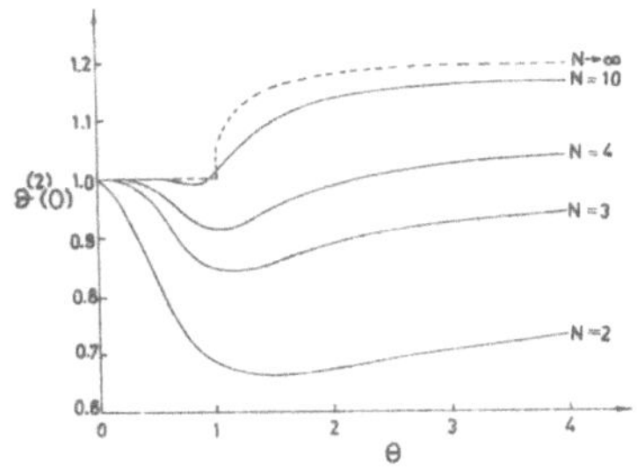

Fig.4. Steady state intensity-intensity correlation function for different values of N.

Similar analysis[6,13] can be carried out for the case when the driving field is off-resonant. Both numerical results for finite N and exact analysis in the thermodynamic limit $N \to \infty$ indicate that there is no critical point phase transition for the atomic system when the detuning parameter $\phi = 2\delta_0/\gamma_0 N \neq 0$. It is also found that as N increases, the quantum fluctuations for a given θ and ϕ go on decreasing approaching a zero value in the limit $N \to \infty$. This is in contrast with the resonance case $(\phi=0)$ where $\sigma_{zz} = 0$ for $\theta < 1$ while for $\theta > 1$, σ_{zz} increases from its value zero (at $\theta=1$) approaching a value 1/12 $(\theta \to \infty)$.

We might also mention that inclusion of phase fluctuations in the driving field[18] also tends to smooth the discontinuities in the atomic observables and their correlations found in the non-fluctuating resonant case. Quantum fluctuations are also reduced due to the phase fluctuation effects.

First Order Phase Transition

There is an interesting way[15] of predicting the conventional bistable behaviour with the Dicke model. It is known that an extended system of atoms placed in a cavity predicts bistability with hysteresis, a typical first order phase transition. This behaviour arises from a competition between the cooperative and individual atom decay processes in presence of the driving field. In contrast, in the driven Dicke model for which $\hat{S}^2$ is conserved, the atomic decay is only cooperative and bistability is not predicted. However, bistable behaviour is predicted if the Dicke system is simply placed in a cavity[15] tuned to the frequency ω_0 of the atomic transition. The cavity mode with which atoms interact is excited by an external field of frequency ω_0 and is coupled to an output mode. In the limit of large N, the cavity field operator A may be replaced by its expectation value $\bar{A}$. The equation for the reduced atomic density operator ρ_a and that for $\bar{A}$ (in the mean field approximation) are given by[15]

$$\frac{d\rho_a}{dt} = - i\bar{A}[S_+ + S_-, \rho_a] + \gamma_0([S_-, \rho S_+] + [S_-\rho_a, S_+]) \tag{27}$$

$$\frac{d\bar{A}}{dt} = - K_A(\bar{A}-A_0) - K^2\bar{S} \tag{28}$$

where K_A is an effective decay constant describing coupling to the external field A_0 and the output mode; $\bar{S} = |\langle S_+\rangle| = |\langle S_-\rangle|$ and K^2 is a coupling constant.

Defining $x = 2\bar{A}/N\gamma_0$, $y = 2A_0/N\gamma_0$, which are proportional to the output and input fields respectively, one obtains in the steady state the equation

$$y = x + 2C|\langle S_+\rangle|/N \tag{29}$$

where according to Eq.(22)

$$\frac{\langle S_+\rangle}{N} = \begin{cases} x/2 & x < 1 \\ (1 - \sqrt{x^2-1}/[x^2\sin^{-1}(1/x)])x/2 & x > 1 \end{cases} \tag{30}$$

in the limit $N \to \infty$ and $C = K^2/K_A\gamma_0$. A plot of x against y for different cooperation numbers C is shown in Fig.5. For $C > 0.1$ we observe the conventional bistable behaviour and hysteresis. Hysteresis arises because curves of negative slope are unstable. Note, however, that the cusps in Fig.5 are due to the second order phase transition contained in the Dicke model. The interesting point here is that the optical bistability arises from the quantum fluctuations and cavity feedback.

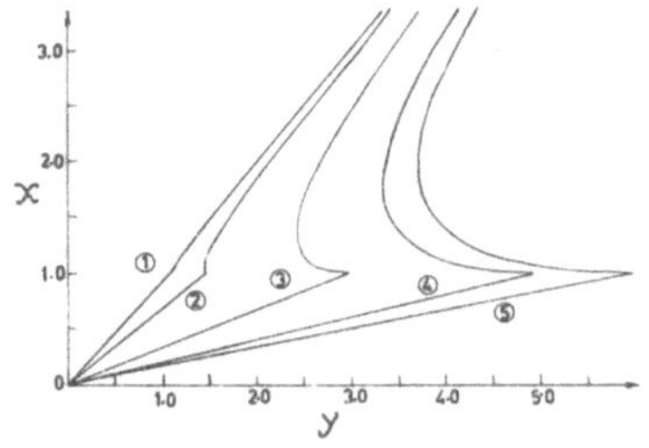

Fig.5. Bistable behaviour from Dicke system placed in a cavity. Curves (1)-(5) correspond to C=0,1,0.5, 2,4,5.

3. DRIVEN DICKE MODEL WITH INTER-ATOMIC INTERACTIONS

We now consider the driven Dicke model taking into account both the frequency detuning and the interaction between the atoms. In the special case for which the interatomic interaction operator $\hat{V}$ has the form[4,16]

$$\hat{V} = \varepsilon S_z + vS_+S_-, \tag{31}$$

the reduced atomic density operator ρ_a satisfies the equation[4,16]

$$\frac{d\rho_a}{dt} = -i\Omega[S_+ + S_-, \rho_a] + i\delta[S_z, \rho_a] - iv[S_+S_-, \rho_a] + \gamma_0([S_-, \rho_a S_+] + [S_-\rho_a, S_+]) \tag{32}$$

where $\delta = \delta_0 - \varepsilon$. This master equation conserves $\langle\hat{S}^2\rangle$. Incidently it describes also an extended system averaged over the space of two-level atoms[5]. This equation also admits an exact steady state solution to be presented later. We shall first discuss the semi-classical approximation.

Semi-classical Solution

The equations of motion obtained from (32) for the mean value $\langle S_\mu\rangle$, $(\mu=\pm,z)$ have the form

$$\frac{d}{dt}\langle S_z\rangle = -2\gamma_0 \langle S_+S_-\rangle - i\Omega[\langle S_+\rangle - \langle S_-\rangle]$$

$$\frac{d}{dt}\langle S_+\rangle = -2i\Omega \langle S_z\rangle - i\delta\langle S_+\rangle - 2(iv-\gamma_0)\langle S_+S_z\rangle = \left(\frac{d}{dt}\langle S_-\rangle\right)^* \tag{33}$$

Neglecting the quantum correlations we can use the factorization $\langle S_\mu S_{\mu'}\rangle = \langle S_\mu\rangle\langle S_{\mu'}\rangle$ to rewrite equations (33) in a convenient form

$$\frac{dm_z}{d\tau} = -2m_+m_- - \frac{i\theta}{2}(m_+ - m_-)$$
$$\frac{dm_+}{d\tau} = - i\theta m_z + \frac{i\phi}{2} m_+ + 2(1-i\xi)m_+m_z \tag{34}$$

where

$$\tau = N\gamma_0 t, \; m_\mu = \frac{<S_\mu>}{N}$$
$$\theta = 2\Omega/N\gamma_0, \; \phi = 2\delta/N\gamma_0, \; \xi = \nu/\gamma_0 \tag{35}$$

We note that within this factorization Eqs.(34) admit that $<\hat{S}^2>=<S_z>^2+<S_+><S_->$ is a constant of motion. The steady state solution of Eqs.(34) is given by

$$m_\pm = \frac{(1-4m_z^2)}{2\theta}[(\phi/4m_z - \theta) \pm i] \tag{36}$$

where m_z is determined from either of the two equivalent relations

$$\theta^2 = h(m_z) = (1 - 4m_z^2)[(\frac{\phi}{4m_z} - \theta)^2 + 1] \tag{37}$$

$$\phi = 4m_z[\xi \pm (\frac{\theta}{1-4m_z^2})^{1/2}] \tag{38}$$

It is clear from Eqs.(37) and (38) that when $\xi=0$, m_z is uniquely defined function of the parameters θ and ϕ. If in addition ϕ is also set equal to zero, we recover the solution for m_z at resonance given in Eq.(21), showing a second order phase transition at the critical point $\theta=1$. On the other hand, when $\theta\neq0$ but $\phi=0$, Eqs.(37) and (38) imply that

$$m_z = \begin{cases} -\frac{1}{2}\sqrt{1 - \frac{\theta^2}{1+\xi^2}} & \theta < \sqrt{1+\xi^2} \\ 0 & \theta > \sqrt{1+\xi^2} \end{cases} \tag{39}$$

which again shows a discontinuity in the derivative (w.r.t.θ) at $\theta = \theta_{cr} = \sqrt{1+\xi^2}$

The most interesting situation arises when both ξ and ϕ are not zero. In this case one expects from Eq.(37) that there will be a range of parameters ξ and ϕ where m_z is a multivalued function of θ. A typical behaviour of m_z is shown in Fig.6. It is seen that there is a range of θ in which each θ corresponds to three values of m_z in the physically relevant interval $[-\frac{1}{2},0]$. Thus one arrives at a bistable behaviour.

In a similar manner, for certain values of the parameters θ and ξ Eq.(38) predicts bistable behaviour of m_z as a function of the detuning parameter ϕ. This is shown in Fig.7. Note that in Figs.6 and 7, the curves with negative slope are unstable and a first order phase transition with hysteresis is seen.

We might mention here that the steady state bistable behaviour was predicted with the Dicke model without taking into account the atom-atom interactions. This prediction was based on the analysis of equations of motion (similar to Eqs.(33) above but with $\nu=0$) subject to a factorization ansatz for operator products. This factorization uses the semiclassical decorrelation as above but retains also the self-

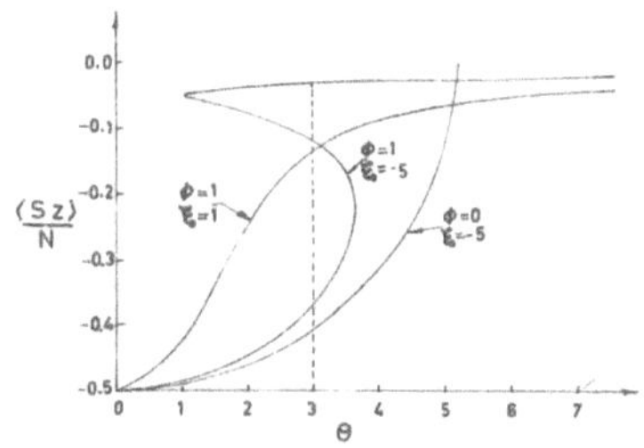

Fig.6. Plot of m_z against θ for different values of the parameters ξ and φ (semi-classical results)

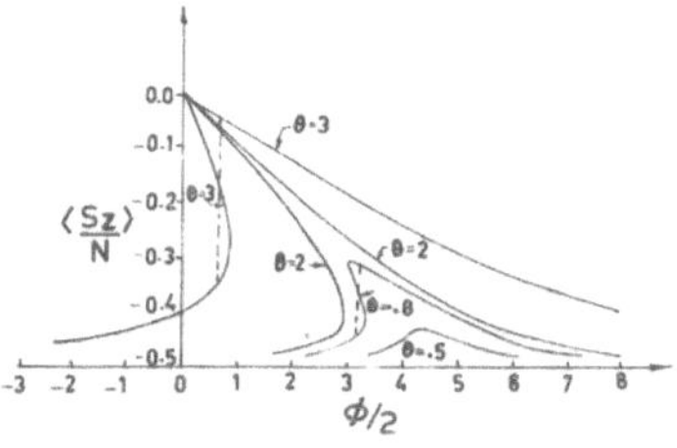

Fig.7. Plot of m_z against φ/2 for several values of θ and ξ = - 5.0

correlation of atoms so that, for example,

$$\langle S_+S_-\rangle \approx (1 - \frac{1}{N}) \langle S_+\rangle\langle S_-\rangle + \langle S_z\rangle + \frac{N}{2} \tag{40}$$

This approximation thus admits independent atomic decay and leads to the usual optical bistability. However, it breaks an important property of the Dicke model, namely that $\langle \hat{S}^2\rangle$ is no longer conserved.

Exact Quantum Mechanical Solution

The steady state solution of the master equation may be easily obtained by writing

$$\rho_a^{ss} = \sum_{m,n=0}^{N} C_{mn} S_-^m S_+^m \tag{41}$$

and determining C_{mn} by direct substitution in the steady state form of Eq.(32). The result is an expression of the same form as in Eqs.(1) with g, Δ replaced by $\tilde{g}$ and $\tilde{\Delta}$ defined as

$$\tilde{g} = (i\Omega/\gamma_o)/(1-i\xi), \quad \tilde{\Delta} = (i\delta_o/\gamma_o)/(1-i\xi) \tag{42}$$

Two limiting cases can be easily inferred from the analysis of the exact ρ_a^{ss} in the non-interacting atoms model[8]. First, it follows that for zero frequency detuning $\tilde{\Delta}$=0 the presence of interatomic interaction leads to an increase of the coherence region with respect to θ. That is, the behaviour of the atomic observables is classical below $\theta = \theta_c = \sqrt{1+\xi^2}$ (instead of θ=1 for $\theta=\delta_o=0$); while above this value quantum effects dominate. The interaction between atoms also leads to a non-zero mean-value $\langle S_x\rangle$ in contrast to the case ξ=0. Secondly, when ξ=0, but δ≠0, the critical behaviour disappears.

More detailed analysis is required to determine the critical behaviour in the regime of parameters ϕ and ξ where the first order phase transition is predicted by semi-classical analysis. Qualitatively one expects a sharp jump in the atomic observables and their fluctuations at the critical values $\theta=\theta_c$(ϕ,ξ fixed) from one stable branch at $\theta<\theta_c$ to another stable branch at $\theta>\theta_c$. A similar critical transition is also expected when the frequency detuning parameter is varied. This is indicated qualitatively in Figs.6 and 7 by dashed lines.

In summary, the Dicke model with atom-atom interactions predicts within $\langle\hat{S}^2\rangle$ conservation a non-equilibrium first order phase transition in the thermodynamic limit.

4. SUMMARY

The driven Dicke model exhibits interesting critical behaviour. Within the conservation of $\langle\hat{S}^2\rangle$, a mathematical condition required for the model, the model predicts a second order phase transition for resonant coherent driving field; bistable behaviour is predicted if the system is placed in a cavity in resonance with the external field as well as with the atomic transition. If interaction between the atoms is included, the system predicts a first order phase transition for certain values of the parameters related to the interaction, field amplitude and frequency detuning. At this stage it is an open question, where these critical phenomena can be observed in an experiment. Experiments with very large wavelength (low frequency) and large dipole moment characteristic of Rydberg atoms may perhaps yield this information. Because of the exactness of the model, it may also be possible to explain the critical behaviour in a more fundamental way. This issue is, however, open and needs further thought.

REFERENCES

1. H. Haken "Synergetics" (Springer-Heidelberg 1977)
2. G. Nicolis and I. Prigogine "Self Organization in Non-equilibrium Systems" (John Wiley: New York 1977)
3. R.H. Dicke, Phys. Rev. 93, 99 (1954)
4. G.S. Agarwal, "Springer Tracts in Modern Physics", Vol.70 (Springer, 1974)
5. S.S. Hassan and R.K. Bullough, "Optical Bistability" eds. C.M. Bowden, M. Ciftan and H. Robl, pp.307-404 (Plenum: New York, 1981)
6. S.V. Lawande, R.R. Puri and S.S. Hassan, J. Phys. B: Atom Mol. 15, 1029 (1982)
7. S.S. Hassan, G.P. Hildred, R.R. Puri and S.V. Lawande, J.Phys.B: Atom Mol.15, 1029 (1982)
8. R.R. Puri and S.V. Lawande, Phys. Lett. A72, 200 (1977), Physica A101, 599 (1980)
9. S. Ya Kilin, Sov. Phys. (JETP) 51, 1081 (1980)
10. P.D. Drummond, Phys. Rev. A22, 1179 (1980)
11. H.J. Carmichael, J. Phys. B13, 3551 (1980)
12. S.S. Hassan, R.K. Bullough, R.R. Puri and S.V. Lawande, Physica A103, 213 (1980)
13. R.R. Puri, S.V. Lawande and S.S. Hassan, Opt. Comm. 35, 179 (1980)
14. L.M. Narducci, D.H. Feng, R. Gilmore and G.S. Agarwal, Phys. Rev. A18, 1751 (1978)
15. R.R. Puri, R.K. Bullough and S.S. Hassan, Applied Physics B, 174 (1982)
16. S. Ya. Kilin, Sov. Phys. JETP 55, 38 (1982)
17. F.T. Arecchi, E. Courtens, R. Gilmore and H. Thomas, Phys. Rev. A6, 2211 (1973)
18. R.R. Puri and S.V. Lawande, to appear in Physica (1983).

DISORDERED SYSTEMS AND RANDOM MEDIA

LOCALIZATION AND DIFFUSION

T.V. Ramakrishnan
Department of Physics
Indian Institute of Science
Bangalore - 560012

The concept of electron localization in a random potential was introduced by Anderson[1] in a paper with the title "Absence of Diffusion in Certain Random Lattices". The intimate connection between localization and quantum diffusion has been explored recently in some detail[2]. I shall briefly summarize this work here after a short introduction.

1. Introduction :

The physical question of interest is the nature of motion (or eigenstates) of a quantum particle in a static random potential. A classical particle would be spatially localized if the allowed regions do not touch each other. One might think naively that since these disparate classically allowed regions are connected by quantum mechanical tunneling across forbidden barriers, a quantum particle always diffuses no matter what the configuration of potentials is. However, the probability for a quantum particle to go from one point to another is the sum of squares of terms which depend on the possible paths. In a strongly random system, the interference between these terms greatly reduces the probability of diffusion. The probability vanishes for sufficiently large separation between the points if the system is highly disordered. Thus, roughly, quantum interference effects inhibit diffusion and eventually lead to localization.

The model considered by Anderson[1] was of an electron hopping from one site of a lattice to its nearest neighbour with an amplitude V. The site energies are random, and are taken to be distributed with equal probability in the range W to -W. Anderson considered the on site Green's function $G_{jj}(E+is)$, i.e. the amplitude for the electron to reside at the site j. By considering the probability distribution of this random number, Anderson showed that the long-time residence probability is non vanishing for disorder greater than a critical value, i.e. for $(W/zV) > n_c$ where z is the number of nearest neighbours, and n_c is a number depending on electron energy E and spatial dimensionality. The best numerical estimates for three dimensions and E=0 (band centre) give $n_c \simeq 2.3$. Thus for stronger disorder, an electron does not diffuse but is spatially localized. This is a general consequence of static disorder, not confined to this particular model for it.

The consequences of localization for physical behaviour of real systems were first explored by Mott[3]. Consider a system with a given static disorder. For this disorder, electronic states with energies lying in some range are localized, others being extended. The boundary energy E_c is called a mobility edge, since states on one side of this edge are mobile and states on the other side are immobile (Fig.1). At T=0, the lowest energy states are occupied till the Fermi energy E_F. If $E_F < E_c$ (see Fig.1) then the low energy excitations are localized, and the system is an insulator. On the other hand if $E_F > E_c$, one has a disordered metal. So, as electron density or disorder is changed, there is a transition from metal to insulator due to disorder (called Anderson transition by Mott). There is a change from diffusion which becomes more and more sluggish to an absence of diffusion. We now outline the ideas of Mott on the nature of this transition.

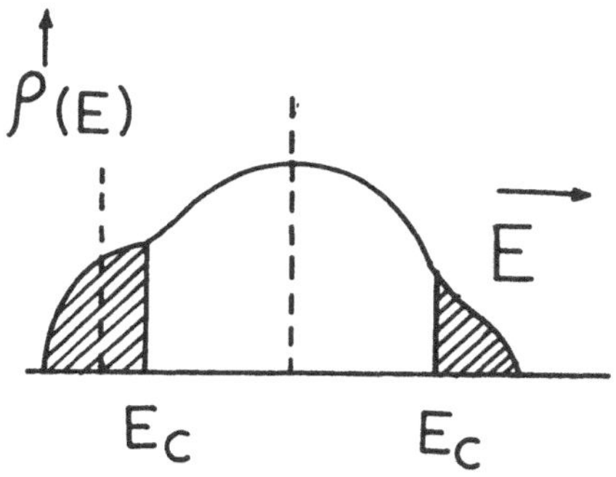

Fig.1 A schematic plot of density of states ρ(E) as a function of energy, localized states being shaded. Dotted lines show two situations, one where $E_F < E_c$ so that the system is an insulator and the other where $E_F > E_c$ so that the system is a metal.

Consider a metal in which electrons move in a relatively weak random potential. Their motion is diffusive because of random changes in direction due to scattering. The cross-over to diffusive behaviour occurs on a length scale of the order of electronic mean free path ℓ for collisions. The diffusion constant D is given by

$$D = \frac{v_F \ell}{d} \tag{1}$$

where d is the spatial dimensionality of the system. Here v_F is the Fermi velocity. The d.c. conductivity is similarly given by

$$\sigma = \frac{e^2}{\hbar} \quad \frac{n\ell}{k_F} \tag{2}$$

where n is the electron density, and k_F the Fermi wavevector. As disorder increases, ℓ becomes shorter. Mott argues following Ioffe and Regel, that the mean free path cannot get shorter than interatomic spacing; the electron has to collide with something. Thus for this mode of diffusive transport, $\ell \gtrsim a$ or since $a \sim k_F^{-1}$, $k_F \ell \gtrsim 1$. Assuming that the expressions (1) and (2) are still valid in this strong collision regime, one has

$$D \gtrsim D_{min.} = \frac{\hbar}{m\, d} \tag{3}$$

and

$$\sigma \gtrsim \sigma_{min.} = \frac{e^2}{\hbar \pi^2} \; k_F \qquad (3d) \tag{4a}$$

$$= \frac{e^2}{2\pi\hbar} \qquad (2d) \qquad (4b)$$

Thus in the metallic state, there is a nonzero minimum diffusion constant with a discontinuous drop to zero at the Anderson transition. The minimum diffusion constant depends on $\hbar$; it is a quantum effect. The numerical value is approximately 1 cm^2/sec The minimum metallic conductivity (Eq.4) again involves Planck's constant. It depends weakly on material properties in three dimensions through k_F, whereas in two dimensions it is universal, with the physical dimensions of conductance. Typically, for a metal with $k_F \sim 10^8 \, A^{o\,-1}$, $\rho_{max} = \sigma_{min}^{-1}$ is about 1000 $\mu\Omega$ cm while in two dimensions $R_{max} = \sigma_{min}^{-1}$ is approximately 30,000 Ω . There is a large body of experimental evidence indicating a change in nature of transport at this resistivity or resistance. For example, near the metal insulator transition in many transition metal oxides and perovskites, the resistivity is in the range $10^3 \mu\Omega$ - cm. In strongly disordered transition metal alloys, the slope of the resistivity vs. temperature curve changes from positive to negative as the resistivity increases beyond 150 - 200 $\mu\Omega$ - cm. In thin metal oxide films there seems to be a resistance 'gap' at R $\simeq$ 30,000 Ω per square.

Approaching the transition from the localized side, the localization length becomes longer and longer as disorder decreases, with the Fermi energy E_F and the mobility edge E_c approaching each other. The divergence can be characterized by an exponent ν i.e. $\xi_{loc} \sim | E_F - E_c |^{-\nu}$. The static dielectric constant which depends on the square of the size of the state also diverges, with an exponent 2ν .

Viewed as a phase transition, localization thus appears rather peculiar. There is a discontinuity in conductivity as in a sudden or first order transition, but from the other side, the localization length diverges as in a second order transition. The source of this asymmetry lies in the neglect of quantum interference effects on the metallic side. The effect of interference between waves scattered back and forth in the random potential is so strong as to localize states for $E_F < E_c$, and to lead to a d.c. conductivity which vanishes exponentially with system size L, provided $L \gg \xi_{loc}$ i.e. $\sigma(L) \sim \sigma_o \exp(-L/\xi_{loc})$. This latter dependence is because in a finite system with localized states near the Fermi energy, electron transport from one end to another takes place by latching on to exponentially small tails of the localized state. Just beyond the transition, $E_F = E_c^+$, however, the random potential is assumed to affect electron motion incoherently, its effect being described in a one collision approximation by a collision time τ . There is no quantum interference, and no dependence of conductivity on scale size. This seems implausible. Abrahams, Anderson, Licciardello and Ramakrishnan[4] showed that quantum interference evolves continuously as a function of scale size. Its effect is described by a scaling function $\beta(g)$ for conductance g in terms of the physical size of the system. From the form for $\beta(g)$ in three dimension, the conductivity is seen to go to zero at the localization transition with the same exponent ν that characterizes the divergence of the localization length. The localization transition is smooth. Near critical disorder, quantum diffusion is strongly scale dependent and nonclassical. We discuss these results below.

In two dimensions, it turns out that the diffusion constant goes to zero at large length scales, so that there is no truly metallic state. This surprising conclusion can be reached by other arguments directly connecting localization and diffusion, and due to Hodges[5] to Maldague[5] and to Mott and Kaveh[5]. In the next section this work is briefly summarized and the scaling theory of localization is then discussed (Section 3).

2. Dimensionality, Diffusion and Localization :

Fluctuations in the local density of states $\rho(\vec{r},E)$ are singularly large for localized states while they are not so if the states are extended. The quantity $\rho(\vec{r},E)$ is defined as

$$\rho(\vec{r},E) = \sum_{\alpha} |\langle\vec{r}|\alpha\rangle|^2 \; \delta(E-E_\alpha) \tag{5}$$

where $|\alpha\rangle$ are the exact eigenstates of the system. For a given potential configuration, if states in the energy range E are localized, $\rho(\vec{r},E)$ will be zero unless a states of energy E is roughly within a localization length of $\vec{r}$. Thus $\rho(\vec{r},E)$ consists of a set of δ - functions, with a strength equal to $|\langle\vec{r}|\alpha\rangle|^2 = |\phi_\alpha(\vec{r})|^2$. Typically the energy spacing between these delta functions is of order $\{N(E)\}^{-1}$ $\{\xi(E)\}^{-d}$ where N(E) is the density of states per unit volume and $\xi(E)$ is the localization length at energy E. Now if states at energy E are extended, each has an amplitude $(1/L^d)$ at the point r , and the spacing between delta function peaks is $\{L^d N(E)\}^{-1}$ which vanishes in the thermodynamic limit so that the local density has a continuous spectrum. On averaging over configurations, however, the distinction is lost. But consider the configuration average of the square of $\rho(\vec{r},E)$. This is expected to be divergent if states of energy E are localized and finite if they are extended. Now for extended states this configuration average can be related to a time integral over the diffusion probability, i.e.

$$\langle \rho(\vec{r},E) \;\; \rho(\vec{r},E)\rangle = \frac{2N(E)}{h} \;\; \frac{1}{[4\pi D_\infty(E)]^{d/2}} \;\; \int_0^\infty \frac{e^{-|\vec{r}-\vec{r}|^2/4Dt}}{t^{d/2}} \, dt \; . \tag{6}$$

clearly the integral diverges for $d \leqslant 2$, as it would for localized states. The assumption that electrons under the influence of a random potential, no matter how weak, diffuse is not internally consistent for two and one dimensions. In one dimension it is easy to visualize localization as an inevitable consequence of repeated backscattering; the localization length is approximately the backscattering mean free path. In two dimensions, the logarithmic divergence of local density of states fluctuations implies that localization takes place on an exponentially large length scale, $\xi_{loc} \simeq k_F^{-1} \exp(k_F \ell)$. Two is the marginal dimension for localization.

3. Scaling Theory of Localization :

As mentioned earlier, the scaling theory of localization focusses on the evolution of quantum interference as a function of scale size L. At any given scale size L, suppose the conductance of the system is g(L). This conductance is itself a measure of interference effects. If they are small electrons move relatively freely and g is large. Conversely, if interference effects are large, g is small. If the scale of the system is changed a little the change in g(L) depends on the size of interference effects at the previous length scale, and on the small change in length[6] . One thus expects that the logarithmic derivative

$$\frac{L}{g} \frac{dg}{dL} = \beta(g)$$

is a function $\beta(g)$ of conductance alone. Now quantum conductance has a natural universal scale, namely $(e^2/\hbar)$ (see for example Eq. (4b). We thus expect that $\beta(g)$ is a universal function in units of $(e^2/\hbar)$ dependent only on the dimensionality of the system.

One can relate, using the scaling curve, the microscopic condition of the system (conductance g_o at microscopic length scale ℓ_o) to its macroscopic conductance (g(L) at large length scales L).

Two limiting forms of $\beta(g)$ are well known. In the weakly disordered regime, the system is a metal characterized by a conductivity σ , and putting together bits of it changes the conductance according to Ohm's law, i.e. $g = \sigma L^{d-2}$ for a hypercube of linear extent L in d dimensions. In this limit, i.e. for $g \gg g_c \simeq e^2/\hbar$, $\beta(g) = (d-2)$. Now for very strong disorder, so that states are localized, $g \sim \exp(-L/\xi_{loc})$ as discussed earlier, and $\beta(g) \simeq \ln g$ $(g \ll g_c \simeq e^2/\hbar)$, Since $\beta(g)$ describes the evolution of quantum interference effects as <u>finite</u> size blocks are scaled up, it is a smooth, monotonic curve. For large values of g, one can do perturbation theory with (1/g) or inverse mean free path as the expansion parameter [4] . It turns out that there are significant scale dependent corrections, and that to leading order,

$$\beta(g) = (d-2) - \frac{a_d}{g} \tag{7}$$

where $a_d = (e^2/\hbar\,\pi^d)$ for a noninteracting electron gas. Calculations using a path integral description of electron motion in a random potential and thence a field theory for the quantum diffusion problem confirm Eq.(7), and show that there are no new terms upto order $(1/g^3)$. Using the limiting forms, the perturbative result Eq.(7) and the continuity as well as monotonicity of $\beta(g)$, we can construct the scaling curve (Fig.2). The scaling curve describes in a succint way the scale dependence of quantum diffusion in a random system. Some of its implications are the following :-

(i) Three dimensions : $\beta(g)$ intersects the x axis at some $g=g_3$ ($g_3 = \frac{e^2}{\hbar\,\pi^3}$ according to the perturbative estimate above). If the system is at the point A at the

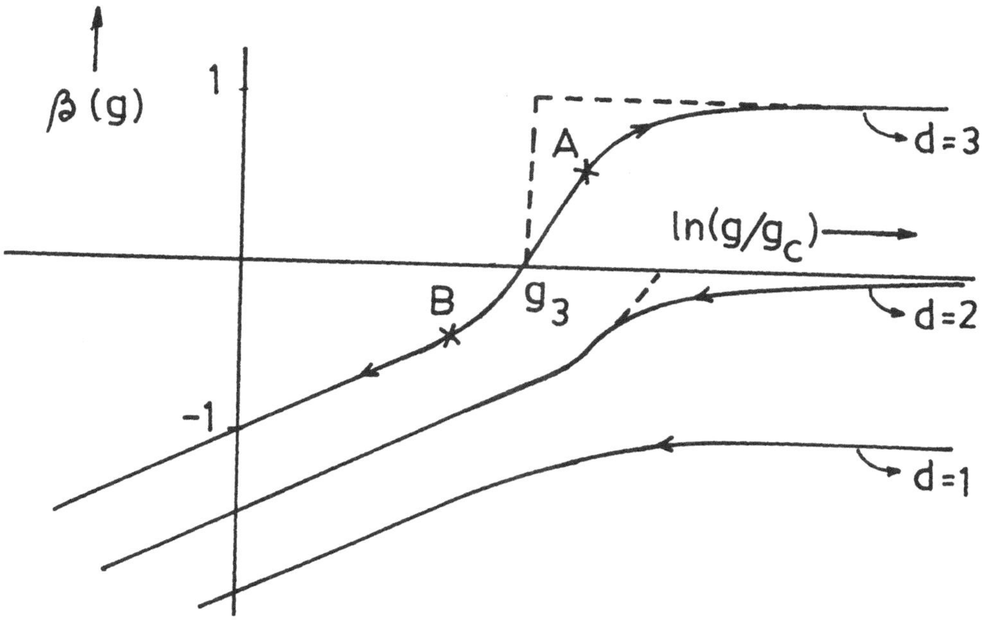

Fig. 2 The scaling curves for conductance in three, two and one dimensions. The direction of change of conductance with increase in scale size is shown by arrows. The dotted curve describes the prediction based on the minimum metallic conductivity idea. The symbols used are explained in the text.

beginning of scaling, i.e. in a state of disorder such that the conductance g_o at the microscopic scale ℓ_o is greater than g_3 one has $\beta(g) > o$ so that g increases with size, finally tending to Ohm's law behaviour. The microscopic conductivity for g_o close to g_3 can be calculated from the scaling curve to be

$$\sigma = (Cg_3/\ell_o) \left\{ (g_o - g_3)/g_3 \right\}^{\nu} \tag{8}$$

where ν^{-1} is the slope of $\beta(g)$ near g_3. C is a numerical constant of order unity. The conductivity goes to zero as critical disorder is approached, with a universal exponent ν. The perturbative estimate for this exponent is $\nu=1$. The Mott minimum metallic conductivity (Cg_3/ℓ_o) is the short length scale conductivity in the critical regime. It thus indicates the state of microscopic disorder at which the localization transition takes place and for which interference effects not included in the conventional transport theory formula Eq. (2) reduce the conductivity according to Eq. (8).

In perturbation theory, an explicit form for conductivity or diffusion constant as a function of scale size can be obtained. In three dimensions, one has

$$\sigma(L) = \sigma_o \left[1 - \frac{3}{\pi} \left(\frac{1}{k_F \ell} \right)^2 \right] + \frac{e^2}{\hbar \pi^3} \frac{1}{L} \tag{9}$$

where σ_o is the transport theory or short length scale conductivity. We see that at $(k_F \ell) = \sqrt{3/\pi}$, $\sigma(\infty) = 0$. This is the condition of critical disorder. For critical disorder, diffusion is entirely nonclassical with $\sigma(L) \propto L^{-1}$. Close to critical disorder, diffusion is slow and depends inversely on scale size upto approximately $L = \xi \sim \ell \left\{ 1 - \frac{3}{\pi} \left(\frac{1}{k_F \ell} \right)^2 \right\}^{-1}$. This is the divergent correlation length in the problem. Beyond ξ, diffusion is classical, with a reduced diffusion constant $D = D_o\,(\ell/\xi)$. Anomalously slow scale dependent diffusion near critical disorder has many consequences, e.g. negative temperature coefficient of resistivity, enhancement of Coulomb repulsion between electrons and consequent suppression of T_c in superconductors.

If the state of microscopic disorder is such that $g_o < g_3$ (e.g. the system is at the point B for a length ℓ_o) then $\beta(g) < o$ always, and as length scale increases conductance decreases. At large enough length scales one has $g(L) \sim g_3 \exp(-L/\xi_{loc})$ where $\xi_{loc} = \frac{\ell_o}{A} \left(\frac{g_3 - g_o}{g_3} \right)^{-\nu}$ for $g_3 \gtrsim g_o$. Near the mobility edge, the localization length diverges with the same exponent as the resistivity. This is what one expects in a theory with a single correlation length.

(ii) Two Dimensions : The $\beta(g)$ curve is negative so that no matter what the initial conductance at microscopic scale ℓ_o, the large length scale behaviour is always that corresponding to localized states. There is no truly metallic state in two dimensions. To estimate the characteristic localization length scale, we use the perturbative form of $\beta(g)$ and integrate it to find

$$g(L) = g_o - \frac{e^2}{\hbar \pi^2} \ln \left(\frac{L}{\pi \ell_o} \right) \tag{10}$$

where $g_o = (e^2/2\pi\hbar)\,(k_F\,\ell_o)$. We see that the conductance is continuously and logarithmically reduced with L, the reduction being significant at a length scale $\xi_{loc}^{2d} \simeq \pi \ell_o \exp(\pi k_F\,\ell_o/2)$.

There is now a large amount of theoretical work on the transition to localization, e.g. a field theory of the transition due to Wegner and others, the effect of a magnetic field, numerical studies of the scaling function $\beta(g)$. Many of these results are reviewed and discussed in reference[2]. We conclude this lecture by outlining briefly the effect of inelastic collisions, of interaction between electrons in disordered systems, and the experimental situation.

4. Inelastic Effects, Interactions and Comparison with Experiment :

(a) Inelastic effects:

At any nonzero temperature, electrons undergo random inelastic collisions which change their energy and thus their phase evolution. The interference effect discussed above is thus terminated at a length scale of the order of the distance an electron diffuses between inelastic collisions (Thouless). Since this increases with decreasing temperature, there are characteristic localization effects in resistivity as a function of temperature (e.g. a term going as ln T in two dimensions).

(b) Interactions:

As electrons slow down in a strongly disordered system, they interact more strongly. Since the motion is diffusive, i.e. slower and slower for longer and longer times ($r \propto \sqrt{t}$ rather than t), the effective interaction is strongly time dependent. The effects of this were worked out first in perturbation theory by Altshuler and Aronov[7]. They showed that there is a $\sqrt{E}$ dip or cusp in the density of states due to interaction effects, and also a $\sqrt{T}$ term in the conductivity. In two dimensions, as pointed out by Altshuler, Aronov, and Lee[7], there is a ln T term in the resistivity due to interactions. From these characteristic temperature dependences it is clear that interaction in a disordered system leads to similar scale dependences as localization. However, interactions affect other quantities besides conductivity, e.g. density of states and screening so that there are in general several (coupled) significantly scale dependent properties. The understanding of these effects is still at an early stage, especially for the most interesting regime, i.e. near the metal insulator transition in disordered three dimensional systems.

(c) Experiment:

The scaling theory, with inelastic cutoff effects and the perturbative theory of interaction effects mentioned above, make explicit experimentally verifiable predictions for resistivity, magnetoresistance, density of states etc. Many of these have been verified in quantitative detail for two dimensional systems (e.g. Si inversion layers, thin films) and for bulk systems (e.g. phosphorus doped silicon). The relative contributions of localization and interaction terms can be disentangled by magnetoresistance measured over a wide range of field and temperature. In the regime where these effects are very strong, e.g. very close to the metal insulator transition, there are two very good experiments. Both find conductivity to go to zero, but with an exponent close to 0.6 in Si:P [8] and close to 1.0 in $Nb_x Si_{1-x}$[9].Interaction and localization effects together give rise to a variety of anomalous diffusion behaviour in real systems which is just beginning to be explored theoretically and experimentally.

References:

1. P.W. Anderson, Phys. Rev. 109, 1492 (1958).
2. See,for example, Anderson Localization (eds. Y. Nagaoka and H. Fukuyama, Springer-Verlag, New York, 1982).
3. See, for example, Electronic Processes in Noncrystalline Materials by N.F. Mott and E.A. Davis (Clarendon; Oxford, 1979).
4. E. Abrahams, P.W.Anderson, D.C.Licciardello and T.V.Ramakrishnan, Phys. Rev. Lett. 42, 673 (1979).
5. C.H. Hodges, P.F. Maldague, Phys. Rev. B. 23, 1719 (1981); M. Kaveh and N.F.Mott, J. Phys. C : Solid State Phys. 14, L177, L183 (1981).
6. The ideas of Thouless on the relation between conductance and sensitivity to change in boundary conditions were very stimulating in this connection. See e.g. D.J. Thouless in Ill-Condensed Matter ed. R. Balian, R. Maynard and G. Toulouse (North-Holland Publishing Co., Amsterdam, 1979) p.1.
7. B.L. Altshuler and A.G. Aronov, Sov. Phys. JETP 50, 968 (1979) ; B.L. Altshuler, A.G. Aronov and P.A. Lee, Phys. Rev. Lett. 44, 1288 (1980).

8. M. Paalanen, T.F. Rosenbaum, G.A. Thomas and R.N. Bhatt, Phys. Rev. Lett. 48, 1284 (1982).
9. G. Hertel, D.J. Bishop, E.G. Spencer, J.M. Rowell and R.C. Dynes (to be published).

CONTINUOUS-TIME RANDOM-WALK IN DISORDERED SYSTEMS

Vipin Srivastava
School of Physics, University of Hyderabad
Hyderabad - 500134, India

Introduction

Diffusion of excitations in a medium of irregular potential has remained a difficult problem. Free transport is the biggest casualty caused in going from an ordered solid to a disordered solid. Environment at different points in a disordered medium is different, consequently a carrier feels different potential as it moves around and experiences scattering due to the fluctuating potential. In situations where the acattering is strong and occurs in random directions, the carrier may get localized [1] in space in presence of an external d.c. field. Such a carrier does not contribute to the electrical conductivity and has a wave function which is peaked around a centre in contrast to the uniformly spread Bloch wave function.

For a carrier of given energy the medium can be divided into 'closed' regions of various spatial extents which can offer enough fluctuations to localize the carrier, as well as 'open' regions through which the carrier can move across and contribute to the conductivity. A trapped carrier can be made to 'hop' to the neighbouring regions by giving it thermally the required activation energy. Thus the transport at non-zero temperature in a disordered medium is to be viewed as hopping along traps and open conducting regions with the traps playing the role of 'immobilizers'. The carrier spends some time in a trap before transition occurs to a neighbouring region, the length of time being decided by the amount of fluctuation offered by the trap. Thus there is a distribution of lengths of time the carrier spends in the traps.

Montroll and coworkers [2,3,4] gave the continuous-time random-walk (CTRW) model to study the above hopping assisted diffusion process. The model assumes a configurationally averaged medium after imbedding all informations of disorder into the distribution function of 'waiting times' (spent in traps), $W(t)$. The carrier, executing random walks in a periodic medium, can abruptly change its position to number of others, each choice weighted by a value of $W(t)$.

It is quite clear that in the CTRW model averaging must have destroyed lot of new and crucial informations expected entirely from the disordered nature of the medium. Experiments [4] have revealed that the transport in disordered solids has a very distinct 'dispersive' characteristic, conceptual visualization of which is not possible in an averaged medium where due to the absence of traps one cannot distinguish 'trapping' from diffusion. Another problem arises when one tries to understand the frequency dependent response to a.c. field of the systems exhibiting dispersive transport [5]. For other criticisms of CTRW model see Schmidlin [6], Ngai and Liu [7] Chekunaeu et al [8] and Rudenko and Ankhipov [9]. All the criticisms emphasize the fact that localized and conducting states should not be treated on equal footing.

So the CTRW model should be generalized to treat the quantum states in the system to be inequivalent. We shall do this in the following [10].

The Generalized CTRW Model

Consider the probability to just arrive at site n in time interval $(t,t+\Delta t)$ in N steps, $R_N(n,t)\Delta t$, which satisfies the recursion formula,

$$R_{N+1}(n,t) \quad = \quad \int_0^t d\tau \sum_{m\neq n} \mathcal{W}_{nm}(t-\tau)R_N(m,\tau) \tag{1}$$

where $\mathcal{W}_{nm}(t)$ is the probability per unit time that transition from a state situated at m occurs to a state situated at n in time t. The states situated at n and m are equivalent in an averaged medium, so $\mathcal{W}_{nm}$ depends only on the separation between n and m. However, in disordered medium the states at n and m shall be different in nature, so this information should also be included in the transition probability. Random walk shall now be represented by the following equation,

$$R_{N+1}(n_\alpha,t) \quad = \quad \int_0^t d\tau \sum_{m_{\alpha'}\neq n_\alpha} W_{n_\alpha m_{\alpha'}}(t-\tau)\; R_N(m_{\alpha'},\tau), \tag{2}$$

where α and α' represent the nature of states at n and m. Following conditions are satisfied by $\mathcal{W}_{nm}$ and $W_{n_\alpha m_{\alpha'}}$:

$$\sum_n \mathcal{W}_{nm}(t) \quad = \quad 0 \; ; \; \sum_{n_\alpha} W_{n_\alpha m_{\alpha'}}(t) \quad = \quad 0, \tag{3a}$$

and

$$\int_0^\infty \sum_{n\neq m} \mathcal{W}_{nm}(\tau)d\tau \quad = 1 \; ; \quad \int_0^\infty \sum_{n_\alpha\neq m_{\alpha'}} W_{n_\alpha m_\alpha},(\tau)d\tau < 1. \tag{3b}$$

Condition (3a) is essential for conservation of probability as will be seen later. Condition (3b) is of central importance — in the case of $\mathcal{W}$ (i.e. averaged medium) it says that the carrier must leave the site m eventually, while in the case of W (i.e. unaveraged medium) the carrier may not leave the site m even if infinite time has elapsed. Some more insight into the difference between the averaged and unaveraged mediums is achieved if we decouple $\mathcal{W}$ and W in space and time parts as,

$$\mathcal{W}_{nm}(t) \quad = \quad \vartheta_{nm}k(t), \tag{4a}$$

and

$$W_{n_\alpha m_{\alpha'}}(t) \quad = \quad V_{n_\alpha m_{\alpha'}} \; k_{m_{\alpha'}}(t), \tag{4b}$$

where ϑ and V cause the transition from m to n and k takes care of the fact that this transition occurs after time t; note the dependence of k on the nature of state at m in the unaveraged medium. For ϑ and V we have the following conditions,

$$\sum_n \vartheta_{nm} = 1 \qquad \text{with} \quad \vartheta_{mm} = 0 \; ; \tag{5a}$$

and

$$\sum_{n_\alpha} V_{n_\alpha m_{\alpha'}} = 1 \text{ with } V_{m_{\alpha'},m_{\alpha'}} > 0 \; . \tag{5b}$$

Also note that, $\vartheta_{mm}(t) = 0$ (6a)

while $W_{m_{\alpha'},m_{\alpha'}}(t) = -(1-V_{m_{\alpha'},m_{\alpha'}})k_{m_{\alpha'}}(t).$ (6b)

Thus,

$$\int_0^\infty \sum_{n_\alpha \neq m_{\alpha'}} W_{n_\alpha m_{\alpha'}}(\tau)d\tau = \underbrace{(1-V_{m_{\alpha'},m_{\alpha'}})}_{\uparrow}\underbrace{\int_0^\infty k_{m_{\alpha'}}(\tau)d\tau}_{\uparrow} \tag{7}$$

$$\begin{array}{lll} = 0 & = 1 & : \text{for averaged medium} \\ > 0 & < 1 & : \text{for unaveraged medium} \end{array}$$

Now we come back to (2) and eliminate the step variable N by summing over all paths of number of steps varying from 0 to ∞ ; using the initial condition that the carrier was at the initial state '0' at time t = 0, we obtain,

$$R(n_\alpha,t) - \int_0^t d\tau \sum_{m_{\alpha'} \neq n_\alpha} W_{n_\alpha m_{\alpha'}}(t-\tau)\, R(m_{\alpha'},\tau) = \delta_{n_\alpha o}\, \delta(t-0) \; . \tag{8}$$

Having considered the probability to just arrive at a given state at a given time we turn to our main object of answering the following question. Suppose the carrier, which was at state '0' at t = 0, reaches the state n_α at time τ , then what is the probability that it can be found at n_α at a later time t? One asks this question because a random walker starting from an arbitrary origin can choose from a large variety of paths to arrive at n_α at different times. We can alternatively put it as follows. A carrier having arrived at n_α at time τ ($<t$) remains unmoved in the remaining time ($t-\tau$). This is easily expressed as follows,

$$P(n_\alpha,t) \equiv P(n_\alpha,t|0,0) = \int_0^t d\tau \;\, \phi_{n_\alpha}(t-\tau)R(n_\alpha,\tau), \tag{9}$$

where $P(n_\alpha,t)$ is the conditional probability that the carrier initially at '0' is found at n_α at time t, and $\phi_{n_\alpha}(t-\tau)$ is the probability for remaining unmoved for time ($t-\tau$). It is important to note that in a random medium the carrier may remain unmoved while at n_α also because n_α may be a trap. The subscript n_α indicates this dependence of ϕ on the nature of state at n. Total probability to remain unmoved in interval (0,t) at n_α is,

$$\phi_{n_\alpha}(t) = 1- \int_0^t d\tau \sum_{g_{\alpha'} \neq n_\alpha} W_{g_{\alpha'},n_\alpha}(\tau) \; . \tag{10}$$

Eliminating R from (9) we obtain,

$$P(n_\alpha,t) = \phi_{n_\alpha}(t)\,\delta_{n_\alpha 0} + \int_0^t d\tau \sum_{m_{\alpha'}\neq n_\alpha} W_{n_\alpha m_{\alpha'}}(t-\tau)P(m_{\alpha'},\tau). \tag{11}$$

This is the generalized CTRW equation. Probability is conserved due to (3a). As discussed earlier $W_{n_\alpha m_{\alpha'}}$ is not a simple function of separation between the positions of the states, although it can be simplified and made function only of separation between positions of states by averaging over the positions of the states (note that this is different from configurational averaging and $V_{m_\alpha, m_{\alpha'}}$ remains non-zero).

We can rearrange (11) and cast it in the form of a rate equation in probability. It is convenient to first take the Laplace transform of (11),

$$\tilde{P}(n_\alpha,z) = \tilde{\phi}_{n_\alpha}(z)\,\delta_{n_\alpha 0} + \sum_{m_{\alpha'}\neq n_\alpha} \tilde{W}_{n_\alpha m_{\alpha'}}(z)\,\tilde{P}(m_{\alpha'},z), \tag{12}$$

and then rearrange it to write,

$$z\,\tilde{P}(n_\alpha,z) - \delta_{n_\alpha 0} = \sum_{m_{\alpha'}} \tilde{M}_{n_\alpha m_{\alpha'}}(z)\,\tilde{P}(m_{\alpha'},z), \tag{13}$$

where, $$\tilde{M}_{n_\alpha m_{\alpha'}}(z) \equiv \Big[(1-\delta_{n_\alpha m_{\alpha'}})\,\tilde{W}_{n_\alpha m_{\alpha'}}(z) - \delta_{n_\alpha m_{\alpha'}} \sum_{g_{\alpha'}\neq n_\alpha} \tilde{W}_{g_{\alpha'},n_\alpha}(z)\Big]/\tilde{\phi}_{n_\alpha}(z). \tag{14}$$

Now taking the Laplace inverse-transform of (13), we get,

$$\dot{P}(n_\alpha,t) = \sum_{m_{\alpha'}} \int_0^t d\tau\; M_{n_\alpha m_{\alpha'}}(t-\tau)P(m_{\alpha'},\tau). \tag{15}$$

This is the familiar generalized master equation. Thus we learn that the GCTRW equation is an exact transport equation and describes a non-Markovian stochastic process [10].

Localization and Dispersive Transport

The transport equation (15) can be useful for understanding certain problems in localization [11] and the problems related with the dispersive nature of transport in disordered systems which, as pointed out in the beginning, can not be explained in the averaged medium - CTRW framework.

As already stated localization is of central importance in disordered solids and is responsible for lots of peculiar physical properties observed in such systems. We will describe the existing localization picture very briefly and then discuss a finer modification introduced in it due to considerations based on (15). Thereafter we will come to dispersive transport which is a consequence of localization and is intimately related with the said modification in the localization picture.

The picture that has emerged from theoretical analyses and is supported by experiments also is the following. In the energy spectrum the energies near the band centre correspond to conducting states and those towards the band edges correspond to trap (or localized) states; the two kinds of states are separated sharply at energies called 'mobility edges', E_c. Localization is defined in terms of a 'stay-put' probability which is same as our $P(n_\alpha,t)$ with n_α as the origin:

$$\lim_{t\to\infty} P(n_\alpha,t) \begin{cases} = 0 \Rightarrow \text{conducting states} \\ > 0 \Rightarrow \text{localized states .} \end{cases} \qquad (16)$$

There are conflicting views about the nature of transition from localized regime to conduction regime. According to one point of view put forward by Mott and coworkers [12], as the mobility edge is approached, conductivity becomes smaller and smaller and takes a minimum value, the 'minimum metallic conductivity', before abruptly becoming zero. The other point of view due to Cohen [13] favours a continuous transition, i.e. the conductivity goes to zero in a continuous manner. This problem can be studied in terms of rate of diffusion, i.e. how $P(n_\alpha,t)$ approaches zero in the vicinity of E_c since $P(n_\alpha,t)$ is related with mobility [3]. For this purpose we introduce the following integral,

$$\mathcal{J} = \int_0^\infty P(n_\alpha,t)\,dt, \qquad (17)$$

which enables us to distinguish between rates of diffusion as follows,

$\lim_{t\to\infty} P(n_\alpha,t)$	$\mathcal{J}$		
$= 0$	$< \infty$	$\Rightarrow$	fast diffusion
$= 0$	$= \infty$	$\Rightarrow$	slow diffusion
> 0	$= \infty$	$\Rightarrow$	no diffusion

The regions of fast and no diffusion are well established. If the existence of the new intermediate regime of 'slow diffusion' could be established, then we could conclude that the conductivity goes to zero in localized regime gradually and there is no minimum metallic conductivity. This, indeed, is possible to show [11] with the help of the rate equation (15). All one has to do is to construct the kernel M which is possible if the waiting-time distribution W can be known. Since our aim is to investigate the vicinity of E_c on conduction side in terms of the divergence of the integral, $\mathcal{J}$, given that the stay-put probability vanishes in the limit $t\to\infty$, we can suitably choose forms for W in conduction and localized regimes so as to satisfy (16) and match them at the E_c. For the matching we make use of experimental results on photoconduction [3,4]. The ansatz for W and the calculational details are discussed in [11].

Now we come to the dispersive nature of transport in presence of localized states. The photoconductivity experiments performed on many semiconducting systems show that a bunch of photo-excited electrons which can be viewed as a Gaussian packet gets deformed and broadened as it moves under an electric field. If the disorder is moderately large the packet disperses into a very wide distribution of carriers with the probability of finding some of the carriers near the starting point remaining considerable even after a large time. The reason is simple. Imagine a packet of carriers starting from an electrode and moving towards the other. On the way they come across traps where they can stop for different durations depending upon the holding strengths of the traps. While some of them remain unmoved, others move ahead. In this manner the packet becomes broader and broader as time elapses.

The nature of transport is not highly dispersive in all regimes. The ansatz for W given in [11] in the GCTRW framework reveals the following picture. In the small disorder-metallic regime transport occurs in the form of mildly broadened Gaussian packet. In the moderately high disorder-slow diffusion regime near E_c, transport occurs in highly dispersive non-Gaussian manner. In high disorder insulating regime the carriers move about among traps in the form of a broadened Gaussian packet. Thus the non-Gaussian transport is an attribute of the new intermediate regime of slow diffusion [14].

An interesting account of thermal equilibrium between mobile and localized carrier fractions in a system with traps having arbitrary energy distribution is given in a series of papers by Rudenko and Arkhipov [9]. They have found that the state of non-equilibrium persists for rather a long time, however, when the equilibrium is attained the carriers form a Gaussian packet which moves with a constant velocity. The dispersion of the packet grows with time as $t^{\frac{1}{2}}$ and the ratio between the dispersion and the displacement of the packet mean decreases with time as $t^{-\frac{1}{2}}$.

Finally we will make a comment on the observed frequency dependent response to a.c. field. To explain this, it is not sufficient to consider the quantum states to be inequivalent and simple traps scattered here and there in the system, instead one must also include the possibility of complex traps or clusters of closely lying traps. In the former situation, when the carrier is free, it drifts in phase with the field and when it occassionally stops in a trap it does not contribute to the current at all, while in the latter situation an occupied complex trap acts as a 'relaxation dipole' — the carrier can oscillate with the a.c. field but out of phase by an amount given by the relaxation time of the dipole. It is thus clear that besides configurational averaging it is also essential that no assumptions should be made regarding the number and arrangement of traps (unlike the work by Scher and Wu [15] who assume a periodic distribution of traps).

References

1. P.W. Anderson, Phys. Rev. 109, 1492 (1958)
2. E.W. Montroll and G.H. Weiss, J. Math. Phys. 6, 165 (1965)
3. H. Scher and M. Lax, Phys. Rev. B7, 4491 and 4502 (1973)
4. H. Scher and E.W. Montroll, Phys. Rev. B12, 2455 (1975)
5. J.K.E. Tunaley, Phys. Rev. Lett. 33, 1037 (1974)
6. F.W. Schmidlin, Phil. Mag. B41, 535 (1980)
7. K.L. Ngai and Fu-Sui Liu, Phys. Rev. B24, 1049 (1981)
8. N.I. Chekunaev, Yu A. Berlin and V.N. Fleurov, J. Phys. C15, 1219 (1982)
9. A.I. Rudenko and V.I. Arkhipov, Phil. Mag. B45, 177, 189 and 209 (1982)
10. V. Srivastava and M. Chaturvedi, J. Phys. C14, L671 (1981); and Z. Phys. B48, 351 (1982)

11. V. Srivastava and M. Chaturvedi, to be published (1983)
12. See e.g. N.F. Mott and E.A. Davis, Electronic Processes in Non-Crystalline Material (Clarendon Press, Oxford, 1971)
13. M.H. Cohen, Physics Today (May 1971) p.26, and the references therein
14. V. Srivastava and M. Chaturvedi, J. Non-Cryst. Sol. (1983)
15. H. Scher and C.H. Wu, Proc. Natl. Acad. Sci. USA 78, 22 (1981)

RANDOM MATRICES IN CONDENSED MATTER PHYSICS

C.K. Majumdar
Magnetism Department
Indian Association for the Cultivation of Science
Jadavpur, Calcutta - 700032, India

The theory of ensembles of random matrices was pioneered by Wigner [1] and has been of considerable utility in the analysis of nuclear spectra [2,3,4]. There are some problems in condensed matter physics where the theory has been applied with some measure of success (see Sec. IX of [4]). It is likely that the use of random matrix ensembles in solid state physics will increase, particularly because the major ensemble types can be realized in practice in solid state problems.

1. New kind of Statistical Mechanics

It has been emphasised by Dyson [5] that the approach through random matrices constitutes a new kind of statistical mechanics characterized by the hypothesis of total ignorance. In ordinary statistical mechanics, the Hamiltonian of the system is well defined, but the system is so large that exact computation is neither feasible nor meaningful, and observation of all the details is not possible. We consider an ensemble of systems described by this fixed Hamiltonian, and assume that all states of the Hamiltonian are equally likely for the ensemble. This leads to useful characterization of the system.

In the new kind of statistical mechanics we renounce knowledge of the exact nature of the system itself. For instance, a nucleus may be pictured as a "black box" containing a large number of particles interacting strongly according to imperfectly understood laws. In the study of phonons in glass, a glass is an irregular network of several types of atoms which interact in a linear harmonic fashion with spring constants varying randomly. The problem is to define in a mathematically precise way an ensemble in which various likely Hamiltonians are equally probable.

A Hamiltonian is represented by a hermitian matrix, usually infinite. For simplicity, we restrict ourselves to large but finite matrices. Certain conservation laws are known; they are associated with some symmetry principles which are supposed to be valid for our system, whatever be its Hamiltonian. There are, for example, rotational invariance, reflection invariance and time reversal invariance. These imply certain symmetries of the matrices and special features of the matrix elements. Thus time reversal invariance implies that the matrix element can be taken to be real. Apart from these general symmetry requirements, the matrix elements are regarded as independent random variables. A common, simple assumption is that the matrix element M_{ij} is normally distributed about a mean value with some variance. Next we ask for the average spectrum of such a matrix ensemble and try to compare the calculated spectrum with some experimental result on the excited states of a nucleus, the phonon density of states in glass, etc.

One interesting feature of this comparison is that we are often more interested in the deviations from the calculated results. If, indeed, there is perfect agreement between experiment and theory, the hypothesis of total ignorance is justified. This may be a happy situation but is devoid of any intellectual challenge. A great deal of work in nuclear physics has been done to confront the random matrix calculations with experimental data; see Reference [4].

In ordinary statistical mechanics we come across several ensembles, uniform, microcanonical, canonical, grand canonical, and so on. It has been found by Dyson [6] that matrix ensembles can be classified into three types - orthogonal, unitary and symplectic - and that all irreducible ensembles belong to one of these three types. Dyson has traced the origin of this classification to the theorem of Frobenius: over the real number field, there exist precisely three associative division algebras, namely the real numbers, the complex numbers, and the real quaternions.

It appears likely that all the three ensembles can be practically realized in solid state problems. Consider small particles of free-electron-like metals at low temperatures. The energy levels are discrete, and around $10^{o}K$ with particle radius about 10^{-6} cm, the level spacing is much larger than the average thermal energy. The system of metal particles will then show "quantum size effect". Now in an actual experiment the shape and size of the small particles may not be precisely controlled. The spectrum will show fluctuations. In fact, the electronic energy levels can be thought to be the eigenvalues of a fixed Hamiltonian with random boundary conditions, which may be incorporated into a random matrix by using fictitious potentials. The appropriate ensembles are then as follows:

(i) if the number of electrons is even and there is no magnetic field, the orthogonal ensemble is applicable;

(ii) if the number of electrons is odd and there is no magnetic field, the symplectic ensemble applies; and

(iii) when the magnetic field is present and the Hamiltonian is no longer time reversal invariant, the unitary ensemble is applicable.

Brody et al have reviewed the comparison between experiment and theory in this field. There are some unsolved problems; in particular, the recent advances in the "Weyl problem" [7] can be confronted with experimental work.

We shall give below a simple derivation of the Wigner semicircular law for the asymptotic distribution of the eigenvalues of a random matrix. A paper by Wigner [8] on bordered matrices contains some other interesting results.

2. Simple Derivation of the Semicircular Law

The derivation of the semicurcular law by Edwards and Jones [9] uses nothing beyond the properties of the ordinary Gaussian integrals, which are first collected here:

$$\int_{-\infty}^{\infty} e^{-ax^2}\,dx = \pi^{\frac{1}{2}} a^{-\frac{1}{2}} \tag{2.1}$$

If a has a negative imaginary part, we can write

$$\int_{-\infty}^{\infty} e^{-iax^2}\,dx = \pi^{\frac{1}{2}} e^{-i\pi/4} a^{-\frac{1}{2}} \tag{2.2}$$

For a positive definite N X N matrix $A = (a_{ij})$ we have

$$\int_{-\infty}^{\infty} \exp\left(-\sum_{i,j} a_{ij}x_i x_j\right) \prod_i dx_i = \pi^{N/2} \det^{-\frac{1}{2}} A \tag{2.3}$$

$\det^{\frac{1}{2}} A$ denotes the square root of the determinant of A. Hence we also write

$$\int_{-\infty}^{\infty} \exp\left(-i\sum_{j,k} a_{jk}x_j x_k\right) \prod_j dx_j = \pi^{N/2} e^{-i\pi N/4} \det^{-\frac{1}{2}} A \tag{2.4}$$

By completing the square one proves

$$\int_{-\infty}^{\infty} e^{-\frac{x^2}{2b} \pm ax}\,dx = e^{\frac{1}{2}a^2 b} (2\pi b)^{\frac{1}{2}} \tag{2.5}$$

Hence we get the representation

$$e^{ia^2/2} = \frac{e^{-i\pi/4}}{(2\pi)^{\frac{1}{2}}} \int_{-\infty}^{\infty} e^{\frac{i}{2}x^2 - ax}\,dx \tag{2.6}$$

The other ingradient is the elementary limit formula of calculus

$$\lim_{n\to 0} \frac{x^n - 1}{n} = \ln x. \tag{2.7}$$

Consider now the problem of calculating the average eigenvalue spectrum of a large N X N real symmetric matrix M. Each matrix element M_{ij} has a gaussian probability density function with zero mean and fixed variance σ :

$$p(M_{ij}) = (2\pi\sigma^2)^{-\frac{1}{2}} \exp\left[-M_{ij}^2/2\sigma^2\right] \tag{2.8}$$

Let the eigenvalues of M be M_i. The density of eigenvalues is defined by

$$\nu(\lambda) = \frac{1}{N} \sum_j \delta(\lambda - M_j) \tag{2.9}$$

where $\nu(\lambda)$ is normalized to unity. Now we give λ a small negative imaginary part and use

$$\frac{1}{\lambda - i\epsilon - M_j} = P\frac{1}{\lambda - M_j} + \pi i\,\delta(\lambda - M_j) \tag{2.10}$$

to obtain

$$\nu(\lambda) = \frac{1}{\pi N} \sum_j \operatorname{Im} \frac{1}{\lambda - i\epsilon - M_j} \tag{2.11}$$

But

$$\det(\lambda I - M) = \prod_j (\lambda - M_j)$$

Hence

$$\ln \det (\lambda I-M) = \sum_j \ln(\lambda - M_j),$$

and

$$\frac{\partial}{\partial x} \ln \det (\lambda I-M) = \sum_j \frac{1}{\lambda - M_j} \tag{2.12}$$

Henceforth λ is supposed to have a small negative imaginary part, then (2.11) can be written as

$$\nu(\lambda) = \frac{1}{\pi N} \operatorname{Im} \frac{\partial}{\partial \lambda} \ln \det (\lambda I-M) \tag{2.13}$$

$$= -\frac{2}{\pi N} \operatorname{Im} \frac{\partial}{\partial \lambda} \ln \det^{-\frac{1}{2}} (\lambda I-M) \tag{2.14}$$

$$= -\frac{2}{\pi N} \operatorname{Im} \frac{\partial}{\partial \lambda} \lim_{n \to 0} \frac{1}{n} \left[(\det^{-\frac{1}{2}}(\lambda I-M))^n - 1 \right] \tag{2.15}$$

where in the last line we use (2.7). With (2.4) we can write

$$\nu(\lambda) = -\frac{2}{\pi N} \operatorname{Im} \frac{\partial}{\partial \lambda} \lim_{n \to 0} \frac{1}{n} \Bigg\{ \left(\frac{e^{i\pi/4}}{\pi^{\frac{1}{2}}} \right)^{Nn}$$

$$\times \int_{-\infty}^{\infty} \prod_{\substack{i,j=i,N \\ \alpha=1,n}} dx_i^\alpha \exp\left[-i \sum_{i,j,\alpha} x_i^\alpha (\lambda \delta_{ij} - M_{ij}) x_j^\alpha \right] - 1 \Bigg\} \tag{2.16}$$

It is here assumed that n is an integer. Equation (2.7) has no such stipulation. Assume that in the result of interest the continuation to $n \to 0$ is still allowed after we get a formal answer of the integral with n an integer. This step makes the derivation nonrigorous. Such a continuation to $n \to 0$ from the initial assumption of only integral n has worked in several problems in solid state physics (this is the replica trick of Edwards and Anderson). The integration is over the Nn variables x_i^α.

It is good to check that in simple cases the formal $n \to 0$ limit goes through. Consider an N X N matrix with all matrix elements M_{ij} equal to M_o/N, where $M_o \sim O(1)$. It is well known that this has one eigenvalue M_o and (N-1) eigenvalues 0. Let us check if (2.16) leads to the correct result.

Equation (2.16) becomes

$$\nu(\lambda) = -\frac{2}{\pi N} \operatorname{Im} \frac{\partial}{\partial \lambda} \lim_{n \to 0} \Bigg[\left(\frac{e^{i\pi/4}}{\pi^{\frac{1}{2}}} \right)^{Nn}$$

$$\times \int_{-\infty}^{\infty} \prod_{i,\alpha} dx_i^\alpha \exp\left(-i\lambda \sum_{i,\alpha} (x_i^\alpha)^2 + i \frac{M_o}{N} \sum_\alpha (\sum_i x_i^\alpha)^2 \right] \tag{2.17}$$

Now we use (2.6) in the form known as the "auxiliary field identity":

$$\exp\left[i\,\frac{M_o}{N}\,(\sum_i x_i^{\alpha})^2\right] = \left[\frac{e^{-i\pi/4}}{(2\pi)^{1/2}}\right] \times \int_{-\infty}^{\infty} dq\; e^{iq^2/2} \exp\left[-\left(\frac{2M_o}{N}\right)^{1/2} q \sum_i x_i^{\alpha}\right] \tag{2.18}$$

The integral in (2.17) becomes

$$J_1 = \prod_{\alpha} \frac{e^{-i\pi/4}}{(2\pi)^{1/2}} \int_{-\infty}^{\infty} dq \prod_i dx_i^{\alpha}\; \exp\left[i\lambda \sum_i (x_i^{\alpha})^2\right] \times \exp\left[-\left(\frac{2M_o}{N}\right)^{1/2} q \sum_i x_i^{\alpha} + \frac{1}{2} iq^2\right] \tag{2.19}$$

The integral over each x_i^{α} is easily performed by completing the square, (2.5), and we find

$$J_1 = \prod_{\alpha} \frac{e^{-i\pi/4}}{(2\pi)^{1/2}}\; e^{-iN\pi/4} \left(\frac{\pi}{\lambda}\right)^{\frac{N}{2}} \int_{-\infty}^{\infty} dq\; \exp\left[-\frac{iq^2}{2}\left(\frac{M_o}{\lambda} - 1\right)\right] \tag{2.20}$$

With (2.2) we get

$$J_1 = \prod_{\alpha} e^{-iN\pi/4}\, \pi^{\frac{N}{2}}\, \lambda^{-\frac{N-1}{2}}\, (\lambda - M_o)^{-1/2} \tag{2.21}$$

Thus (2.17) becomes

$$\nu(\lambda) = -\frac{2}{N\pi} \mathrm{Im} \frac{\partial}{\partial\lambda} \lim_{n\to 0} \frac{1}{n}\left[\lambda^{-\frac{(N-1)n}{2}} (\lambda - M_o)^{-\frac{n}{2}} - 1\right] \tag{2.22}$$

The formula allows continuation in n to fractional values and we take (2.7) as true

$$\nu(\lambda) = -\frac{2}{N\pi} \mathrm{Im} \frac{\partial}{\partial\lambda} \ln\left[\lambda^{-\frac{N-1}{2}} (\lambda - M_o)^{-1/2}\right]$$

$$= \frac{1}{\pi N} \mathrm{Im} \frac{\partial}{\partial\lambda} \left[(N-1)\ln\lambda + \ln(\lambda - M_o)\right]$$

$$= \frac{1}{\pi N} \mathrm{Im} \left[\frac{N-1}{\lambda} + \frac{1}{\lambda - M_o}\right]. \tag{2.23}$$

Recall λ has a small negative imaginary part and use (2.10):

$$\nu(\lambda) = \frac{N-1}{N}\,\delta(\lambda) + \frac{1}{N}\,\delta(\lambda - M_o) \tag{2.24}$$

We now go back to the real symmetric matrix M with $M_{ij} = M_{ji}$. From (2.8) we define J by

$$\sigma^2 = J^2/N \tag{2.25}$$

with J of order unity. The averaged density of eigenvalues $\rho(\lambda)$ is obtained by averaging $\nu(\lambda;\{M_{ij}\})$ of (2.16) over all configurations of the M_{ij} given by (2.8):

$$\rho(\lambda) = \int \nu(\lambda;\{M_{ij}\}) \prod p(M_{ij})\, d M_{ij} \tag{2.26}$$

Putting (2.8) with (2.26), we carry out all the gaussian integrations over M_{ij}:

$$\rho(\lambda) = -\frac{2}{N\pi}\,\mathrm{Im}\,\frac{\partial}{\partial\lambda}\lim_{n\to 0}\frac{1}{n}\Big[\Big\{\Big(\frac{e^{i\pi/4}}{\pi^{1/2}}\Big)^{Nn}$$

$$\times\int_{-\infty}^{\infty}\prod dx_i^{\alpha}\exp\Big[-i\lambda\sum_i (x_i^{\alpha})^2\Big]\exp\Big[-\frac{J^2}{N}\sum_{i,j}\Big(\sum_{\alpha} x_i^{\alpha}x_j^{\alpha}\Big)^2\Big]$$

$$\times\exp\Big[\frac{J^2}{2N}\sum_i\Big(\sum_{\alpha}(x_i^{\alpha})^2\Big)^2\Big]\Big\} - 1\Big] \tag{2.27}$$

We want to retain the leading terms in N as $N \to \infty$ and the term linear in n as $n \to o$. We can simplify the calculation of (2.27) by estimating order of magnitude of the second term of the exponentials.

$$\frac{J^2}{N}\sum_{i,j}\Big(\sum_{\alpha} x_i^{\alpha}x_j^{\alpha}\Big)^2 = \frac{J^2}{N}\sum_{i,j}\sum_{\alpha\beta} x_i^{\alpha}x_j^{\alpha}x_i^{\beta}x_j^{\beta}$$

$$= \frac{J^2}{N}\sum_{\alpha}\Big(\sum_i (x_i^{\alpha})^2\Big)^2 + \frac{J^2}{N}\sum_{\alpha\neq\beta}\sum_{i\neq j} x_i^{\alpha}x_j^{\alpha}x_i^{\beta}x_j^{\beta} \tag{2.28}$$

The first term is of order Nn. The second has a zero mean, but its square is of order n. The third exponential term in (2.27) is of order n^2. Hence it is enough to keep the terms

$$\frac{J^2}{N}\sum_{\alpha}\Big(\sum_i (x_i^{\alpha})^2\Big)^2$$

Thus for large N,

$$\rho(\lambda) = -\frac{2}{\pi N}\,\mathrm{Im}\,\frac{\partial}{\partial\lambda}\lim_{n\to 0}\frac{1}{n}\Big[\Big(\frac{e^{i\pi/4}}{\pi^{1/2}}\Big)^{Nn}$$

$$\Big\{\int_{-\infty}^{\infty}\prod dx_i\,\exp\Big[-i\lambda\sum_i x_i^2 - \frac{J^2}{N}\Big(\sum_i x_i^2\Big)^2\Big]\Big\}^n - 1\Big] \tag{2.29}$$

Use again an auxiliary field identity

$$e^{-\frac{J^2}{N}\left(\sum_i x_i^2\right)^2} = \Big(\frac{N}{2\pi}\Big)^{1/2}\frac{\lambda}{(2J^2)^{1/2}}\int_{-\infty}^{\infty} ds\,\exp\Big(-i\lambda s\sum_i x_i^2\Big)\,\exp\Big(-\frac{\lambda^2}{4J^2}Ns^2\Big) \tag{2.30}$$

The integral in (2.29) becomes

$$J_2 = \Big[\int_{-\infty}^{\infty} ds\prod_i dx_i\Big(\frac{N}{2\pi}\Big)^{1/2}\frac{\lambda}{(2J^2)^{1/2}}\exp\Big(-i\lambda(1+s)\sum_i x_i^2\Big)\exp\Big(-\frac{N\lambda^2 s^2}{4J^2}\Big)\Big]^n \tag{2.31}$$

The integrals over the $\{x_i\}$ are straightforward, but the convergence of this integral depends on the small negative imaginary part of λ. We could make λ real and maintain convergence by putting a small negative imaginary part to s. Hence the result is

$$J_2 = \left\{ \left(\frac{N}{2\pi}\right)^{\frac{1}{2}} \left(\frac{\pi^N}{2J^2}\right)^{\frac{1}{2}} \lambda\, e^{-\frac{N}{2}\ln\lambda} \int_{-\infty}^{\infty} ds \exp\left[-Ng(s)\right]\right\}^n \tag{2.32}$$

where

$$g(s) = \frac{\lambda^2 s^2}{4J^2} + \frac{1}{2}\ln\left[\, i(1+s)\,\right] \tag{2.33}$$

The negative imaginary part of s implies that the branch point in ln (1+s) lies slightly above the real axis in the upper half plane at -1. We cut the complex s-plane by a line running parallel to but above the real axis from -1 to $-\infty$. The contour of integration in (2.32) lies along the real axis. Since we are interested in the result as $N \to \infty$, we can now do a simple saddle point integration.

Now $g'(s) = 0$ has the roots

$$s_0^{\pm} = \frac{1}{2}\left[-1 \pm i\left(\frac{4J^2}{\lambda^2} - 1\right)^{\frac{1}{2}}\right] \tag{2.34}$$

for $|\lambda| < 2J$ at complex conjugate points $s_0^{\pm}$. For $|\lambda| > 2J$, the roots are on the real axis

$$s_-^+ = \frac{1}{2}\left[-1 \pm \left(1 - \frac{4J^2}{\lambda^2}\right)^{\frac{1}{2}}\right] \tag{2.35}$$

In the case $|\lambda| < 2J$, the contour must be chosen such that Re g(s) is minimum at the saddle point. If the contour integration is deformed downwards to follow the line

$$s = x - \frac{i}{2}\left(\frac{4J^2}{\lambda^2} - 1\right)^{\frac{1}{2}} \tag{2.36}$$

Re g(s) has a minimum at $x = -\frac{1}{2}$ corresponding to the saddle point s_0^-. Along Re $s = -\frac{1}{2}$ we find Re g(s) has maxima at $s_0^{\pm}$ and minimum at $s = -\frac{1}{2}$. Hence the contour runs through s_0^- and, to leading order in N as $N \to \infty$, we get

$$\int_{-\infty}^{\infty} ds\ e^{-Ng(s)} \simeq e^{-Ng(s_0^-)} \tag{2.37}$$

Hence we get

$$\rho(\lambda) = -\frac{2}{N\pi}\,\mathrm{Im}\frac{\partial}{\partial\lambda}\lim_{n\to 0}\frac{1}{n}\left[\left(\frac{e^{i\pi/4}}{\pi^{\frac{1}{2}}}\right)^{Nn}\left(\frac{N}{2\pi}\right)^{\frac{n}{2}}\left(\frac{\pi}{2J^2}\right)^{\frac{Nn}{2}}\right.$$

$$\left. \times\ \lambda^n e^{-\frac{1}{2}Nn\ln\lambda} e^{-Nng(s_0^-)} - 1\right]$$

$$= -\frac{2}{N\pi}\,\mathrm{Im}\frac{\partial}{\partial\lambda}\ln\left\{\frac{e^{iN\pi/4}}{(2J^2)^{N/2}}\left(\frac{N}{2\pi}\right)^{\frac{1}{2}}\lambda\, e^{-\frac{N}{2}\ln\lambda - Ng(s_0^-)}\right\}$$

$$\simeq -\frac{2}{N\pi}\,\mathrm{Im}\frac{\partial}{\partial\lambda}\left\{-\frac{N}{2}\ln\lambda - Ng(s_0^-)\right\}$$

to leading order in N. Recall that λ is now real. After elementary differentiation, we get

$$\rho(\lambda) = \frac{1}{2\pi J^2}\left(4J^2 - \lambda^2\right)^{\frac{1}{2}} \quad (\text{for } |\lambda| < 2J) \tag{2.38}$$

for $|\lambda| > 2J$, the integral turns out to be real and $\rho(\lambda) = 0$.

Equation (2.38) represents the well-known semicircular law of Wigner.

If the matrix elements are distributed about a fixed mean value M_o/N, we have

$$p(M_{ij}) = (2\pi\sigma^2)^{-\frac{1}{2}} \exp\left[-\frac{1}{2\sigma^2}(M_{ij}-(M_o/N))^2\right] \tag{2.39}$$

As before we shall define $J^2 = N\sigma^2$, where J is of order unity. The calculation is very similar and involves only rather detailed manipulations of the contour integration at the end. The details are available in the paper of Edwards and Jones [9]. We shall simply quote the result.

For a large N x N random symmetric matrix, the elements of which are independent gaussian random variables with mean M_o/N and variance J^2/N, the average density of states in the limit $N \longrightarrow \infty$ is

$$\rho_o(\lambda) = \begin{cases} (4J^2 - \lambda^2)^{\frac{1}{2}}/2\pi J^2, & M_o = 0, |\lambda| < 2J \\ 0 & , M_o = 0, |\lambda| > 2J \end{cases}$$

(this is (2.38)), and

$$\rho(\lambda) = \begin{cases} \rho_o(\lambda) + \frac{1}{N}\delta\left\{\lambda - (M_o + \frac{J^2}{M_o})\right\}, & |M_o| > J \\ \rho_o(\lambda) & , |M_o| < J \end{cases} \tag{2.40}$$

Equation (2.40) has some relevance to the eigenvalue spectrum of a strongly coupled localized perturbation in a solid such as a substitutional impurity coupled to the phonons. For certain values of the coupling constant of the system a state may be split off from the band of extended states and contribute a delta function outside the band of continuum of states.

The other interesting question relates to "band-tailing". Equation (2.40) gives a sharp cut-off for the averaged density of states. This is only true for $N \longrightarrow \infty$. For large and finite N, the spectrum has an exponential tail of states with a finite number of eigenvalues concentrated in a region of order $N^{-1/6}$ beyond 2J [3].

3. Random Matrix in Glass

The random matrix appeared in solid state physics in connection with phonons in glass. The solution of this problem in one dimension was given by Dyson [10].

Consider a chain of N masses, each coupled to its nearest neighbours by elastic forces obeying Hooke's law. Only motion in one dimension is envisaged, so each mass is described by a single coordinate. Let the mass of the particle number j in the chain be m_j and its displacement from equilibrium position x_j. The spring constant between particles j and j+1 is k_j. Thus the equations of motion of the system are

$$m_j \ddot{x}_j = k_j(x_{j+1}-x_j) + k_{j-1}(x_{j-1}-x_i) \tag{3.1}$$

with appropriate changes for the end masses. There is one trivial zero frequency mode with all displacements equal. The problem is to calculate the remaining (N-1) eigenmodes as N becomes large.

When the masses and the spring constants are all equal, the calculation of the frequency spectrum is elementary. In the interesting case of glass, we have several species of atoms and the positions are irregular. One could assume that the masses m_j and spring constants k_j are arranged along the chain in a random fashion.

Let us put

$$y_j = m_j^{\frac{1}{2}} x_j \tag{3.2}$$

and introduce new constants $\lambda_1, \lambda_2, \ldots\ldots, \lambda_{2N-1}$ by $\lambda_{2j-1} = k_j/m_j$, $\lambda_{2j} = k_j/m_{j+1}$. Then

$$\ddot{y}_j = (\lambda_{2j-1}\lambda_{2j})^{\frac{1}{2}} y_{j+1} + (\lambda_{2j-1}\lambda_{2j-2})^{\frac{1}{2}} y_{j-1} - (\lambda_{2j-1} + \lambda_{2j-2})\, y_j \tag{3.3}$$

The coefficient matrix is now symmetric. Now we difine new variables $z_1, z_2, \ldots\ldots, z_{N-1}$ by

$$z_j = \lambda_{2j}^{\frac{1}{2}} y_{j+1} - \lambda_{2j-1}^{\frac{1}{2}} y_j \,, \tag{3.4}$$

so (3.3) becomes

$$\dot{y}_j = \lambda_{2j-1}^{\frac{1}{2}} z_j - \lambda_{2j-2}^{\frac{1}{2}} z_{j-1} \tag{3.5}$$

Finally, we introduce the variables $u_1, u_2, \ldots, u_{2N-1}$ by

$$u_{2j-1} = y_j, \quad u_{2j} = z_j, \quad j = 1,2,\ldots,N. \tag{3.6}$$

Then (3.4) and (3.5) are combined into a set of (2N-1) linear equations

$$\dot{u}_j = \lambda_j^{\frac{1}{2}} u_{j+1} - \lambda_{j-1}^{\frac{1}{2}} u_{j-1} \tag{3.7}$$

The eigenfrequencies ω_j of the chain are therefore the characteristic roots of the (2N-1) x (2N-1) matrix Λ whose elements are given by

$$\Lambda_{j+1,j} = -\Lambda_{j,j+1} = i\lambda_j^{\frac{1}{2}} \tag{3.8}$$

All other elements are zero. There is one zero root corresponding to the degenerate mode in which all displacements are equal. The remaining roots occur in (N-1) pairs, ω_j and $-\omega_j$.

The spectrum of eigenfrequencies is given by the function $M(\mu)$ which is defined as the proportion of the roots ω_j for which $\omega_j^2 \leqslant \mu$. As $N \longrightarrow \infty$, the function $M(\mu)$ will become a smooth differentiable function and then a density of eigenfrequencies can be defined by $D(\mu) = dM(\mu)/d\mu$. Corresponding to given $\{\lambda_j\}$ we have to determine M and D.

There are several different ways to introduce randomization. The masses may be independent random variables, while spring constants are all equal. The masses may all be the same, while the spring constants are random. Or we may consider the λ_j's, which are combinations of masses and spring constants to be random.

Dyson studies the function

$$\Omega(x) = \lim_{N\to\infty} \frac{1}{2N-1} \sum_j \ln (1+x\,\omega_j^2) \quad . \tag{3.9}$$

$$= \int_0^\infty \ln (1+x\mu)\, D(\mu)\, d\mu \tag{3.10}$$

as a function of the complex variable x. That branch of the logarithm is taken which is real for real, positive x. The spectral density functions D and M are determined by the limiting values of $\Omega(x)$ on the negative real axis approached from above.

By a rather intricate analysis, Dyson shows that an alternative expression for $\Omega(x)$ is

$$\Omega(x) = \lim_{N\to\infty} \frac{1}{N} \sum_{a=1}^{2N-1} \ln (1+\xi(a)) \tag{3.11}$$

where $\xi(a)$ is a continued fraction

$$\xi(a) = x\lambda_a/(1+x\lambda_{a+1}/(1+x\lambda_{a+2}/(\ldots. \tag{3.12}$$

A simpler derivation of this equation is due to Bellman [11]. When all the λ_j's are the same, $\lambda = k/m$, ξ satisfies

$$\xi = \cfrac{x\lambda}{1+\cfrac{x\lambda}{1+\cdots}} = \frac{x\lambda}{1+\xi} \tag{3.13}$$

and $\xi \longrightarrow 0$ as $x \longrightarrow 0$. Hence

$$\xi = \frac{1}{2}[(1+4x\lambda)^{\frac{1}{2}}-1] \quad , \tag{3.14}$$

$$\Omega(x) = 2 \ln [\frac{1}{2}(1+4x\lambda)^{\frac{1}{2}} + \frac{1}{2}] \tag{3.15}$$

and

$$D(\mu) = \begin{cases} \frac{1}{\pi}(4\lambda\mu - \mu^2)^{\frac{1}{2}} \quad , & \mu < 4\lambda \\ 0 \quad , & \mu > 4\lambda \end{cases} \tag{3.16}$$

Equation (3.16) can be checked by direct calculation. In general solvable cases are rare, but Dyson found one analytically tractable. Each of the parameters λ_j of (3.8) is an independent random variable with the probability distribution

$$G_n(\lambda) = \frac{n^n e^{-n\lambda} \lambda^{n-1}}{(n-1)!} \tag{3.17}$$

the integer n taking the values 1,2, ... The distribution has mean 1 and standard deviation $n^{-1/2}$ and in the limit of large n, $G_n(\lambda)$ becomes a gaussian

$$G_n(\lambda) \sim \left(\frac{n}{2\pi}\right)^{1/2} e^{-\frac{n}{2}(n-1)^2} \qquad (3.18)$$

Notice the probability distribution is defined for a combination of spring constant and mass, and not of each separately.

The details of the solution are complicated. For small z, the function M(z) is given by

$$M_n(z) \sim \left[\frac{\pi^2}{6} - t_{n-1}\right] / \left[\pi^2 + \ln nz + S_{n-1} + \gamma\right] \qquad (3.19)$$

with

$$s_j = \sum_{l=1}^{j} \frac{1}{l}, \qquad t_j = \sum_{l=1}^{j} \frac{1}{l^2} \qquad (3.20)$$

and γ is Euler's constant = 0.5772,.... For large z,

$$M_n(z) = 1 - 2(\ln nz - s_{n-1} + \gamma)\, e^{-nz} \, (nz)^{2n-1} \left[(n-1)!\right]^{-2} \qquad (3.21)$$

For comparison, the uniform chain result (3.16) leads to

$$\begin{aligned} M(z) &= \frac{1}{\pi}\cos^{-1}(1-\tfrac{1}{2}z), \quad && z<4, \\ &= 1, && z>4 \end{aligned} \qquad (3.22)$$

Follows a very important conclusion, which is probably valid generally : A disordered chain has a much greater proportion of very low characteristic frequencies than a uniform ordered chain.

Several points worth further study may be mentioned.

(i) Dyson's calculation is in one dimension and does not include topological disorder which is possible in three dimensions. Very little has been done in this regard.

(ii) The distribution function G_n does not correspond to any physical chain. One could consider problems of equal coupling constants but random masses. A formulation was given by H. Schmidt [12], but Schmidt's equation has been characterized by Lieb and Mattis [13] as one of the most difficult equations in mathematical physics.

(iii) Dyson's formulation, though elegant, requires analytic continuation of $\Omega(x)$ to the negative real axis through the upper half plane. This makes the formulation difficult for numerical computation. The numerical work done in this field has been reviewed by P. Dean [14] . Whether one can reformulate Dyson's method in a form suitable for computation has not been studied much. Neither has any other analytical solution, except (3.17), been discussed in the literature.

Stress Relaxation in Glass

We shall now turn to another problem in glass which is again not clearly understood. It is known that glass is a linear solid and exhibits "delayed elasticity" (see a lucid exposition by Douglas [15]). Apart from the normal instantaneous elastic response, common soda-lime-silicate glass exhibits linear viscoelastic flow. Experimentalists have established that creep and stress relaxation in glass are non-exponential. With the strain kept constant, the stress S in glass relaxes because of delayed elasticity according to a law

$$S(t) \quad = \quad S_o \exp\left[-(t\,(\bar{\tau})^{\alpha})\right] \tag{3.23}$$

S_o is the initial stress. The mean relaxation time $\bar{\tau} \sim 4 \times 10^4$ sec and is directly proportional to viscosity, showing the same temperature dependence. The viscosity here is around 10^{14} poise . The index α changes with time; for $t \ll \bar{\tau}$, $\alpha \sim 0.3$; for $t \approx \bar{\tau}$, α is around 0.5 to 0.6; and for $t \gg \bar{\tau}$, $\alpha \rightarrow 1$. The problem is to find an explanation for the form (3.23), the existence of large relaxation times and the values of α [16].

It is natural to start with the Navier Stokes equation for viscous flow:

$$\rho\left[\frac{\partial \vec{v}}{\partial t} + (\vec{v}.\vec{\nabla})\vec{v}\,\right] \quad = \quad -\nabla p + \eta\nabla^2\vec{v} + \tfrac{1}{3}\eta\nabla(\nabla.\vec{v}) \tag{3.24}$$

We may consider glass as an incompressible fluid. Then $\nabla.\vec{v} = 0$ and the pressure term ∇p also drops out (that is, we ignore sound propagation). As we are in the linear regime we also drop the non-linear term$(\vec{v}.\nabla)\vec{v}$. We then end up with

$$\rho\,\frac{\partial \vec{v}}{\partial t} = \eta\,\nabla^2\vec{v} \tag{3.25}$$

which we may write as a diffusion equation

$$\frac{\partial \vec{v}}{\partial t} \quad = \quad D\,\nabla^2\vec{v} \tag{3.26}$$

where $D = \eta/\rho$.

Equation (3.25) is a classical continuum equation. Suppose we try to write a discrete version and take into account the different atomic masses. Consider a 1-dimensional case first.

$$\rho\,\frac{\partial u}{\partial t} \quad = \quad \eta\,\frac{\partial^2 u}{\partial x^2} \tag{3.27}$$

After simple manipulations this can be replaced by

$$\frac{\partial u_i}{\partial t} \quad = \quad \eta\left[\frac{a_i}{m_i}\,(u_{i+1}-u_i) + \frac{a_{i-1}}{m_i}\,(u_{i-1}-u_i)\right] \tag{3.28}$$

where m_i is the ith mass and a_i the distance between the i and (i+1) th masses. a_i and m_i are random variables. This equation has a form similar to Dyson's equation (3.1), with ω^2 replaced by τ^{-1}, where τ denotes a relaxation time. From the theory of random matrices we can carry over the qualitative conclusion : the disordered system will have larger proportion of longer relaxation times than ordered systems. So the delayed elasticity is expected.

The trouble is that the starting point (3.24) is not quite right. Each relaxation time τ is inversely rather than directly proportional to viscosity and the relaxation time does not increase with viscosity as observed. If we use the kinetic theory, the diffusion coefficient $D = \eta/\rho = \frac{1}{3}c\lambda = \lambda^2/3\tau$ where λ is the mean free path, c the mean velocity and τ the time between collisions. The mean path picture breaks down when the viscosity is high and the Navier-Stokes equation becomes inapplicable.

It has been agreed [17] that the equation (3.26) being a macroscopic equation can be retained in the region of high viscosity, but the diffusion coefficient should be something like the Einstein diffusion coefficient in the Brownian motion

$$D = \frac{k_B T}{6\pi\eta a} \tag{3.29}$$

a is a characteristic average length of the network in glass. The indices α could be produced with suitable mode distribution [17]. The value of $\bar{\tau}$ comes out right,

$$\bar{\tau} \simeq \frac{\bar{\lambda}^2}{4D} \tag{3.30}$$

where $\bar{\lambda}$ is the short range order in glass. By (3.29) $\bar{\tau}$ is directly proportional to η. The problem of establishing the flow equation at large viscosities has not been satisfactorily solved. What seems to be happening is epitomized in a stochastic model of Kramers [18] on the escape of particles over potential barriers. When the viscosity is small, the rate of escape depends very little on viscosity (the Navier-Stokes situation), but when the viscosity is large the rate is inversely proportional to viscosity (the glass problem).

4. Conclusion

We have indicated several areas in solid state physics where random matrices appear. Another closely related area not touched above is that of spin waves in disordered Heisenberg magnets [19]. Analytical and numerical techniques used in the problems mentioned above and the technique of diagonalization of sparse matrices are often useful for such problems.

References

1. E.P. Wigner, Ann. Math. 53, 36 (1951), 62, 145 (1955).
2. C.E. Porter, Statistical Theories of Spectra: Fluctuations (Academic Press, New York and London, 1965).
3. M.L. Mehta, Random Matrices and the Statistical Theory of Energy Levels (Academic Press, New York and London, 1967).
4. T.A. Brody, J. Flores, J.B. French, P.A. Mello, A. Pandey and S.S.M. Wong, Rev. Mod. Phys. 53, 385 (1981).
5. F.J. Dyson, J. Math. Phys. 3 140 (1962).
6. F.J. Dyson, J. Math. Phys. 3, 1200 (1962).
7. R. Balian and C. Bloch, Ann. Phys. (N.Y.) 85, 514 (1974).
8. E.P. Wigner, Ann. Math. 62, 548 (1955).
9. S.F. Edwards and R.C. Jones, J. Phys. A: Math Gen. 9, 1595 (1976).
10. F.J. Dyson, Phys. Rev. 92, 1331(1951).
11. R. Bellman, Phys. Rev. 101, 19 (1956).
12. H. Schmidt, Phys. Rev. 105, 425 (1957).
13. E. Lieb and D.C. Mattis, Mathematical Physics in One Dimension (Academic Press, New York and London, 1966).

14. P. Dean, Rev. Mod. Phys. 44, 127 (1972).
15. R.W. Douglas, Br. J. Appl. Physics 17, 435 (1966).
16. R.W. Douglas, P. J. Duke and O.V. Mazurin, Phys. Chem. Glasses 9, 169 (1968).
17. C.K. Majumdar, Sol. St. Commun. 9, 1087 (1971).
18. H.A.Kramers Physica 7, 284 (1940).
19. S.M. Bose and E.Ni Foo, J. Phys. C. Sol. St. Phys. 5, 1082 (1972).

STOCHASTIC EVOLUTION IN ISING MODELS

Deepak Dhar
Tata Institute of Fundamental Research
Homi Bhabha Road, Bombay 400005, INDIA

I. Introduction and Outline

Stochastic evolution models are often used to study time-dependent properties of many-body systems. Some examples are the Brownian motion of particles, tracks of nuclear particles through dense media, kinetics of nucleation in super-heated liquids etc. These models provide the basic framework for studying time-dependent phenomena in statistical mechanics, such as the approach to thermal equilibrium from an arbitrarily prepared initial state, or of non-equilibrium steady states in dissipative systems, or the appearance of co-operative long-time correlations in the neighbourhood of second-order phase-transitions. In the following, the relaxation properties of some kinetic Ising models are briefly discussed.

The plan of these lectures is as follows: In section II, we discuss briefly how the probabilistic description of evolution of many-particle systems can be reconciled with the deterministic (microscopic) mechanical evolution. In section III, the rate-equation for the Markovian evolution, and the condition of detailed balance are described. In section IV, we introduce the single-spin-flip kinetic Ising model and general crystal-growth model (of which the kinetic Ising model is a special case). The dynamical scaling hypothesis, and some of its consequences are discussed in section V. Sections VI and VII contain brief discussions of long-time relaxation in a disordered Ising model in one and higher dimensions respectively. It is shown that in the disordered Ising model with broken bonds, the relaxation of magnetization to the equilibrium-value is slower than exponential for all temperatures below the critical temperature of the model without disorder.

II. Probabilistic Versus Deterministic Evolution

The understanding of the coexistence of thermodynamic irreversibility (as canonized in the second law of thermodynamics) with microscopic mechanical reversibility has been the central theme in non-equilibrium statistical mechanics. The fact that the time-evolution of a gas undergoing free expansion, and its time-reversed evolution are both consistent with the laws of mechanics, clearly shows that it is not possible to 'prove' the approach to equilibrium in isolated systems without making additional assumptions about the evolution of macroscopic systems. These assumptions are necessarily extra-mechanical. They may be very plausible (e.g. the unprovable assertion that initial states corresponding to sets of measure zero in phase space are unlikely to occur in real experiments) or much less obvious ones (these also are usually preceded by the qualifier 'almost always'). These may be assumptions about the large size of the system (absence of Poincare recurrences), the tendency of the phase space trajectory of the system to diverge (the mixing property), the decay of multiparticle correlations (Boltzmann's collision-number hypothesis) or the presence of weak but uncontrollable interactions of the system with the outside (evolution in the presence of weak noise). In the following, we shall adopt the position that the macroscopic relaxation behavior of large systems is very well modelled by a probabilistic evolution law, in particular by Markovian dynamics, and shall side-step the question of deriving the master-equation from more elementary principles. (For a discussion of these issues, see [1-3]). Note that macroscopic deterministic evolution (e.g. the Navier-Stokes equations) is a special case of Markovian dynamics, and corresponds to the case when fluctuations in the macroscopic variables are small.

A simple illustrative example of a system which undergoes deterministic evolution, but may be equivalently described by a probabilistic law, is the following: Consider a particle performing a walk on the points of a linear chain. The internal state of the particle is described by an angle-variable θ ($0 < \theta < 2\pi$), which undergoes a (discrete-time) deterministic evolution according to the law

$$\theta_{t+1} = 2\theta_t \quad (\text{modulo } 2\pi). \tag{1}$$

The particle starts at time t=0. At subsequent times t=1,2,3,..., it takes a step of unit length to the right, or to the left, according as $\cos\theta_t$ is positive or negative. If the value $\theta_{t=0}$ is known, the motion of the particle is completely determined. Assume, however, that the value $\theta_{t=0}$ is known only to a finite accuracy; the a priori probability density of $\theta_{t=0}$ being constant for $|\theta_{t=0}| < 2^{-N}\pi$, and zero elsewhere. It is easy to see that under these assumptions, the steps of the particle for $t > N$ are perfectly random and uncorrelated. As far as the motion of the particle on the chain is concerned, after the decay of initial state correlations (for $t > N$), it can be described as a simple unbiassed random walk.

In this example, the stochastic evolution is a result of the exponential growth of the initial-state uncertainty with time. Note that the stochastic characterization of the random walk does not involve the 'internal degree of freedom' θ_t. In general, the number of variables in terms of which the 'mesoscopic' state of the system is characterized need not be as large as the number required for a full microscopic characterization.

III. The Rate Equations

Consider a finite system which at any time t may exist in any one of a denumerable number of states labelled by integers 1,2,3... The system is in contact with a heat-reservoir, and the interaction causes transitions between these states, the transition rate from state m to state n being W_{mn}. Let $P_m(t)$ be the probability that the system exists in the state m at time t. From the general theory of Markov chains [see e.g. [4]] it follows that under very weak conditions on the transition rates W_{mn}'s (each state must be reachable from every other), as the time t tends to infinity, $P_m(t)$ tends to a limiting value $P_m(\infty)$ independent of the initial state. The time evolution of $P_m(t)$ is governed by the equation

$$\frac{d}{dt} P_m(t) = \sum_{n \neq m} [P_n(t) W_{nm} - P_m(t) W_{mn}] \qquad (2)$$

In problems of physical interest, P_m^{eq} is the well known equilibrium distribution

$$P_m^{eq} = \exp(-\beta E_m)/[\sum_n \exp(-\beta E_n)]; \qquad (3)$$

where E_m is the energy of the state m and β is the inverse temperature characteristic of the heat bath. The requirement that a system at large times should tend to thermal equilibrium constrains the physically admissible transition rates W_{mn}'s, but it does not determine them uniquely. Many different choices of W_{mn}'s would be consistent with a given limiting distribution. $\{P_m^{eq}\}$ is a time-invariant probability distribution for the rate equation (2) if

$$\sum_n [W_{mn} P_m^{eq} - W_{nm} P_n^{eq}] = 0, \text{ for all m.} \qquad (4)$$

These conditions are clearly satisfied, if W_{mn}'s satisfy the detailed balance condition

$$W_{mn} = W_{nm} \exp[-\beta E_n + \beta E_m]. \qquad (5)$$

Since the limiting distribution is the unique time-invariant distribution (for finite systems), the assumptions of Markovian evolution, the detailed balance condition, and non-existence of other conserved quantities guarantee that the system will relax to thermal equilibrium with time.

Define $\bar{P}_m(t) = P_m(t)\exp(\beta E_m/2)$. (6)

In terms of $\bar{P}_m(t)$'s, eq.(2) may be rewritten as

$$\frac{d}{dt}\bar{P}_m(t) = \sum_n \bar{P}_n(t)\bar{W}_{nm} \tag{7}$$

where $\bar{W}_{mn} = W_{mn}\exp[\beta(E_n - E_m)/2]$ for $m \neq n$ (8)

and $\bar{W}_{mm} = -\sum_{n\neq m} W_{mn}$ (9)

If the transition rates W_{mn} satisfy the detailed balance condition, then $\bar{W}$ is a real symmetrical matrix. Hence all its eigenvalues are real. The rate of growth of $\bar{P}_m(t)$ with time is related to the largest eigenvalue of $\bar{W}$. Since $\bar{P}_m(t)$ tends to a constant value for large t, the largest eigenvalue of $\bar{W}$ is zero.

The problem of integration of Eq.(7) is reduced to that of diagonalizing the matrix $\bar{W}$. This is quite hard, and has not been solved yet for any nontrivial model so far. Some specific models of interest are described in the next section.

IV. Stochastic Evolution Models

Our progress in understanding irreversible phenomena in statistical mechanics is severely hampered by the lack of simple illustrative models which are exactly soluble for their nonequilibrium properties, and thus serve as testing-grounds for theories or guides to intuition. An important motivation for the study of the kinetic Ising model and the crystal-growth model discussed below is that the powerful techniques of equilibrium statistical mechanics can be used to study these models, and thus they help us bridge the gap between our understanding of these disciplines.

(a) The Kinetic Ising Model: Consider an Ising ferromagnet in which the spins σ_i (i=1 to N), taking values ±1, are located at the sites of a d-dimensional hypercubical lattice. The Hamiltonian of the system is given by

$$H = -\sum_{ij} J_{ij}\sigma_i\sigma_j \ . \tag{10}$$

We denote a configuration of spins by $\{\sigma\}$. There are 2^N possible configurations. In the Glauber model [5] the evolution is assumed to be due to single-spin-flips. The probability that the ith spin flips between the times t and t+dt, when the configuration of spins at time t is $\{\sigma\}$ is assumed to be equal to $W_i(\{\sigma\})dt$, where

$$W_i(\{\sigma\}) = \frac{1}{2}\,[1 - \tanh \beta h_i(t)], \tag{11}$$

and

$$h_i(t) = \sum_{j\neq i} J_{ij}\sigma_j(t). \tag{12}$$

It is easy to see that the choice (11) satisfies the detailed balance condition, and hence in the long-time limit the equal-time correlation functions of this model agree with those calculated using equilibrium ensembles. Glauber studied the relaxation of a one-dimensional chain with nearest-neighbour couplings in zero external field. An exact solution for a linear chain in the presence of an external field, or for a higher dimensional lattice has not yet been found.

The single-spin-flip Glauber model is not applicable to Ising systems like binary alloys, because the assumed dynamics does not conserve magnetization (particle-number in alloys). However, a similar rate matrix W, which describes a simultaneous exchange of spins between neighbouring sites i and j is easy to write down such that it satisfies the condition of detailed balance. The evolution equations for this model are similar to the single-spin-flip case [6].

Much insight into the behavior of these models near phase transitions has been obtained by Monte-Carlo simulations. These have been comprehensively reviewed in ref.[7]. A review of the theoretical aspects may be found in Kawasaki [6].

(b) Crystal-Growth Models: Consider a model of crystal-growth shown in Fig.1. The lattice is a square lattice. At time t=0, each of the sites (x,y) is unoccupied if $x+y > 0$, and if $x+y \leqslant 0$, it is occupied by one (and only one) of two kinds of atoms A and B. The state of an occupied site (ij) is characterized by an Ising variable $\sigma_{i,j}$ which is +1 or -1 according as the site is occupied by A or B. At time t=1, each of the sites (x,y) lying on the line x+y=1 is filled by a particle A or B taken randomly from an external source of particles. We assume that sites on this new 'layer' are occupied independently of each other, and that the probability that the site (i,j) is filled by an A atom (or B) depends on the configuration of neighbouring earlier-filled

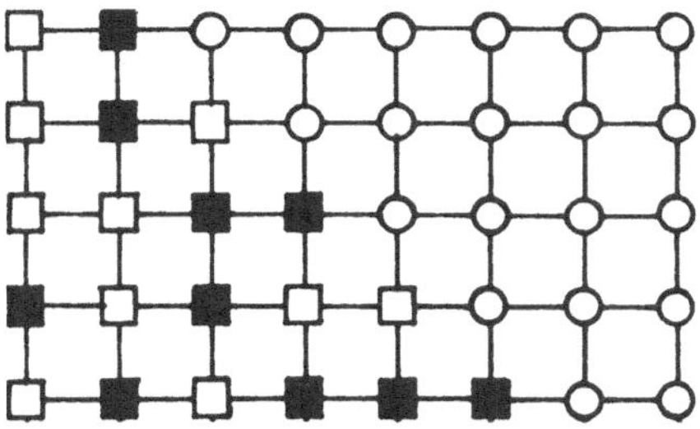

Fig.1: The crystal-growth model. The full and open squares denote sites occupied by A and B atoms respectively. Unoccupied sites are denoted by open circles.

sites (i-1,j), (i,j-1) and (i-1,j-1). Once a site is occupied, it stays occupied by the same atom at all subsequent times. At times t=2,3,..., successive layers of sites x+y=2,3,... are filled and the boundary of the 'crystal' (occupied sites) moves outwards. The model simulates the growth of mixed crystals from solution, under conditions when particle diffusion in the solid phase is negligible [8-9].

The full specification of the model requires a specification of the conditional probabilities $\mathrm{Prob}(\sigma_{i,j}|\sigma_{i-1,j},\sigma_{i,j-1},\sigma_{i-1,j-1})$. Since $\mathrm{Prob}(+1|\alpha,\beta,\gamma) + \mathrm{Prob}(-1|\alpha,\beta,\gamma)$ must be 1 for all α,β,γ, we may write

$$\mathrm{Prob}(\sigma_{i,j}|\sigma_{i-1,j},\sigma_{i,j-1},\sigma_{i-1,j-1}) = \frac{1}{2}\,[1+\sigma_{i,j}\, f(\sigma_{i-1,j},\sigma_{i,j-1},\sigma_{i-1,j-1})] \qquad (13)$$

The model is thus defined in terms of 8 parameters, the values $f(\pm1,\pm1,\pm1)$.

We may treat t=x+y as the 'time' coordinate, and the configurations of $\sigma_{i,j}$'s on the lattice as time-histories of a discrete-time evolution kinetic Ising model on a linear chain in which the spins on odd-numbered sites are allowed to flip at odd values of the discrete time-parameter, and the even numbered ones when the time-parameter is even.

The general model involving 8 parameters is quite difficult to analyse. Of special interest is the case when the function f in Eq. (13) is independent of the spin $\sigma_{i-1,j-1}$. If we also assume mirror symmetry about the line x=y, the model can be characterized in terms of 3 parameters only. We write

$$\begin{aligned} f(+1,+1,\pm1) &= a; \\ f(+1,-1,\pm1) &= f(-1,+1,\pm1) = b; \\ f(-1,-1,\pm1) &= c. \end{aligned} \qquad (14)$$

It is easy to see that the case

$$a = 4p-2p^2-1; \; b = 2p-1; \; c = -1 \tag{15}$$

corresponds to problem of directed bond percolation with bond-concentration p. The case

$$a = b = 2p-1; \; c = -1 \tag{16}$$

corresponds to the directed site-percolation problem. These problems have attracted much attention recently [10-15]. The problem with a=b=-1 is exactly soluble and is related to the directed animals problem [16,17]. Various other special soluble cases of the general crystal-growth model have been discussed by Enting [18]. Crystal-growth models in higher dimensions, or on other lattices are easy to define, but are usually quite difficult to solve exactly.

V. Dynamical Scaling Theory

Near phase-transitions, the spatial and temporal correlations in the Ising model became long-ranged. The singular behavior of these correlation functions is described in terms of critical exponents, which are expected to show universal characteristics, as in the case of equilibrium phase transitions. In fact, from the discussion in the previous section, it is clear that there is no essential difference between the statistical-mechanical descriptions of static and dynamical critical phenomena. Calculating time-dependent correlation functions in a d-dimensional system involves averaging over histories of configurations, which is just like a (d+1)-dimensional static calculation. Dynamical scaling theory [19] is a natural generalization of the static scaling theory used to describe equilibrium phase-transitions. Using this theory, the time-dependent critical behavior of a wide class of systems (including kinetic Ising models) can be described in terms of only 3 independent critical exponents.

Consider a d-dimensional kinetic Ising model at a temperature $T = T_c(1+\varepsilon)$, where T_c is the critical temperature of the model, and ε is small. We consider only the case, when there is no external magnetic field present. Inclusion of magnetic field requires only a straightforward extension of the formalism. The time-dependent two-point correlation function in equilibrium is given by

$$G(\bar{r},t,\varepsilon) \overset{\text{def}}{=\!=} \langle \sigma_{\bar{r}+\bar{x}}(t+t_o)\sigma_{\bar{x}}(t_o) \rangle \tag{17}$$

Note that $G(\bar{r},t,\varepsilon)$ described the time-dependent correlations <u>in</u> the equilibrium state and hence does not depend on $\bar{x}$ or t_o.

According to the dynamical scaling hypothesis, the function $G(\bar{r},t,\varepsilon)$ in the limit of large spatial or temporal separations and small ε, equals a generalized homogenous function of $\bar{r}$, t and ε. The difference between $G(\bar{r},t,\varepsilon)$ and the homogeneous function is a non-universal correction-to-scaling function which is asymptotically negligible in the critical region. Since we are not going to discuss corrections to scaling here, the generalized homogeneous function will be denoted by the same symbol $G(\bar{r},t,\varepsilon)$. A function $G(\bar{r},t,\varepsilon)$ is said to be a generalized homogeneous function if there exist constants β, ν and z such that

$$G(\bar{r},t,\varepsilon) = \lambda^{+2\beta}G(\lambda^{+\nu}\bar{r},\ \lambda^{\nu z}t,\ \lambda^{-1}\varepsilon) \quad \text{for all } \lambda > 0. \tag{18}$$

A similar equation describes the behavior of more general n-point correlation function. The n-point function is scaled by $\lambda^{n\beta}$ if all distances, times and ε scaled as in Eq.(18). From the scaling equation (18), the singular behavior of various physical quantities can be deduced in terms of the three critical exponents ν, z and β. In the presence of external magnetic field, an exponent Δ characterizing the scaling of the external field has to be included. However, if we assume the hyperscaling relation ($d\nu = 2-\alpha$, where α is a known function of β, ν and Δ), then the number of independent exponents is 3 again.

On putting t=0 in the above equation, it reduces to the static scaling equation. It follows that β and ν appearing in Eq.(18) are the conventional magnetization and correlation length exponents. Consider now, the autocorrelation function $G(0,t,-\varepsilon)$. By the scaling hypothesis, it can be written in the form

$$G(0,t,-\varepsilon) = \varepsilon^{2\beta}f(t\varepsilon^{\nu z}) \tag{19}$$

where f(x) is a function of a single variable x. As x tends to infinity, f(x) tends to a constant. Since t scales as $\varepsilon^{-\nu z}$ in the above equation, the relaxation time must diverge as $\varepsilon^{-\nu z}$ for small ε. If we fix t, and let $\varepsilon \to 0$, then $G(0,t,-\varepsilon)$ has a smooth limit only if f(x) varies as $x^{-2\beta/\nu z}$ for small x. This then implies that

$$G(0,t,0) \sim t^{-2\beta/\nu z} \tag{20}$$

In the analysis of the decay of magnetization from an initially aligned state, we cannot use time-translational invariance. However, if we assume that an equation analogous to Eq.(19) can be written down for M(t), we get

$$M(t) \simeq \varepsilon^{\beta} g(t\varepsilon^{\nu z}), \tag{21}$$

where, since $M(t) \to \varepsilon^{\beta}$ as $t \to \infty$; $g(x)$ should tend to a constant as x tends to infinity. Then arguing as before, $g(x)$ must vary as $x^{-\beta/\nu z}$ for small x, giving

$$M(t) \sim t^{-\beta/\nu z}, \quad \text{for } \varepsilon = 0. \tag{22}$$

This is slower than the decay of the auto-correlation function, and is due to multi-spin correlations. The relative simplicity of such results using scaling theory becomes especially valuable, as a complete solution is not available for any of the kinetic Ising models showing phase transitions. In the finite-size scaling method, similar scaling techniques are used to determine the values of the exponents β, ν and z by extrapolating the behavior of infinite systems from a sequence of finite-size realizations which are solved exactly (numerically). The procedure yields very good numerical estimates of static and dynamic critical exponents [13,20].

VI. Relaxation in Disordered Ising Chains

The Ising model in one dimension does not undergo any phase transition, and hence relaxation in Ising chains does not show all the features of critical slowing down near phase transitions. However, it is the only non-trivial model showing approach to thermal equilibrium for which the time-dependent correlation functions can be calculated without too much trouble, and hence serves as a useful test case for various approximations. In the following we study the relaxation of magnetization in a chain from an initially aligned state. The relaxation is found to be exponential in the homogeneous case. On introducing a quenched disorder in the band strengths, the relaxation is not exponential any longer, and we study its long-time behavior.

The Hamiltonian of a linear chain of Ising spins σ_i (i=1 to N) is given by

$$H = \sum_{i=1}^{N-1} J_{i+1/2}\, \sigma_i \sigma_{i+1}. \tag{23}$$

The coupling constants $J_{i+1/2}$'s are assumed to be quenched, independent, identically distributed random variables taking values J_1 and J_o ($0 \leq J_1 < J_o$) with probabilities (1-p) and p respectively. The time-evolution of the probability $\mathcal{P}$ of the configuration $\{\sigma_1, \sigma_2, \ldots \sigma_N\}$ in the single-spin-flip Glauber model is governed by the equation

$$\frac{d}{dt}\,\mathcal{P}(\{\sigma_1,\sigma_2...\sigma_N\}) = \sum_i [W_i^+\,\mathcal{P}(\{\sigma_1...-\sigma_i..\sigma_N\})-W_i^-\,\mathcal{P}(\{\sigma_1\sigma_2...\sigma_N\})]. \tag{24}$$

where $W_i^{\pm}$ are the spin-flip probabilities per unit time for the spin σ_i. The most general expression for $W_i^{\pm}(\{\sigma\})$ which depends only on σ_{i-1},σ_i and σ_{i+1}, and satisfies the detailed balance condition is

$$W_i^{\pm} = \frac{1}{2}\,[1 \pm \sigma_i\,\tanh\beta h_i(t)][A+B(\sigma_{i-1}+\sigma_{i+1}) + C\sigma_{i-1}\sigma_{i+1}], \tag{25}$$

where $h_i(t)$ is the effective field at site i given by

$$h_i(t) = J_{i-1/2}\sigma_{i-1}(t) + J_{i+1/2}\sigma_{i+1}(t). \tag{26}$$

In the special case A=1, B=C=0, $S_i(t)$ defined as the expectation value of $\sigma_i(t)$ evolves according to the equation

$$(1 + \frac{d}{dt})S_i(t) = C_i^- S_{i-1}(t) + C_i^+ S_{i+1}(t) \tag{27}$$

where

$$C_i^{\pm} = \frac{1}{2}\,[\tanh\beta(J_{i+1/2} + J_{i-1/2}) \pm \tanh\beta(J_{i+1/2} - J_{i-1/2})] \tag{28}$$

We assume that at time t=0, all spins are +1. The average magnetization at time t is given by

$$M(t) = \frac{1}{N}\sum_{i=1}^{N} S_i(t) \tag{29}$$

Eq.(27) may be written as the matrix equation

$$\frac{d}{dt}\,|S(t)\,> = -\Lambda|S(t)\,> \tag{30}$$

where Λ is an NxN tridiagonal matrix, independent of time. The equation may be solved formally to give

$$M(t) = \int_0^{\infty} d\lambda D(\lambda)\exp(-\lambda t) \tag{31}$$

where

$$D(\lambda)d\lambda = <\sum_{\lambda<\lambda'<\lambda+d\lambda} <S(0)|\lambda'><\lambda'|S(0)>>_c \tag{32}$$

where $|\lambda'>$ and $<\lambda'|$ are the right-and left-eigenvectors of Λ with eigenvalue λ', and $<\,>_c$ denotes configuration averaging over the quenched variables $J_{i+1/2}$'s. If p=1, and $N \to \infty$, straightforward diagonalization of Λ shows that

$$M(t) = \exp(-\lambda_0 t) \tag{33}$$

with

$$\lambda_o = 1 - \tanh 2\beta J_o. \quad (34)$$

If $p \neq 1$, the arguments of Lifshitz [21] show that (details of these calculations may be found in Dhar and Barma [22]) $D(\lambda)$ is non-zero if $|\lambda-1| < \tanh 2\beta J_o$, and as λ tends to λ_o from above, $D(\lambda)$ varies as $\exp[-(\lambda-\lambda_o)^{-1/2}C]$, where C is a known constant. Substituting this asymptotic form of $D(\lambda)$ in Eq.(31), the behavior of $M(t)$ for large t is easily determined. We find

$$M(t) \sim \mathrm{Exp}(-\lambda_o t - at^{1/3}), \text{ as } t \to \infty. \quad (35)$$

If the temperature is very low so that the thermal correlation length $\xi_T = \exp(2\beta J_o)$ is much larger than the percolation correlation length $\xi_p = 1/(1-p)$, then in the time domain $\xi_T \xi_p << t << \xi_T^3/\xi_p$, the magnetization is approximately given by the formula

$$M(t) \sim \exp(-\lambda_o t - bt^{1/2}). \quad (36)$$

While this is an admittedly simplified model, the existence of non-exponential relaxation in the presence of disorder is gratifying. Many experimental disordered materials show such a behavior.

VII. Higher Dimensions

Consider now the kinetic Ising model on a d-dimensional hyper-cubical lattice ($d > 1$) with single-spin-flip dynamics given by Eq.(11). We restrict ourselves to the case of nearest-neighbour ferromagnetic couplings and $J_1=0$.

In this case, the equation of evolution for $S_i(t)$ involves multi-spin correlation functions, and a rigorous analysis of the problem is difficult. Even in the limit of high temperatures, when higher order correlations may be neglected, the resulting equations are difficult to solve in the disordered problem. However, it is easy to see that the relaxation would be non-exponential in general.

If p is less than the critical percolation probability p_c on the lattice, all clusters are finite, and the equilibrium state is paramagnetic with no spontaneous magnetization. The density of clusters of size n varies as $\exp(-An)$, where A is a p-dependent constant. Because of the absence of mutual interactions, different clusters evolve independently of each other. For each finite cluster, the relaxation problem is in principle soluble, involving a diagonalization of $2^n \times 2^n$ matrix. Let τ_n be the relaxation time (actually this is the longest of the spectrum of relaxation times) of a typical cluster of size n. [A more careful argument would take into account the fact that τ_n depends on the shape of the cluster also.] If the temperature T is greater than T_c, the critical temperature for the pure case (p=1), then all clusters have finite relaxation times. We may write for large n

$$\tau_n^{-1} \simeq B(T,p) + cn^{-x}. \tag{37}$$

Here we have used the fact that τ_n^{-1} is a bounded decreasing function of n. We expect B(T,p) to vanish as T tends to T_c from above. In the special case p=1, $B(T,p) \simeq (T-T_c)^{\nu z}$ for T near T_c, where z is standard dynamical critical exponent. The extrapolation form (37) is only one of several equivalent forms showing the correct n dependence in the limit of large n. Using a Lifshitz-like argument, we would expect that $x = 2/d$. The average magnetization at time t is given by

$$M(t) \simeq \sum_n P_n \exp(-t/\tau_n) \tag{38}$$

For large times t, Eqs.(37) and (38) together imply that

$$M(t) \sim \exp[-B(T,p)t - Dt^{\frac{1}{1+x}}] \tag{39}$$

where D is some function of T and p.

If $T < T_c$, there will be exceptionally large clusters for which there are two metastable states of opposite magnetization with very infrequent transitions between them. For these clusters, the relaxation times are expected to be a strongly rising function of n

$$\tau_n \sim \exp[n^{x'}A'(T,p)] \tag{40}$$

Again, we expect A'(T,p) to vanish as T tends to T_c from below. Also x' is equal to (1-1/d). This is because the free energy barrier between up and down magnetized states is the surface energy of a domain wall spanning the cluster. Such a wall would be of size $n^{1-1/d}$ for compact clusters having n sites.

Using Eqs.(38) and (40) we get for $T < T_c$ and $p < p_c$

$$M(t) \sim \exp[-A''(\log t)^{1/x'}], \text{ as } t \to \infty; \tag{41}$$

where A" is some constant which depends on T and p.

If $p > p_c$, there is an infinite cluster, and the possibility of a ferromagnetic-paramagnetic phase transition. However, the analysis given above would still hold for the finite clusters, which still would relax to zero magnetization very slowly. Hence the overall relaxation of magnetization would be non-exponential. Note that for all $T < T_c$, whatever the value of p, the relaxation of magnetization is slower than exponential. This is essentially a rigorous result and depends only on the finite probability of occurrence of very large compact clusters with no holes. These clusters have large relaxation times and their contribution to the average magnetization of the sample dominates the long-time relaxation. This result is the non-equilibrium counterpart of Griffiths singularities (Griffiths [23]). For intermediate time regimes, the relaxation of noncompact clusters, as well as of the infinite cluster have to be considered.

References

1) O. Penrose, Foundations of Statistical Mechanics: A Deductive Treatment, (Pergamon Press, Oxford, 1970).
2) R. Jancel, Foundations of Classical and Quantum Statistical Mechanics, (Pergamon Press, Oxford, 1963).
3) R. Balescu, Equilibrium and Non-equilibrium Statistical Mechanics, (John Wiley, New York, 1975).
4) W. Feller, An Introduction to Probability Theory and its Applications, (John Wiley, New York, 1968).
5) R.J. Glauber, J. Math. Phys. 4, 294 (1963).
6) K. Kawasaki, in Phase Transitions and Critical Phenomena, Vol.II, eds. C. Domb and M.S. Green (Academic, London, 1972).
7) K. Binder (editor), Monte Carlo Methods in Statistical Physics (Springer-Verlag, Berlin, 1979).
8) T.R. Welberry and R. Galbraith, J. Appl. Crystallogr. 6, 87 (1973).
9) I.G. Enting, J. Phys. C10, 1379 (1977).
10) J. Blease, J. Phys. C10, 925 (1977).
11) J.L. Cardy and R.L. Sugar, J. Phys. A13, L423 (1980).
12) D. Dhar and M. Barma, J. Phys. C14, L1 (1981).
13) W. Kinzel and J.M. Yeomans, J. Phys. A14, L405 (1981).
14) S. Redner, in Percolation Structures and Processes, in Ann. Isr. Phys. Soc., Vol.3, eds. G. Deutscher, R. Zallen and J. Adler (Adam-Hilger, Bristol, 1982).
15) R. Durrett, Oriented Percolation in Two Dimensions, Univ. of California (Los Angeles) preprint.
16) A.M.W. Verhagen, J. Stat. Phys. 15, 219 (1976).
17) D. Dhar, Phys. Rev. Lett. 49, 959 (1982).

18) I.G. Enting, J. Phys. A11, 2001 (1978).

19) P.C. Hohenberg and B.I. Halperin, Rev. Mod. Phys. 35, 1678 (1975).

20) M.P. Nightingale, Physica A83, 561 (1976).

21) I.M. Lifshitz, Adv. Phys. 13, 483 (1964).

22) D. Dhar and M. Barma, J. Stat. Phys. 22, 259 (1980).

23) R.B. Griffiths, Phys. Rev. Lett. 23, 17 (1969).

RELAXATIONAL DYNAMICS OF SPIN-GLASSES NEAR TRANSITION TEMPERATURE

Deepak Kumar
Department of Physics, University of Roorkee
Roorkee, India

The dynamic behaviour of spin-glasses is of interest from several points of view. (i) At low temperatures, the relaxation of magnetisation and other associated properties show very slow, non-exponential behaviour. (ii) In order to explain the low temperature thermodynamic properties, viz., magnetic specific heat, one has to study the spectrum of low energy excitations. (iii) On higher temperature side, one is interested in understanding how the spin-motion freezes to give the spin-glass phase as the temperature is lowered. It is with the last aspect that we shall be concerned with in this lecture.

The seminal paper which has given us first understanding of how spin-motion freezes to cause a cooperative phase transition at a well defined temperature is due to Edwards and Anderson [1]. Since this work, a vast amount of papers have been written on the subject, and it is now well recognised that the question of phase transition in spin-glasses is a very complex one, involving very delicate competition between randomness and critical fluctuations. Due to glassy nature of freezing in spin-glasses, it is believed that the key to understanding the spin-glass state is via the dynamics. For the discussion of the current theoretical situation, we refer to some recent review articles [2-5]. In this lecture, we shall describe only the work of Edwards and Anderson (EA) in detail.

The starting point of the discussion is the Hamiltonian given by

$$H = \sum_{ij} J_{ij} \vec{S}_i . \vec{S}_j \tag{1}$$

where J_{ij}'s are the exchange constants, the sum is over the pairs of sites and S_i are taken to be two dimentional unit vectors. J_{ij}'s are taken to be random quantities with zero average and a non-zero mean square average:

$$\langle J_{ij}^2 \rangle_R = \bar{J}_{ij}^2 \tag{2}$$

where $\langle\ \rangle_R$ denotes the averaging over the distribution of J_{ij}'s. Since the idea here is to merely illustrate concepts, we shall take J_{ij}'s to be non-zero only for nearest neighbours. Due to zero mean value of J_{ij}'s, unlike ordered magnetic phases, the quantity $\langle\langle \vec{S}_i \rangle_T \rangle_R$ (where $\langle\ \rangle_T$ denotes the averaging with respect to the thermal ensemble) remains zero at all temperatures. So EA introduced a rather novel way to describe the frozen state [6]. They considered the average $\langle\langle \vec{S}_i(t) . \vec{S}_i(0) \rangle_T \rangle_R$, and argued that even though there is no magnetic long range order of any kind in the frozen state, the frozen state is different from, say the paramagnetic state, in the sense that the spins retain infinitely a memory of their directions. Mathematically

$$\lim_{t\to\infty} \langle\langle \vec{S}_i(t) \cdot \vec{S}_i(0) \rangle_T \rangle_R = q_i \neq o \quad \text{for} \quad T \leqslant T_g \tag{3}$$

In the mean field description, q_i is independent of the site index i and plays the role of order parameter for the second order phase transition. Thus the EA picture of the phase transition is : as the temperature decreases the spin-motion slows down and at a well defined temperature T_g, the various spins freeze in different directions depending upon their local environment in a somewhat collective manner.

Since $\vec{S}_i$'s are fixed length vectors, we rewrite (1) in terms of angles in the plane, and add a kinetic energy term for planar rotators

$$H = \sum_{ij} J_{ij} \operatorname{Cos}(\theta_i - \theta_j) + \frac{I}{2} \sum_i \dot{\theta}_i^2 \tag{4}$$

where I is the moment of inertia of the rotators. The equations of motion are

$$I \ddot{\theta}_i + \nu \dot{\theta}_i = - \sum_j J_{ij} \operatorname{Sin}(\theta_i - \theta_j) + f_i(t) \tag{5}$$

In writing down (5), we have allowed the spins to interact with a heat bath in the usual way. Here ν is the coefficient of viscosity which causes a drag on the motion of the rotators. The force $f_i(t)$ is a random one which helps maintain the system in equilibrium with the heat bath at temperature T. As is customary, $f_i(t)$ is taken to be a Gaussian, delta correlated random force satisfying the relation

$$\langle f_i(t) f_j(t') \rangle = \nu kT \delta_{ij} \delta(t-t') \tag{6}$$

In the long time limit, i.e., $t \gg \nu^{-1}$, the inertial effects on the velocity are washed out, and it is sufficient to consider the equations

$$\nu \dot{\theta}_i = - \sum_j J_{ij} \operatorname{Sin}(\theta_i - \theta_j) + f_i(t) \tag{7}$$

One can also look at the corresponding Fokker-Planck equation which reads

$$\frac{\partial}{\partial t} P(\{\theta_i\}, t) = \frac{kT}{\nu} \sum_i \left\{ \frac{1}{kT} \frac{\partial}{\partial \theta_i} \left[-\sum_j J_{ij} \operatorname{Sin}(\theta_i - \theta_j) P \right] + \frac{\partial^2 P}{\partial \theta^2} \right\} \tag{8}$$

It is easy to see that the stationary solution of this equation is the canonical distribution is

$$P_o = N \exp\left[- \frac{1}{kT} \sum_{ij} J_{ij} \operatorname{Cos}(\theta_i - \theta_j) \right] . \tag{9}$$

In order to understand the physics implied by (7), let us first consider the behaviour, when there are no interactions

$$\nu \dot{\theta}(t) = f(t) \tag{10}$$

$$\theta(t) - \theta(o) = \frac{1}{\nu} \int_o^t f(t')\, dt' \tag{11}$$

$$\langle (\theta(t)-\theta(o))^2 \rangle = \frac{2kT}{\nu} t \tag{12}$$

$$\langle \mathrm{Cos}(\theta(t)-\theta(o)) \rangle = \mathrm{Re}[\exp i(\theta(t)-\theta(o))] \tag{13}$$

To evaluate (13), we use cumulant expansion and note that (6) and (11) imply that $(\theta(o)-\theta(o))$ is a Gaussian variable, with zero average. Thus

$$\langle \mathrm{Cos}(\theta(t)-\theta(o)) \rangle = \langle \vec{S}_i(t).\vec{S}_i(o) \rangle = \exp[-\frac{kT}{\nu} t] \tag{14}$$

Now let us see the role of interactions. The latter restrict the motion of each spin, so there must be some slowing down in general. The actual behaviour is indeed quite complex, but as a first level description EA proposed that the autocorrelation function $\langle\langle \vec{S}_i(t).\vec{S}_i(o) \rangle\rangle$ has the same form as (14) but with a renormalised coefficient of viscosity η in place of ν. Thus the entire complexity introduced by interactions is tackled by calculating selfconsistently a renormalised coefficient of viscosity, which now becomes a temperature dependent object.

The motion of a given spin θ_i depends upon its nearest neighbours θ_j's, whose motion in turn depend upon θ_i. This back reaction of the medium i.e. the other spins, is the principle factor which renormalises the self viscosity η. To isolate this, we write the equation of motion of one of the neighbours of θ_j in the following way

$$\nu\dot{\theta}_j + J_{ij} \mathrm{Sin}(\theta_j-\theta_i) = f_j - \sum_{k(\neq i)} J_{jk} \mathrm{Sin}(\theta_j-\theta_k) \tag{15}$$

Now the renormalised quantities are introduced through the definitions

$$\eta\dot{\theta}_j + J_{ij} \mathrm{Sin}(\theta_j-\theta_i) = F_j \tag{16}$$

In (16) the interactions on θ_j, apart from that due to θ_i have been lumped into the new coefficient of viscosity η, and a new fluctuating force F. In this way the major interdependence between θ_i and θ_j is identified and rest of the terms are regarded as part of the heat bath. Equation (16) is now integrated to give

$$\theta_j(t)-\theta_j(-\infty) = \frac{1}{\eta}[\int_{-\infty}^{t} F_j(\tau)d\tau - J_{ij}\int_{-\infty}^{t} \mathrm{Sin}[\theta_j(\tau)-\theta_i(\tau)] d\tau$$

$$\theta_j(t) = \tilde{\theta}_j(t) - \frac{J_{ij}}{\eta}\int_{-\infty}^{t} \mathrm{Sin}(\theta_j(\tau) - \theta_i(\tau))d\tau \tag{17}$$

Note that $\tilde{\theta}_j(t)$ can be taken to be independent of θ_i. Putting this solution back into (16) gives

$$\nu\dot{\theta}_i + \sum_j J_{ij} \mathrm{Sin}[\theta_i-\tilde{\theta}_j + \frac{J_{ij}}{\eta}\int_{-\infty}^{t}\mathrm{Sin}(\theta_j(\tau) - \theta_i(\tau))d\tau] = f_i \tag{18}$$

The second term in (17), being the contribution of a single spin, can be treated as small. Using this fact, (18) can be written as

$$\nu\dot{\theta}_i + \sum_j J_{ij}\mathrm{Sin}(\theta_i-\tilde{\theta}_j) + \frac{1}{\eta}\sum_j J_{ij}^2 \mathrm{Cos}(\theta_i-\tilde{\theta}_j)\int_{-\infty}^{t}\mathrm{Sin}(\theta_j(\tau) - \theta_j(\tau))d\tau = f_i \tag{19}$$

The second term in the left of (19) is now brought together with f_i to define a new fluctuating force

$$F_i = f_i - \sum_j J_{ij} \operatorname{Sin}(\theta_i - \tilde{\theta}_j) \tag{20}$$

and the third term is manipulated in the following manner

$$2\operatorname{Cos}(\theta_i - \tilde{\theta}_j)\operatorname{Sin}(\theta_j(\tau) - \theta_i(\tau)) = \operatorname{Sin}[\theta_i(t) - \theta_i(\tau) - \tilde{\theta}_j(t) + \theta_j(\tau)]$$

$$+\operatorname{Sin}[\theta_i(t) + \theta_i(\tau) - \tilde{\theta}_j(t) - \theta_j(\tau)] \tag{21}$$

The second of these terms can be ignored as being small when averages with respect to J_{ij}'s are taken, and the first term can again be decomposed to write

$$\operatorname{Sin}[\theta_i(t) - \theta_i(\tau) - \tilde{\theta}_j(t) + \theta_j(\tau)] = \operatorname{Sin}(\theta_i(t) - \theta_i(\tau))\operatorname{Cos}(\tilde{\theta}_j(t) - \theta_j(\tau))$$

$$+\operatorname{Cos}(\theta_i(t) - \theta_i(\tau))\operatorname{Sin}(\tilde{\theta}_j(t) - \theta_j(\tau)) \tag{22}$$

Now, following the mean field philosophy,we replace the j-dependent terms by the averaged quantities

$$\langle \operatorname{Cos}(\tilde{\theta}_j(t) - \theta_j(\tau)) \rangle = q(t-\tau) \tag{23}$$

$$\langle \operatorname{Sin}(\tilde{\theta}_j(t) - \theta_j(\tau)) \rangle \approx 0 \tag{24}$$

Substituting the result of these manipulations in (19), one obtains

$$\nu\dot{\theta}_i + \frac{1}{2\eta}\sum_j \bar{J}^2 \int_{-\infty}^{t} q(t-\tau)\operatorname{Sin}(\theta_i(t) - \theta_i(\tau))\, d\tau = F_i(t) \tag{25}$$

The self-consistency is now imposed by requiring that the final equation (25) is linear in θ_i and $F_i(t)$ satisfies the fluctuation-dissipation relation

$$\langle F_i(t)\, F_i(t')\rangle = \eta kT\ \delta(t-t') \tag{26}$$

Thus

$$\nu\dot{\theta}_i + \frac{J_o^2}{2\eta}\int_{-\infty}^{t} q(t-\tau)\,(\theta_i(t) - \theta_i(\tau))\, d\tau = F_i(t) \tag{27}$$

where $J_o^2 = z\,\bar{J}^2$. Further, one can approximate

$$\theta_i(t) - \theta_i(\tau) = (t-\tau)\,\dot{\theta}_i + \ldots \tag{28}$$

and write for $q(t-\tau)$ $+\exp\left[-\frac{kT}{\eta}(t-\tau)\right]$, to get

$$\left[\nu + \frac{J_o^2}{2\eta}\int_{-\infty}^{t} (t-\tau)\, e^{-(kT/\eta)(t-\tau)}\, d\tau\right]\dot{\theta}_i = F_i(t). \tag{29}$$

Use of (26) finally yields the following self-consistent equation for η

$$\eta = \nu + \frac{J_o^2}{2\eta} \int_o^\infty t' \, e^{-kT/\eta t'} \, dt' \tag{30}$$

$$\eta = \nu [1-\frac{1}{2}(J_o/kT)^2]^{-1} \tag{31}$$

The consistency of this approximation can be further checked by using (20) to evaluate (26). Equation (31) shows that at as $T \rightarrow T_g = \frac{J_o}{\sqrt{2}k}$, $\eta \rightarrow \infty$, which means that at this temperature the spin motion slows down to the extent of having a strong memory even for infinite time.

As mentioned in the beginning, EA picture outlines above has come under much scrutiny in recent work, but its idea remains essential for an understanding of spin-glass physics.

References

1. S.F. Edwards and P.W. Anderson, J. Phys. F6, 1927 (1976)
2. P.W. Anderson in "Ill-condensed Matter" p.162, Eds. R. Balian, R. Maynard and G. Toulouse (North-Holland Publishing Co., 1979)
3. A. Blandin, J.de.Physique 39, C6-1499 (1978)
4. A.P. Murani, J. Mag. Materials 5, 95 (1977)
5. D. Kumar in "Current Trends in Magnetism" p.388 Ed. N.S. Satyamurthy and L. Madhav Rao (Published by Indian Physics Association, 1981)
6. S.F. Edwards and P.W. Anderson, J. Phys. F5, 965 (1975)

WAVE PROPAGATION IN RANDOM MEDIA

G.S. Agarwal
School of Physics, University of Hyderabad
Hyderabad - 500134, India

In this lecture we will apply some of the techniques developed in this volume to the study of the wave propagation in a random medium. This is a subject on which enormous body of the literature [1] exists and hence we will cover only certain basic stochastic aspects of the system. In the specific context of this talk, wave propagation refers to the propagation of the electromagnetic waves through a medium whose refractive index is a random function. However one could very well also consider the propagation of acoustic waves or even Schrödinger equation in a random potential, the latter being covered by other speakers at this school.

Let us consider the steady state phenomena and let $\epsilon(\vec{r})$ be the spatially inhomogeneous dielectric function of the medium. The Maxwell equations in such a case lead to an equation for the electric field $\vec{E}$ alone

$$\vec{\nabla}\times\vec{\nabla}\times\vec{E} - k_m^2(\vec{r})\,\vec{E} = 0,\quad k_m = (\omega/c)\sqrt{\epsilon(\vec{r})} \tag{1}$$

which can also be written as

$$\nabla^2\vec{E} + k^2\epsilon(\vec{r})\,\vec{E} + \vec{\nabla}(\vec{E}\cdot\vec{\nabla}\ln\epsilon(\vec{r})) = 0,\quad k = \omega/c \tag{2}$$

Since $\epsilon(\vec{r})$ is a random function whose statistical properties are supposedly known (2) represents a very complex type of Langevin equation. No exact solutions for such an equation appear to be known even if $n(\vec{r})$ is a Gaussian random process and hence one needs to use approximate methods in order to obtain information on the characteristics of $\vec{E}$ for such a medium.

(1) Weak Refractive Index Fluctuations

We first consider the case when the refractive index fluctuations are weak i.e. $n(\vec{r}) = 1+n_1(r)$; $n_1(r) \ll 1$. We can then integrate equation (1) in a perturbative manner. Let us introduce the Green's dyadic defined by

$$\vec{\nabla}\times\vec{\nabla}\times\overset{\Rightarrow}{G}(\vec{r},\vec{r}',\omega) - k^2\overset{\Rightarrow}{G}(\vec{r},\vec{r}',\omega) = 4\pi\overset{\Rightarrow}{I}\,\delta(\vec{r}-\vec{r}'). \tag{3}$$

Assuming outgoing boundary conditions at infinity, the solution of (3) is

$$\overset{\Rightarrow}{G}(\vec{r},\vec{r}',\omega) = \left(\overset{\Rightarrow}{I} + \frac{\vec{\nabla}\vec{\nabla}}{k^2}\right)\frac{e^{ik|\vec{r}-\vec{r}'|}}{|\vec{r}-\vec{r}'|} \tag{4}$$

This represents the free space Green's function. Using (4) and $\epsilon(r) = 1+\epsilon_1(r)$, one can convert (1) into an integral equation

$$\vec{E}(\vec{r},\omega) = \vec{E}^{(i)}(\vec{r},\omega) + \frac{k^2}{4\pi}\int d^3r'\,\overset{\Rightarrow}{G}(\vec{r},\vec{r}',\omega)\cdot\epsilon_1(\vec{r}')\vec{E}(\vec{r}',\omega) \tag{5}$$

where $\vec{E}^{(i)}(\vec{r},\omega)$ represents the incident field on the medium. Note that $\epsilon_1(\vec{r}) \sim 2n_1(\vec{r})$. The scattered field can now be evaluated to different orders in ϵ_1. In Born approximation, the scattered field will be

$$\vec{E}_{Sc}(\vec{r},\omega) = \frac{k^2}{4\pi}\int d^3r' \,\vec{\vec{G}}(\vec{r},\vec{r}',\omega)\,.\,\epsilon_1(\vec{r}')\,.\vec{E}^{(i)}(\vec{r}',\omega)\,. \tag{6}$$

If the fluctuations in ϵ_1 are Gaussian, then the scattered field will also be Gaussian provided that the input field $\vec{E}^{(i)}$ is non-stochastic in nature. It may be of interest to note that $\vec{E}_{Sc}$ will not be Gaussian even if both ϵ_1 and $\vec{E}^{(i)}$ are Gaussian random processes. The expression for the scattered field can be simplified in the far zone limit whence

$$\vec{\vec{G}}(\vec{r},\vec{r}',\omega) \longrightarrow (\vec{\vec{I}}-\vec{n}\,\vec{n})\,\frac{e^{ikr-i\vec{n}.\vec{r}'k}}{r}\quad, \tag{7}$$

where $\vec{n}$ is the unit vector in the direction $\vec{r}$. The scattering amplitude then becomes

$$\vec{E}_{Sc} \sim -\frac{k^2}{4\pi}\left(\frac{e^{ikr}}{r}\right)\int d^3r'\,\epsilon_1(\vec{r}')e^{-i\vec{n}.\vec{r}'k}\,\vec{n}\times(\vec{n}\times\vec{E}^{(i)}(\vec{r}')) \tag{8}$$

which on assuming $\vec{E}^{(i)}(\vec{r}) = \vec{\mathcal{E}}\,e^{i\,\vec{K}.\vec{r}}$ leads to the following expression for the differential cross section

$$\frac{d\sigma}{d\Omega} = \frac{k^4}{16\,\pi^2}\int d^3r_1\int d^3r_2\,\epsilon_1(\vec{r}_1)\,\epsilon_1^*(\vec{r}_2)e^{i(\vec{K}-\vec{n}k).(\vec{r}_1-\vec{r}_2)}\sin^2\theta\quad, \tag{9}$$

where θ is the angle between $\vec{n}$ and $\vec{\mathcal{E}}$. Expression (9) is the basic formula for the electromagnetic scattering. For our medium ϵ_1 is a random quantity and hence we have to average (9) over the stochastic fluctuations of $\epsilon_1(\vec{r})$. Let us write the correlation function of n as

$$\Gamma(\vec{r}_1,\vec{r}_2) = \langle n_1(\vec{r}_1)n_1^*(\vec{r}_2)\rangle = F\left\{\frac{\vec{r}_1+\vec{r}_2}{2}\right\}g(\vec{r}_1-\vec{r}_2) \tag{10}$$

For a homogeneous medium F is unity . Then the average cross section can be written as

$$\frac{d\sigma}{d\Omega} = \frac{k^4\sin^2\theta}{4\pi^2}\int_V d^3R\;F(\vec{R})\int_V g(\vec{r})e^{i(\vec{k}-k\vec{n}).\vec{r}}d^3r, \tag{11}$$

which is the basic formula relating the scattered flux to the refractive index fluctuations of the random medium. Further progress can be made depending on the structure of F and g. Often Kolmogorov spectrum is used in connection with the atmospheric turbulence

$$\langle(n_1(\vec{r}_1+\vec{r}) - n_1(\vec{r}_1))^2\rangle = D(r) \sim r^{2/3},\; L_o \gg r \gg l_o \qquad \sim r^2\;,\quad r \ll l_o \tag{12}$$

where l_o and L_o are, respectively, the inner and outer scales of turbulence. This form was generalized by Kármán to

$$1 - \frac{2^{2/3}}{\Gamma(\frac{1}{3})} \left(\frac{r}{L_o}\right)^{1/3} K_{1/3} \left(\frac{r}{L_o}\right).$$

More generally we can write

$$g(\vec{r}) = \frac{\bar{n}_1^2}{2^{\nu-1}\Gamma(\nu)} \left(\frac{r}{L_o}\right)^{\nu} K_{\nu} \left(\frac{r}{L_o}\right), \qquad \nu \sim 1/3, \tag{13}$$

whose cosine Fourier transform $g(\vec{k})$ is

$$g(\vec{k}) = \frac{\Gamma(\nu+3/2)}{\pi^{3/2}\Gamma(\nu)} \bar{n}_1^2 L_o^3/(1+k^2L_o^2)^{\nu+3/2} \tag{14}$$

On substituting (14) in (11) and assuming F = 1 we find the result

$$\frac{d\sigma}{d\Omega} = \frac{2k^4 \sin^2\theta \;\; V \;\; \Gamma(\nu+3/2)\; \bar{n}_1^2 \; L_o^3}{\pi^{1/2} \;\; \Gamma(\nu) \;\; \left[1+4L_o^2k^2\sin^2\frac{\theta}{2}\right]^{\nu+3/2}}, \tag{15}$$

which is quite a general expression for the scattering from a random medium. Notice the important frequency dependence coming from the denominator. The refractive index fluctuations are characterized by the parameter $\bar{n}_1^2/L_o^{2\nu}$.

2. Strong Fluctuations : Non Perturbative Methods

When the refractive index fluctuations are strong, then one has to use methods which yield the scattered fields to all orders in the refractive index. This obviously is difficult and it turns out that a good deal of progress can be made by incorporating several physical assumptions. We assume that the wavelength is small compared to the typical scale of inhomogeneities and hence we can use the idea of small angle scattering in the forward direction and ignore any back scattering. We will also, for simplicity, ignore the depolarization effects so that the electromagnetic problem can be treated as a scalar problem. Writing further $E = ue^{ikz}$ and making the slowly varying approximation, we find that the original equation (2) reduces to the well known parabolic equation

$$2ik\frac{\partial u}{\partial z} + \nabla_t^2 u + k^2 \epsilon_1 u = 0; \tag{16}$$

where z is the direction of propagation. This is a stochastic differential equation of the type treated earlier [2]. As noted there, exact results for the moments and the correlation functions of u can be obtained by assuming that ϵ_1 is a Gaussian delta correlated random process

$$\langle \epsilon_1(\vec{r}) \;\; \epsilon_1(\vec{r}\,') \rangle = \delta(z-z') \;\; \Gamma(|\vec{\rho} - \vec{\rho}\,'|), \tag{17}$$

where $\vec{\rho}$ is the coordinate transverse to z. This assumption has been more or less universally used by every one in the field and very many different methods like the ones based on Novikov's theorem [3], diagrams [4] have been developed to obtain the equations for the mean and the correlations. Here we will show how the results of [2] can be used to derive in a systematic manner the various averaged equations. For this purpose we recall the result from [2]: If the basic stochastic differential equation is written in the form (z now playing the role of time)

$$\frac{\partial \psi}{\partial z} = [L_o + L_1 (z, \cdots)] \psi \tag{18}$$

and if

$$L_1 = \sum F_\alpha (z, \cdots) L_{1\alpha} , \tag{19}$$

where F_α's are Gaussian delta correlated process with diffusion coefficient $2 D_{\alpha\beta}$, then the exact equation for the ensemble average of ψ is

$$\frac{\partial \langle \psi \rangle}{\partial z} = (L_o + \sum_{\alpha\beta} D_{\alpha\beta} L_{1\alpha} L_{1\beta}) \langle \psi \rangle . \tag{20}$$

we can now directly apply (20) to (16) by using identification $L_o = \frac{-1}{2ik} \nabla_t^2$, $L_1 = -k^2 \epsilon_1 (z, \vec{\rho})/2ik$ with the result

$$2ik \frac{\partial \langle u \rangle}{\partial z} + \nabla_t^2 \langle u \rangle + \frac{ik^3}{4} \Gamma(o) \langle u \rangle = 0, \tag{21}$$

which has the solution

$$\langle u(\vec{\rho}, z) \rangle = u_o (\vec{\rho}, z) \exp(-\alpha z), \quad \alpha = \frac{k^2 \Gamma(o)}{2} , \tag{22}$$

where u_o is the field in free space and which for a Gaussian beam has the structure

$$u_o = \frac{1}{(1+i\beta z)} \exp \left\{ - \frac{k\beta}{2} \frac{\rho^2}{(1+i\beta z)} \right\}. \tag{23}$$

Thus α can be identified with the absorption coefficient and is related to the spectral density g(k) of fluctuations for $\Gamma(\rho)$ can be written as

$$\Gamma(\rho) = (2\pi)^2 \int_0^\infty kdk \, g(k) \, J_o (k\rho), \tag{24}$$

where we can, for example, choose for g(k), the Kolmogorov form (14).

We next show how the equations for the mutual coherence function $K(\vec{\rho}_1, \vec{\rho}_2, z)$

$$K(\vec{\rho}_1, \vec{\rho}_2, z) = \langle u(\vec{\rho}_1, z) \, u^* (\vec{\rho}_2, z) \rangle \tag{25}$$

can be obtained. Using the parabolic equation (16), we find the equation for $u(\vec{\rho}_1, z) \, u^*(\vec{\rho}_2, z) \equiv u_1 u_2^*$

$$2ik \frac{\partial}{\partial z}(u_1 u_2^*) + (\nabla_{t1}^2 - \nabla_{t2}^2) u_1 u_2^* + k^2 \left\{ \epsilon_1 (z, \vec{\rho}_1) - \epsilon_1^* (z, \vec{\rho}_2) \right\} u_1 u_2^* = 0. \tag{26}$$

The application of (20) to (26) is straight forward if we identify $\psi = u_1 u_2^*$, $L_0 = (-\nabla_{t1}^2 + \nabla_{t2}^2)/2ik$, $L_1 = -k^2 \{\epsilon_1(z, \vec{P}_1) - \epsilon_1^*(z, \vec{P}_2)\}/2ik$ with the result

$$2ik \frac{\partial K}{\partial z} + (\nabla_{t1}^2 - \nabla_{t2}^2) K + \frac{ik^3}{2}[\Gamma(0) - \Gamma(\vec{P}_1 - \vec{P}_2)] K = 0. \tag{27}$$

Note that the mutual coherence function gives information regarding the correlations in a given plane and hence (27) essentially shows how the correlations change in propagation from one plane to another plane. Thus (27) will yield the generalization of the van Zittert-Zernike theorem [5] for propagation in a random medium. Exact solution to (27) is known from the work of Tatarskii. Introducing the sum and difference coordinates $\vec{P} = \vec{P}_1 - \vec{P}_2$, $\vec{R} = (\vec{P}_1 + \vec{P}_1)/2$ (27) becomes

$$2ik \frac{\partial K}{\partial z} + 2 \vec{\nabla}_\rho \cdot \vec{\nabla}_R K + \frac{ik^3}{2} (\Gamma(0) - \Gamma(\vec{P})) K = 0. \tag{28}$$

Equation (28) can be solved by first taking the Fourier transform with respect to the variable $\vec{R}$. This results in a first order equation in $\vec{P}$ and z and hence can be solved by the method of characteristics. The result being

$$K(\vec{R}, \vec{P}, z) = \int d^2K \, m\left(\vec{K}, \vec{P} - \frac{\vec{P} z}{k}, 0\right) \exp\{i\vec{K}.\vec{R} - H(\vec{K}, \vec{P}, z)\},$$

$$H(\vec{K}, \vec{P}, z) = \frac{k^2}{4} \int_0 [\Gamma(o) - \Gamma(\vec{P} - \frac{\vec{K}}{k} z')] dz' \tag{29}$$

where m is the Fourier transform of K in the plane z = 0

$$m(\vec{k}, \vec{P}, 0) = \frac{1}{(2\pi)^2} \int K(\vec{R}, \vec{P}, 0) \, e^{-i\vec{k}.\vec{R}} d^2R. \tag{30}$$

For the Gaussian beam case the function m has a simple form, however the function H and hence K has to be evaluated numerically for the case of Kolmogorov spectrum.

The closed form of equations for the higher oreder moments can also be obtained using (20). However such equations acquire increasingly complex structure, for example the intensity correlation

$$K_2 = \langle I(\vec{P}_1, z) I(\vec{P}_2, z) \rangle, \quad I = uu^* \tag{31}$$

obeys the equation

$$2ik \frac{\partial K_2}{\partial z} + (\nabla_{\rho_1}^2 + \nabla_{\rho_2}^2 - \nabla_{\rho_1'}^2 - \nabla_{\rho_2'}^2) K_2$$
$$+ \frac{ik^3}{2}[2\Gamma(0) + \Gamma(\vec{P}_2 - \vec{P}_1) + \Gamma(\vec{P}_2 - \vec{P}_1') - \Gamma(\vec{P}_1 - \vec{P}_1')$$
$$- \Gamma(\vec{P}_1 - \vec{P}_2') - \Gamma(\vec{P}_2 - \vec{P}_1') - \Gamma(\vec{P}_2 - \vec{P}_2')] K_2 = 0 \tag{32}$$

No exact solutions to (32) appear to be known; approximate solutions can be found for example in [6].

Finally we would like to add that if ϵ_1 were to correspond to a random telegraphic signal, then certain exact results can be obtained following the general formulation given in [2]. Similarly if the fluctuations of ϵ_1 had a finite correlation length, in the direction of propagation, then we have to use more general formula than (20) such as those given in [2] .

References

1. V.I. Tatarskii, "Wave Propagation in a Turbulent Medium" (Dover, N.Y. 1967); J.W. Strohbehn "Laser Beam Propagation in the Atmosphere" (Springer-Verlag, Berlin 1978); U. Frisch in "Probabilistic Methods in Applied Mathematics" edited by A.T. Bharucha-Reid (Academic, N.Y. 1968) Vol.I, p.75, N.G. van Kampen, Physics Reports 24C, 171 (1976)
2. G.S. Agarwal, this volume p.
3. E.A. Novikov, Sov. Phys. JETP 20, 1290 (1965)
4. R.C. Bourret, Nuovo Cimento 26, 1 (1962); Can. J. Phys. 40, 782 (1962); see also J. Molyneux, J. Opt. Soc. Am. 61, 248 (1971); V.I. Tatarskii "The Effects of the Turbulent Atmosphere on Wave Propagation" (NTIS, Springfield, Va 1971)
5. cf.M. Born and E. Wolf, "Principles of Optics" (Pergamon, Oxford 1970, p.508)
6. R.L. Fante, Proc. IEEE 63, 1669 (1975)

Springer Series in

Synergetics

Series Editor: **H. Haken**

Volume 1: **H. Haken**
Synergetics
An Introduction. Nonequilibrium Phase Transitions and Self-Organization in Physics, Chemistry and Biology. 3nd revised and enlarged edition
ISBN 3-540-12356-3

Volume 2:
Synergetics
A Workshop. Proceedings of the International Workshop on Synergetics at Schloß Elmau, Bavaria, May 2–7, 1977
Editor: **H. Haken**
ISBN 3-540-08483-5

Volume 3:
Synergetics
Far from Equilibrium. Proceedings of the Conference Far From Equilibrium: Instabilities and Structures, Bordeaux, France, September 27–29, 1978
Editors: **A. Pacault, C. Vidal**
ISBN 3-540-09304-4

Volume 4:
Structural Stability in Physics
Proceedings of Two International Symposia on Applications of Catastrophe Theory and Topological Concepts in Physics, Tübingen, Federal Republic of Germany, May 2–6 and December 11–14, 1978
Editors: **W. Güttinger, H. Eikemeier**
ISBN 3-540-09463-6

Volume 5:
Pattern Formation by Dynamic Systems and Pattern Recognition
Proceedings of the International Symposium on Synergetics at Schloß Elmau, Bavaria, April 30–May 5, 1979
Editor: **H. Haken**
ISBN 3-540-09770-8

Volume 6:
Dynamics of Synergetic Systems
Proceedings of the International Symposium on Synergetics, Bielefeld, Federal Republic of Germany, September 24–29, 1979
Editor: **H. Haken**
ISBN 3-540-09918-2

Volume 7: **L. A. Blumenfeld**
Problems of Biological Physics
ISBN 3-540-10401-1

Volume 8:
Stochastic Nonlinear Systems
in Physics, Chemistry and Biology. Proceedings of the Workshop, Bielefeld, Federal Republic of Germany, October 5–11, 1980
Editors: **L. Arnold, R. Lefever**
ISBN 3-540-10713-4

Volume 9:
Numerical Methods in the Study of Critical Phenomena
Proceedings of a Colloquium, Carry-le-Rout, France, June 2–4, 1980
Editors: **J. Della Dora, J. Demongeot, B. Lacolle**
ISBN 3-540-11009-7

Volume 10: **Y. L. Klimontovich**
The Kinetic Theory of Electromagnetic Processes
ISBN 3-540-11458-0

Volume 11:
Chaos and Order in Nature
Proceedings of the International Symposium on Synergetics at Schloß Elmau, Bavaria, April 27 - May 2, 1981
Editor: **H. Haken**
ISBN 3-540-11101-8

Volume 12:
Nonlinear Phenomena in Chemical Dynamics
Proceedings of an International Conference, Bordeaux, France, September 7–11, 1981
Editors: **C. Vidal, A. Pacault**
ISBN 3-540-11294-4

Volume 13: **C. W. Gardiner**
Handbook of Stochastic Methods
for Physics, Chemistry and the Natural Sciences
ISBN 3-540-11357-6

Volume 14: **W. Weidlich, G. Haag**
Concepts and Models of a Quantitative Sociology
The Dynamics of Interacting Populations
ISBN 3-540-11358-4

Volume 15: **W. Horsthemke, R. Lefever**
Nonequilibrium Transitions Induced by External Noise
ISBN 3-540-11359-2

Volume 16: **L. A. Blumenfeld**
Physics of Bioenergetic Processes
ISBN 3-540-11417-3

Volume 17:
Evolution of Order and Chaos
in Physics, Chemistry, and Biology
Proceedings of the International Symposium on Synergetics at Schloß Elmau, Bavaria, April 26 - May 1, 1982
Editor: **H. Haken**
ISBN 3-540-11904-3

Springer-Verlag
Berlin
Heidelberg
New York
Tokyo

Vol. 144: Topics in Nuclear Physics I. A Comprehensive Review of Recent Developments. Edited by T.T.S. Kuo and S.S.M. Wong. XX, 567 pages. 1981.

Vol. 145: Topics in Nuclear Physics II. A Comprehensive Review of Recent Developments. Proceedings 1980/81. Edited by T. T. S. Kuo and S. S. M. Wong. VIII, 571-1.082 pages. 1981.

Vol. 146: B. J. West, On the Simpler Aspects of Nonlinear Fluctuating. Deep Gravity Waves. VI, 341 pages. 1981.

Vol. 147: J. Messer, Temperature Dependent Thomas-Fermi Theory. IX, 131 pages. 1981.

Vol. 148: Advances in Fluid Mechanics. Proceedings, 1980. Edited by E. Krause. VII, 361 pages. 1981.

Vol. 149: Disordered Systems and Localization. Proceedings, 1981. Edited by C. Castellani, C. Castro, and L. Peliti. XII, 308 pages. 1981.

Vol. 150: N. Straumann, Allgemeine Relativitätstheorie und relativistische Astrophysik. VII, 418 Seiten. 1981.

Vol. 151: Integrable Quantum Field Theory. Proceedings, 1981. Edited by J. Hietarinta and C. Montonen. V, 251 pages. 1982.

Vol. 152: Physics of Narrow Gap Semiconductors. Proceedings, 1981. Edited by E. Gornik, H. Heinrich and L. Palmetshofer. XIII, 485 pages. 1982.

Vol. 153: Mathematical Problems in Theoretical Physics. Proceedings, 1981. Edited by R. Schrader, R. Seiler, and D.A. Uhlenbrock. XII, 429 pages. 1982.

Vol. 154: Macroscopic Properties of Disordered Media. Proceedings, 1981. Edited by R. Burridge, S. Childress, and G. Papanicolaou. VII, 307 pages. 1982.

Vol. 155: Quantum Optics. Proceedings, 1981. Edited by C.A. Engelbrecht. VIII, 329 pages. 1982.

Vol. 156: Resonances in Heavy Ion Reactions. Proceedings, 1981. Edited by K.A. Eberhard. XII, 448 pages. 1982.

Vol. 157: P. Niyogi, Integral Equation Method in Transonic Flow. XI, 189 pages. 1982.

Vol. 158: Dynamics of Nuclear Fission and Related Collective Phenomena. Proceedings, 1981. Edited by P. David, T. Mayer-Kuckuk, and A. van der Woude. X, 462 pages. 1982.

Vol. 159: E. Seiler, Gauge Theories as a Problem of Constructive Quantum Field Theory and Statistical Mechanics. V, 192 pages. 1982.

Vol. 160: Unified Theories of Elementary Particles. Critical Assessment and Prospects. Proceedings, 1981. Edited by P. Breitenlohner and H.P. Dürr. VI, 217 pages. 1982.

Vol. 161: Interacting Bosons in Nuclei. Proceedings, 1981. Edited by J.S. Dehesa, J.M.G. Gomez, and J. Ros. V, 209 pages. 1982.

Vol. 162: Relativistic Action at a Distance: Classical and Quantum Aspects. Proceedings, 1981. Edited by J. Llosa. X, 263 pages. 1982.

Vol. 163: J. S. Darrozes, C. Francois, Mécanique des Fluides Incompressibles. XIX, 459 pages. 1982.

Vol. 164: Stability of Thermodynamic Systems. Proceedings, 1981. Edited by J. Casas-Vázquez and G. Lebon. VII, 321 pages. 1982.

Vol. 165: N. Mukunda, H. van Dam, L.C. Biedenharn, Relativistic Models of Extended Hadrons Obeying a Mass-Spin Trajectory Constraint. Edited by A. Böhm and J.D. Dollard. VI, 163 pages. 1982.

Vol. 166: Computer Simulation of Solids. Edited by C.R.A. Catlow and W.C. Mackrodt. XII, 320 pages. 1982.

Vol. 167: G. Fieck, Symmetry of Polycentric Systems. VI, 137 pages, 1982.

Vol. 168: Heavy-Ion Collisions. Proceedings, 1982. Edited by G. Madurga and M. Lozano. VI, 429 pages. 1982.

Vol. 169: K. Sundermeyer, Constrained Dynamics. IV, 318 pages. 1982.

Vol. 170: Eighth International Conference on Numerical Methods in Fluid Dynamics. Proceedings, 1982. Edited by E. Krause. X, 569 pages. 1982.

Vol. 171: Time-Dependent Hartree-Fock and Beyond. Proceedings, 1982. Edited by K. Goeke and P.-G. Reinhard. VIII, 426 pages. 1982.

Vol. 172: Ionic Liquids, Molten Salts and Polyelectrolytes. Proceedings, 1982. Edited by K.-H. Bennemann, F. Brouers, and D. Quitmann. VII, 253 pages. 1982.

Vol. 173: Stochastic Processes in Quantum Theory and Statistical Physics. Proceedings, 1981. Edited by S. Albeverio, Ph. Combe, and M. Sirugue-Collin. VIII, 337 pages. 1982.

Vol. 174: A. Kadić, D.G.B. Edelen, A Gauge Theory of Dislocations and Disclinations. VII, 290 pages. 1983.

Vol. 175: Defect Complexes in Semiconductor Structures. Proceedings, 1982. Edited by J. Giber, F. Beleznay, J. C. Szép, and J. László. VI, 308 pages. 1983.

Vol. 176: Gauge Theory and Gravitation. Proceedings, 1982. Edited by K. Kikkawa, N. Nakanishi, and H. Nariai. X, 316 pages. 1983.

Vol. 177: Application of High Magnetic Fields in Semiconductor Physics. Proceedings, 1982. Edited by G. Landwehr. XII, 552 pages. 1983.

Vol. 178: Detectors in Heavy-Ion Reactions. Proceedings, 1982. Edited by W. von Oertzen. VIII, 258 pages. 1983.

Vol. 179: Dynamical Systems and Chaos. Proceedings, 1982. Edited by L. Garrido. XIV, 298 pages. 1983.

Vol. 180: Group Theoretical Methods in Physics. Proceedings, 1982. Edited by M. Serdaroğlu and E. İnönü. XI, 569 pages. 1983.

Vol. 181: Gauge Theories of the Eighties. Proceedings, 1982. Edited by R. Raitio and J. Lindfors. V, 644 pages. 1983.

Vol. 182: Laser Physics. Proceedings, 1983. Edited by J. D. Harvey and D. F. Walls. V, 263 pages. 1983.

Vol. 183: J. D. Gunton, M. Droz, Introduction to the Theory of Metastable and Unstable States. VI, 140 pages. 1983.

Vol. 184: Stochastic Processes – Formalism and Applications. Proceedings, 1982. Edited by G. S. Agarwal and S. Dattagupta. VI, 324 pages. 1983.

Selected Issues from

Lecture Notes in Mathematics